Numerical Simulation: Theory and Analysis

Numerical Simulation: Theory and Analysis

Edited by **Gregory Rago**

CLANRYE
INTERNATIONAL

New Jersey

Published by Clanrye International,
55 Van Reypen Street,
Jersey City, NJ 07306, USA
www.clanryeinternational.com

Numerical Simulation: Theory and Analysis
Edited by Gregory Rago

International Standard Book Number: 978-1-63240-399-5 (Hardback)

Printed in the United States of America.

Contents

Preface

This book encompasses the fundamentals as well as contemporary developments of numerical simulation associated with fluid dynamics in the natural environment and scientific applications. It also discusses numerical simulation in various industrial areas, like metallurgy, power engineering and building. Latest numerical methodologies, as well as software, the most precise and enhanced in treating the physical phenomena, are applied for the purpose of explanation of the investigated processes in terms of numbers. Since it plays a significant role in both industrial and theoretical research, this book regarding simulation of several physical procedures will serve as a useful tool for researchers as well as scientists, industrial engineers, applied mathematicians, and post-graduate students.

After months of intensive research and writing, this book is the end result of all who devoted their time and efforts in the initiation and progress of this book. It will surely be a source of reference in enhancing the required knowledge of the new developments in the area. During the course of developing this book, certain measures such as accuracy, authenticity and research focused analytical studies were given preference in order to produce a comprehensive book in the area of study.

This book would not have been possible without the efforts of the authors and the publisher. I extend my sincere thanks to them. Secondly, I express my gratitude to my family and well-wishers. And most importantly, I thank my students for constantly expressing their willingness and curiosity in enhancing their knowledge in the field, which encourages me to take up further research projects for the advancement of the area.

Editor

Fluid Dynamics

BG Model Based on Bagnold's Concept and Its Application to Analysis of Elongation of Sand Spit and Shore – Normal Sand Bar

Takaaki Uda, Masumi Serizawa and Shiho Miyahara

Additional information is available at the end of the chapter

1. Introduction

The accurate prediction of three-dimensional (3D) beach changes on a coast with a large shoreline curvature, such as a coast with a sand spit, and a wave field that significantly changes in response to topographic changes, has been difficult to achieve in previous studies. As a result, regarding the beach changes around a sand spit, most previous studies have focused on the shoreline changes. Ashton et al. (2001) showed that when the incident angle of deep-water waves to the mean shoreline exceeds 45°, shoreline instability occurs, resulting in the development of sand spits from a small perturbation of the shoreline. They successfully predicted the planar changes in a shoreline containing sand spits using infinitesimal meshes divided in the x- and y-directions. However in Ashton et al.'s model, only the longshore sand transport equation is employed as the sand transport equation instead of a two-dimensional (2D) sand transport equation in which both cross-shore and longshore sand transport are considered. Furthermore, in evaluating the wave field, wave conditions at the breaking point are transformed into deep-water values assuming a bathymetry with straight parallel contours and using Snell's law. Since the sand transport equation is expressed using these deep-water parameters, the effect of large 3D changes in topography on the wave field cannot be accurately evaluated. Furthermore, the finite-difference scheme used to evaluate the breaker angle and the method of calculating the wave field around the wave-shelter zone are altered depending on the calculation conditions, resulting in a complicated calculation method that requires special calculation techniques. Watanabe et al. (2004) developed a model for predicting the shoreline changes of a sand spit under the conditions that the sand spit significantly changes its configuration with changes in the wave field. They selected orthogonal curvilinear coordinates parallel and normal to the shoreline of the sand spit, and the seabed topography after various numbers of time steps was inversely determined from the time-dependent shoreline configuration given by the seabed slope. They also predicted the shoreline changes due to the spatial changes in longshore sand transport. Their model is also not a definitive model for predicting the 3D topographic changes of a sand spit.

A sand spit is often formed by wave action at a location where the direction of the coastline abruptly changes. Uda & Yamamoto (1992) carried out a movable-bed experiment using a plane-wave basin to investigate the development of a sand spit. Two experiments were carried out: sand was deposited (1) on a shallow flat seabed and (2) on a coast with a steep slope. Their results showed that a slender sand spit extends along the marginal line between the shallow sea and offshore steep slope in Case 1, whereas a cuspate foreland is formed owing to the deposition of sand on the steep slope in Case 2, suggesting the importance of the effect caused by the difference in the depth of water where sand is deposited. We have developed a model for predicting beach changes based on Bagnold's concept (Serizawa et al., 2006) by applying the concept of the equilibrium slope introduced by Inman& Bagnold (1963) and the energetics approach of Bagnold (1963). Here, the BG model is used to simulate the extension of a sand spit on a shallow seabed and the formation of a cuspate foreland on a steep coast (Serizawa & Uda, 2011).

As another type of beach change due to waves on a coast with a shallow flat seabed, a tidal flat facing an inland sea is considered. On such a tidal flat subject to the action of waves with significant energy, a sandy beach may develop along the marginal line between the tidal flat and the land, and the sandy beach with a steep slope is clearly separated from the tidal flat along a line with a discontinuous change in the slope. On such beaches developing along the marginal line between the tidal flat and the land, longshore sand transport due to the oblique wave incidence to the shoreline and cross-shore sand transport during storm surges often occur. However, in addition to these sand transport phenomena, as part of the interaction between the tidal flat and the sandy beach, shoreward transport and the landward deposition of sand originally supplied from the offshore zone of the tidal flat, forming a slender sand bar, are often observed. Although this landward sand movement due to waves on the shallow tidal flat is considered to be part of the process by which sand transported offshore by river currents during floods returns to the shore, its mechanism has not yet been studied. These phenomena were observed on the Kutsuo coast, which has a very wide tidal flat and faces the Suo-nada Sea, part of the Seto Inland Sea, Japan. Here, the BG model was also used to predict the extension of a slender shore-normal sand bar observed on this coast (Serizawa et al., 2011). The observed phenomena were successfully explained by the results of the numerical simulation.

2. Numerical model (BG model)

With the elongation of a sand spit or a sand bar, the shape of the wave-shelter zone behind the spit or the sand bar changes, and therefore, the repeated calculation of the wave field and topographic changes is required. We use Cartesian coordinates (x, y) and consider the elevation at a point $Z(x, y, t)$ as a variable to be solved, where t is time. The beach changes are assumed to take place between the depth of closure h_c and the berm height h_R. A modified version of the BG model proposed by Serizawa et al. (2009a) was used to predict the formation of a sand spit. An additional term given by Ozasa & Brampton (1980) was also incorporated into the fundamental equation of the BG model to accurately evaluate the longshore sand transport due to the effect of the longshore gradient of the wave height. The fundamental equation is given by

$$\vec{q} = C_0 \frac{P}{\tan\beta_c} \left\{ \begin{array}{l} K_n \left(\tan\beta_c \vec{e_w} - |\cos\alpha| \ \overline{\nabla Z} \right) \\ + \left\{ (K_s - K_n)\sin\alpha - \frac{K_2}{\tan\beta}\frac{\partial H}{\partial s} \right\} \tan\beta \vec{e_s} \end{array} \right\} \tag{1}$$

$$\left(-h_c \le Z \le h_R \right)$$

$$P = \rho \, u_m^3 \tag{2}$$

$$u_m = \frac{H}{2}\sqrt{\frac{g}{h}} \tag{3}$$

Here, $\vec{q} = (q_r, q_y)$ is the net sand transport flux, Z (x, y, t) is the elevation, n and s are the local coordinates taken along the directions normal (shoreward) and parallel to the contour lines, respectively, $\overline{\nabla Z} = (\partial Z/\partial x, \partial Z/\partial y)$ is the slope vector, $\vec{e_w}$ is a unit vector in the wave direction, $\vec{e_s}$ is a unit vector parallel to the contour lines, α is the angle between the wave direction and the direction normal to the contour lines, $\tan\beta = |\overline{\nabla Z}|$ is the seabed slope, $\tan\beta_c$ is the equilibrium slope, $\tan\beta \, \vec{e_s} = (-\partial Z/\partial y, \partial Z/\partial x)$, K_s and K_n are the coefficients of longshore sand transport and cross-shore sand transport, respectively, K_2 is the coefficient of the term given by Ozasa & Brampton (1980), $\partial H/\partial s = \vec{e_s} \cdot \nabla H$ is the longshore gradient of the wave height H measured parallel to the contour lines and $\tan\overline{\beta}$ is the characteristic slope of the breaker zone. In addition, C_0 is the coefficient transforming the immersed-weight expression into a volumetric expression $(C_0 = 1/\{(\rho_s - \rho)g(1-p)\}$, where ρ is the density of seawater, ρ_s is the specific gravity of sand particles, p is the porosity of sand and g is the acceleration due to gravity), u_m is the amplitude of the seabed velocity due to the orbital motion of waves given by Eq. (3), h_c is the depth of closure, and h_R is the berm height.

The intensity of sand transport P in Eq. (1) is assumed to be proportional to the wave energy dissipation rate ϕ based on the energetics approach of Bagnold (1963). In the model of Serizawa et al. (2006), P was formulated using the wave energy at the breaking point, but in this study, it is combined with the wave characteristics at a local point. Bailard & Inman (1981) used the relationship $\phi_t = \tau u_t = \rho C_f u_t^3$ for the instantaneous wave energy dissipation rate ϕ_t to derive their sand transport equation, where τ is the bottom shear stress, u_t is the instantaneous velocity and C_f is the drag coefficient. We basically follow their study but assume that ϕ is proportional to the third power of the amplitude of the bottom oscillatory velocity u_m due to waves instead of the third power of the instantaneous velocity. The intensity of sand transport P is then given by Eq. (2), and its coefficient is assumed to be included in the coefficients of longshore and cross-shore sand transport, K_s and K_n, respectively. u_m can be calculated by small-amplitude wave theory in shallow water using the wave height H at a local point (Eq. (3)), which can be obtained by the numerical calculation of the plane-wave field. The depth of closure h_c is assumed to be proportional to the wave height H at a local point and is given by Eq. (4), referring to the relationship given by Uda & Kawano (1996).

$$h_c = KH \quad (K=2.5) \tag{4}$$

In the numerical simulation of beach changes, the sand transport equation and the continuity equation are solved on the x-y plane by the explicit finite-difference methodemploying staggered mesh scheme. In the estimation of sand transport near the berm top and the depth of closure, sand transport was linearly reduced to 0 near the berm height or the depth of closure to prevent sand from depositing in the area higher than the berm height and beach erosion in the zone deeper than the depth of closure.

The wave field was calculated using the energy balance equation given by Mase (2001), in which the directional spectrum D (f, θ) of the irregular waves varies with the energy dissipation term due to wave breaking (Dally et al., 1984). Here, f and θ are the frequency and wave direction, respectively. In this method, wave refraction, wave breaking and wave diffraction in the wave-shelter zone can be calculated with a small calculation load. The energy dissipation term due to wave breaking ϕ (Dally et al., 1984), which is incorporated into the energy balance equation (Eq. (5)), is given by

$$\frac{\partial}{\partial x}\left(DV_x\right) + \frac{\partial}{\partial y}\left(DV_y\right) + \frac{\partial}{\partial \theta}\left(DV_\theta\right) = F - \phi \tag{5}$$

$$\phi = \left(K/h\right)DC_g\left[1 - \left(\Gamma/\gamma\right)^2\right] \quad \left(\phi \geq 0\right) \tag{6}$$

Here, D is the directional spectrum, (V_x, V_y, V_θ) is the energy transport velocity in the (x, y, θ) space, F is the wave diffraction term given by Mase (2001), K is the coefficient of the wave-breaking intensity, h is the water depth, C_g is the wave group velocity ($C_g \approx \sqrt{gh}$ in the approximation in shallow-water wave theory), Γ is the ratio of the critical breaker height to the water depth on the horizontal bed and γ is the ratio of the wave height to water depth. To prevent the location where the berm develops from being excessively seaward compared with that observed in the experiment or the field, a lower limit was considered for h in Eq. (6). As a result of this procedure, wave decay near the berm top was reduced, resulting in a higher landward sand transport rate. In the calculation of the wave field on land, the imaginary depth h' between the minimum depth h_0 and berm height h_R was considered, as given by Eq. (7), similarly to in the ordinary 3D model (Shimizu et al., 1996).

$$h' = \left(\frac{h_R - Z}{h_R + h_0}\right)^r h_0 \quad \left(r=1\right) \quad \left(-h_0 \leq Z \leq h_R\right) \tag{7}$$

In addition, at locations whose elevation is higher than the berm height, the wave energy was set to 0. The calculation of the wave field was carried out every 10 steps in the calculation of topographic changes.

Equation (1) shows that the sand transport flux can be expressed as the sum of the component along the wave direction and the components due to the effect of gravity normal to the contours and the effect of longshore currents parallel to the contours. To investigate the physical meaning of Eq. (1), $\vec{q}$ in Eq. (1) is separately expressed as Eq. (8) when neglecting the additional term given by Ozasa & Brampton (1980), and when the inner products of $\vec{e_n}$ and $\vec{q}$ and of $\vec{e_s}$ and $\vec{q}$ are taken, Eqs. (9) and (10) are derived for the cross-shore and longshore components of sand transport, q_n and q_s, respectively. Furthermore, under the condition that the seabed slope is equal to the equilibrium slope, Eq. (10) reduces to Eq. (11).

$$\vec{q} = q_n \vec{e_n} + q_s \vec{e_s} \tag{8}$$

$$q_n = \vec{e_n} \cdot \vec{q} = C_0 K_n P |\cos\alpha| \left(\frac{\cos\alpha}{|\cos\alpha|} - \frac{\tan\beta}{\tan\beta_c} \right) \tag{9}$$

$$q_s = \vec{e_s} \cdot \vec{q} = C_0 K_s P \sin\alpha \left\{ \frac{\tan\beta}{\tan\beta_c} + \frac{K_n}{K_s} \left(1 - \frac{\tan\beta}{\tan\beta_c} \right) \right\} \tag{10}$$

$$q_s \approx C_0 K_s P \sin\alpha \quad (\because \tan\beta \approx \tan\beta_c) \tag{11}$$

In Eq. (9), the cross-shore sand transport q_n becomes 0 when the local seabed slope is equal to the equilibrium slope, and the longshore sand transport q_s becomes 0 when the wave direction coincides with the normal to the contour lines, as shown in Eqs. (10) and (11). When a discrepancy from these conditions arises, sand transport is generated by the same stabilization mechanism as in the contour-line-change model (Uda & Serizawa, 2010).

Taking the above into account, the first term in the parentheses in Eq. (1) gives the sand transport in the case that the rates of longshore and cross-shore sand transport are equal ($K_s = K_n$), and the second term is the additional longshore sand transport in the case that the rates are different ($K_s > K_n$). The physical meaning of the second term is that longshore sand transport is generated by the small angular shift that occurs when the wave direction is incompletely reversed in the oscillatory movement due to waves, and the second term also models the additional longshore sand transport due to the effect of longshore currents, the effect of which is only partially included in the first term.

Although the applicability of the contour-line-change model to the prediction of beach changes is limited when the shape of coastal structures is complicated because it tracks the movement of lines with specific characteristics, the BG model can be applied to the prediction of topographic changes under all structural conditions, because the depth changes in the x-y plane are calculated, similarly to in the ordinary model for predicting 3D beach changes, and therefore the calculation can be carried out systematically. This is an advantage of the BG model.

3. Movable-bed experiment on elongation of a sand spit

A movable-bed experiment was carried out using a plane-wave tank of 16 m width and 21 m length (Uda & Yamamoto, 1992). A model beach was made of sand with d_{50} = 0.28 mm. A sandy beach was established as the source of sand in the right half of the plane basin and conditions were set up such that leftward longshore sand transport developed. In Case 1, a shallow seabed with a water depth of 5 cm was formed in the left half of the wave basin and an offshore bed was formed with a steep slope of 1/5. In Case 2, a steep slope of 1/5 was produced instead of a shallow sea where the sand spit was formed. The angle between the direction normal to initial shoreline and the wave direction was 20° in order for sufficient longshore sand transport to occur. The elevation of the flat surface on the land was assumed to be 10 cm above mean sea level. Regular waves with H_0' = 4.6 cm and T = 1.27 s incident to the model beach were generated for 8 hr. When a shallow sea exists, incident waves break immediately offshore of the shallow seabed, resulting in the rapid decay of waves on the shallow seabed. Because of this effect, sand is deposited near the marginal line between the shallow flat seabed and the steep offshore slope, resulting in the rapid elongation of a sand spit.

Figures 1(a)-1(c) show the initial bathymetry and the beach topography after wave generation for 1 and 8 hr in Case 1, respectively. Here, the arrows in Figs. 1(a) and 1(d) show the breaking point (the tip of the arrows), the breaker height (the length of the arrows), and the wave directionat the breaking point (the direction of the arrows) measured immediately after the start of wave generation. Because the shoreline had a discontinuity due to a sudden change in the shoreline direction between the sand supply zone and the shallow seabedwhere sand was deposited, a straight sand spit extended from the boundary, and a slender sand spit extended along the marginal line between the shallow seabed and the steepoffshore slope over time. After 8 hr, the sand spit had reached the left boundary while forming a barrier island, and the width of the barrier island expanded upcoast from the left boundary because of the continuous sand supply.

Figures 1(d) and 1(e) show the initial bathymetry and the beach topography after wave generation for 8 hr in Case 2 with a steep offshore slope. The water depth in the zone where sand was deposited was large; thus, sand was deposited while forming a steep slope.

Because this steep slope reaches a great depth, a cuspate foreland was formed without the development of a sand spit. These experimental results were used for validating the improved BG model.

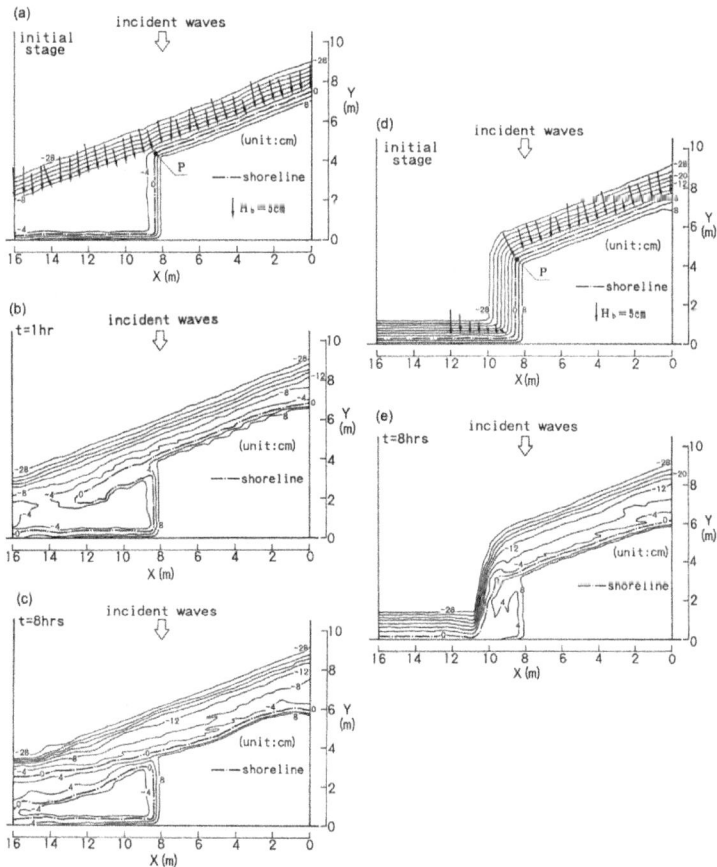

Figure 1. Experimental results for development of sand spit on a coast with abrupt change in coastline orientation (Uda & Yamamoto, 1992).

4. Field observation of formation of slender shore-normal sand bar

4.1. General conditions

The study area is the Kutsuo coast facing the Suo-nada Sea, part of the Seto Inland Sea, as shown in Fig. 2. Figure 3 shows an aerial photograph of the study area taken in 1999. The Harai River flows into the coast, which has a very wide tidal flat of approximately 1.5 km width offshore of the river. Although a river mouth bar extends on the north side of the Harai River, another slender sand bar has developed along the north side of the channel extending between the river mouth and the offshore tidal mud flat, and it intersects the river mouth bar perpendicular to the shoreline. The sand source for this slender sand bar is assumed to be the Harai River; sand transported offshore by flood currents is deposited along both sides of the channel, and is then transported shoreward owing to the wave action.In this area, the slender sand bar as shown in Fig. 3 has been continuously developing. Figure 4 shows the bathymetry around the slender sand bar relative to the reference level (0.1m below mean sealevel) in 2008. Although the slender sand bar has moved slightly north compared with its position in Fig. 3, it extends in the cross-shore direction at $X = 120$ m in the central part of the river mouth bar, and the shoreline slightly protrudes near the connection point. The foreshore has been developing in the zone with elevation between 1.0 and 3.0 m. This foreshore is composed of coarse and medium-size sand, and its slope is as steep as 1/10. A tidal flat extends in the offshore zone, the elevation of which is lower than 1.0 m. In addition, there is a difference in the elevation of the tidal flat on both sides of the slender sand bar extending in the cross-shore direction with the ground elevation on the south side slightly higher than that on the north side.

Figure 2. Location of Kutsuo coast facing Suo-nada Sea, part of Seto Inland Sea.

Figure 3. Aerial photograph of Kutsuo coast.

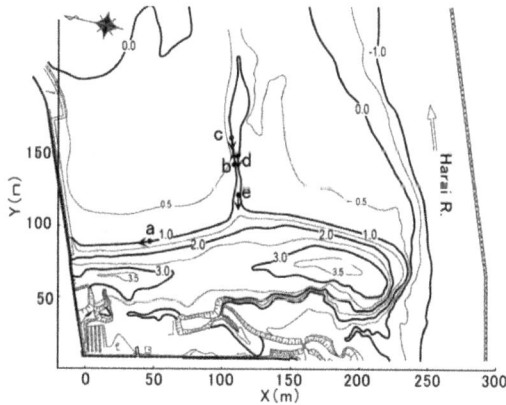

Figure 4. Bathymetry around slender sand bar on tidal flat offshore of Kutsuo coast.

4.2. Field observation

During low tide on December 27, 2009, a field observation of the tidal flat offshore of the coast was carried out. The north end of the coast is separated by a vertical seawall protecting a park, as shown in Fig. 5. A sandy beach abruptly begins with a steep slope from the mud flat covered by cohesive materials, and the mud flat and sandy beach are clearly separated along a line where the slope changes abruptly. The tidal flat is composed of cohesive materials, whereas the sandy beach is composed of coarse sand and is well compacted.

Figure 6 shows a view of the entire slender sand bar. Although the sand bar is submerged during high tide, it is completely exposed during low tide, as shown in Fig. 6. The sand bar extends in the direction normal to the mean coastline. Comparing the elevations of the mud flat on both sides of the sand bar, the elevation on the right (south) side, which is next to theoffshore channel, is higher than that on the left (north) side. On the south side of the slender sand bar, the Harai River flows into the sea, and sand supplied to the tidal flat from the river mouth is considered to be transported and deposited on the surface of the tidal flat owing to the shoreward sand transport due to waves. In this case, sand supply to the area north of the slender sand bar is obstructed by the sand bar itself, and this is assumed to be the cause of the difference in ground elevation on both sides of the sand bar.

Figure 7 shows the tip of a branch separated from the offshore part of the main slender sand bar, as shown in Fig. 3. On this sand bar, decomposed granite, which was considered to be transported offshore by flood currents of the river, has been deposited. The elevation of the sand bar gradually increases shoreward, then at the landward end suddenly drops to the tidal flat with a steep slope approximately equal to the angle of repose of the sand, implying the occurrence of shoreward sand transport.

Figure 8 shows the coastal conditions, looking landward from the tip of the slender sand bar. Although sand was deposited with a foreshore slope of 1/10 along the north side of the slender sand bar, the foreshore and flat tidal flat were clearly separated along the abrupt

Figure 5. Beach separated by vertical seawall (December 27, 2009).

Figure 6. View of entire slender sand bar (December 27, 2009).

change in the slope passing through point B. By a sieve analysis of the beach material sampled at point A, the median diameter of the beach material was determined to be d_{50} = 1.50 mm at point A. Figure 9 shows the narrow neck of the slender sand bar connected to the land. Several lines showing high tide marks extend in the cross-shore direction on the surface of the sand bar, implying that the sand bar stably exists during tidal changes. Figure 10 shows the beach condition near the point connecting the land and the slender sand bar. The triangular high tide lines show that the contour lines are parallel to these high tide lines. Therefore, if waves are incident from the direction normal to the coastline, large shoreward sand transport may occur along these contour lines because of the large incident wave angle. However, the slender sand bar is stable without rapid beach changes. Taking into account the continuity condition of sand and the fact that the slender sand bar is stable, the materials forming the sand bar are considered to have been carried from the Harai River during floods. The observation results indicate that the sand transported offshore by flood currents returns to the beach owing to shoreward sand transport due to waves. On an exposed beach, the formation of a stable sand bar extending normal to the coastline, as observed on this coast, is difficult, and such a sand bar rapidly deforms under wave action. Taking this into account, the formation of a slender sand bar observed in this study is

BG Model Based on Bagnold's Concept and Its Application to Analysis of Elongation of
Sand Spit and Shore – Normal Sand Bar

13

considered to be due to the typical sand movement observed on only a very shallow tidal flat.

Figure 7. Tip of branch separated from offshore part of main slender sand bar (December 27, 2009).

Figure 8. Coastal condition, while looking landward from tip of slender sand bar (December 27, 2009).

Figure 9. Narrow neck of slender sand bar connected to land (December 27, 2009).

Figure 10. Beach condition near point connecting land and slender sand bar (December 27, 2009).

5. Application to movable-bed experiment

At locations with elevations higher than the berm height, the wave energy was set to 0. For convenience, the space scale in the calculation was set to 100-fold that in the experiment, and then the calculated results were reduced 100-fold. Although the movable-bed experiment was carried out under regular wave conditions, the wave field was calculated using the energy balance equation for irregular waves while regarding regular waves in the experiment as irregular waves, because repeated calculations were necessary owing to the bathymetric changes in this calculation.

5.1. Formation of barrier island on flat shallow seabed

Given the same initial topography and wave conditions as those in the experiment (regular waves with $H_0' = 4.6$ cm and $T = 1.27$ s obliquely incident to the model beach with an angle of 20°), the beach changes after 8 hr were predicted. The depth of closure was given by $h_c = 2.5H$ (H: wave height at a local point). The berm height was assumed to be 5 cm, and the angles of the equilibrium slope and repose slope were set as 1/5 and 1/2, respectively, on the basis of the experimental results. The calculation domain was divided by $\Delta x = \Delta y = 20$ cm intervals in the cross-shore and longshore directions, respectively, and the calculation for up to 8 hr (8×10^4 steps) was carried out using time intervals of $\Delta t = 1 \times 10^{-3}$ hr. Table 1 shows the calculation conditions.

5.1.1. Bathymetric changes

Figures 11(a)-11(f) show the results for the predicted development of a sand spit on a shallow flat seabed given the same conditions as those in the experiment. A slender sand spit with a length of approximately 2 m was formed until 0.5 hr because of the deposition of sand supplied from upcoast along the marginal line between the flat shallow seabed and the steep offshore bottom, as shown in Fig. 11(b). Rapid shoreward sand transport also occurred owing to the restoration effect of the beach slope corresponding to the deviation from the

BG Model Based on Bagnold's Concept and Its Application to Analysis of Elongation of
Sand Spit and Shore – Normal Sand Bar

15

Wave conditions	Incident waves: H_I= 4.6 m (4.6 cm), T = 12.7 s (1.27 s), wave direction θ=20° relative to normal to initial shoreline
Berm height	h_R= 5 m (5 cm)
Depth of closure	h_c = 2.5H (H: wave height)
Equilibrium slope	$\tan\beta_c$=1/5
Angle of repose slope	$\tan\beta_x$=1/2
Coefficients of sand transport	Coefficient of longshore sand transport K_s=0.045 Coefficient of Ozasa & Brampton (1980) term K_2 = 1.62K_s Coefficient of cross-shore sand transport K_n = 0.1K_s
Mesh size	$\Delta x = \Delta y$ = 20 m
Time intervals	Δt = 0.001 hr (0.01 hr)
Duration of calculation	80 hr (8×10⁴ steps) (8 hr)
Boundary conditions	Shoreward and landward ends: q_x = 0, left and right boundaries: q_y = 0
Calculation of wave field	Energy balance equation (Mase, 2001) • Term of wave dissipation due to wave breaking: Dally et al. (1984) model • Wave spectrum of incident waves: directional wave spectrum density obtained by Goda (1985) • Total number of frequency components N_F = 1 and number of directional subdivisions N_θ= 8 • Directional spreading parameter S_{max} = 75 • Coefficient of wave breaking K = 0.17 and Γ= 0.3 • Imaginary depth between minimum depth h_0 and berm height h_R : h_0= 2 m (2 cm) • Wave energy = 0 where $Z \geq h_R$ • Lower limit of h in terms of wave decay due to breaking: 0.7 m (0.7 cm)
Remarks	Numbers in parentheses show experimental values. Space and time scales in the calculation are 100- and 10-fold those in the experiment, respectively.

Table 1. Calculation conditions (numbers in parentheses: experimental conditions).

equilibrium slope, because the seabed had an abrupt change in the slope along this marginal line along which sand was deposited, whereas the intervals between the contours became large in the offshore zone shallower than h_c. The sand spit further extended along the marginal line with increasing time, and the length of the spit reached 3.5 m after 1 hr, as shown in Fig. 11(c). After 2 hr, the tip of the spit was connected to the left boundary and a barrier island had formed, enclosing a lagoon behind the barrier island (Fig. 11(d)). Although a slender, straight sand spit extended along the marginal line until 2 hr after the start of wave generation, sand started to be deposited upcoast of the left boundary after 4 hr because of the blockage of longshore sand transport by the left boundary. Offshore sand transport in the deep zone also occurred, as shown in Fig. 11(e). During this process, the shoreline advanced and the width of the barrier island formed by the extension of the sand spit gradually increased from the left boundary. After 8 hr, the effect of the blockage of longshore sand transport by the left boundary had reached the upcoast and the width of the barrier island had also increased at $X = 9$ m, where the sand spit first developed, as shown in Fig. 11(f).

We compared the experimental results after 1 hr with the calculated results, as shown in Figs. 1(b) and 11(c), respectively. Both sets of results indicated that a sand spit extended from the location with a sudden change in the coastline orientation along the marginal line on the flat shallow seabed and were in good agreement. However, there was some discrepancy in the location of the tip of the sand spit. Similarly, both experimental and calculated results after 8 hr, shown in Figs. 1(c) and 11(f), respectively, are in good agreement in that the width of the barrier island was increased by the blockage of longshore sand transport by the left boundary and that a gentle slope was formed at a depth of approximately 8 cm owing to erosion along with the formation of a scarp near the right boundary. With regard to the experimental results for the extension of the sand spit reported by Uda & Yamamoto (1992), Watanabe et al. (2004) successfully predicted the shoreline changes related to the extension of the sand spit using a one-line model with a curvilinear coordinate system. However, in the present study, we were able to predict the development of a barrier island after the extension of the sand spit.

Figures 12(a)-12(f) show bird's-eye view of the extension of the barrier island in Case 1, looking upcoast from above the downcoast. Although a simple sand spit extended from the boundary between the sand supply and accretion zones on the flat shallow seabed, sand had already been transported shoreward, forming a subsurface sand bar along the marginal line between the steep offshore slope and flat shallow seabed, until 0.5 hr before the extension of the sand spit, implying the generation of rapid shoreward sand transport at the abrupt change in the slope. Furthermore, sand was deposited over the steep offshore slope between 4 and 8 hr after the start of wave generation and an extremely steep slope was formed near the downcoast boundary. In contrast, a wave-cut gentle slope was formed offshore of the erosion zone located upcoast.

5.1.2. Changes in wave field

Significant changes in the wave field occurred on the shallow flat seabed with the extension of the barrier island, as shown in Fig. 13. Initially, the wave height was reduced by up to approximately 1.5 cm because of wave breaking along the marginal line of the shallow flat seabed. However, the extension of the sand spit owing to the shoreward sand transport was very rapid, and the wave height was markedly reduced on the shallow flat seabed after wave generation for 1 hr. After 8 hr, a calm wave zone extended in the entire area behind the barrier because of the rapid development of the barrier, and the wave height had a uniform distribution.

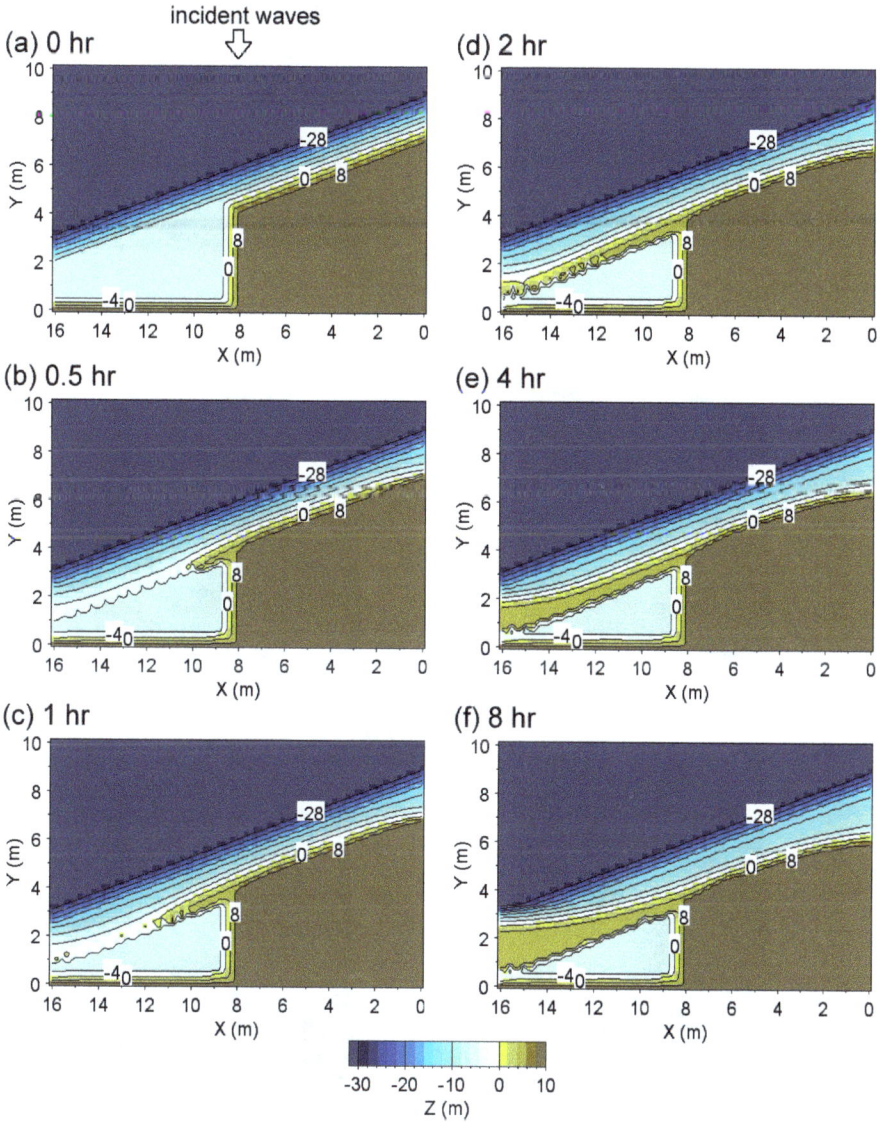

Figure 11. Results for predicted development of sand spit on a coast with abrupt change in coastline orientation (Case 1: flat shallow seabed).

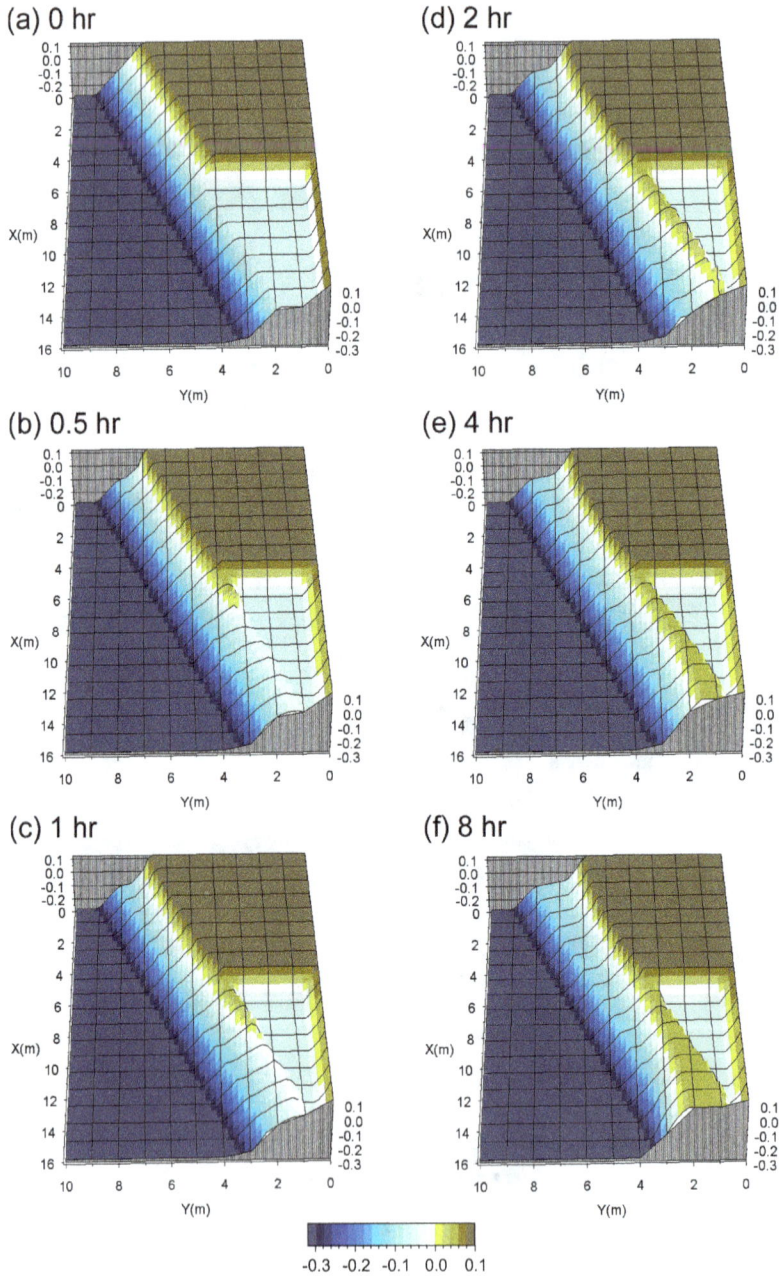

Figure 12. Bird's-eye view of topographic changes (Case 1: flat shallow seabed).

BG Model Based on Bagnold's Concept and Its Application to Analysis of Elongation of
Sand Spit and Shore – Normal Sand Bar

19

5.1.3. Sand transport flux

Figure 14 shows the sand transport flux 0, 1 and 8 hr after the start of wave generation. Although the initial sand transport flux was large in the area to the right of $X = 8$ m, where the sand source was located, the area with a large sand transport flux had moved left with the extension of the sand spit after 1 hr. In contrast, the sand transport flux decreased in magnitude near the right boundary, because the angle between the direction normal to the contour lines and the direction of incident waves was reduced. After 8 hr, the sand spit had reached the left boundary and the sand transport flux had significantly decreased, and the area with a large sand transport flux had become small.

5.1.4. Changes in longitudinal profiles

Figures 15(a)-15(d) show the experimental and predicted changes in longitudinal profiles along transect $X = 0$ m located at the right boundary, and transects $X = 9$, 12 and 14 m crossing the flat shallow seabed, respectively. Along transect $X = 0$ m, although the experimental and predicted results, which indicated that the depth of closure was -12 cm, are in agreement, as shown in Fig. 15(a), the eroded volume in the calculation was overestimated in the nearshore zone, where there was less scarp erosion. However, the sand budget in the cross section was approximately maintained and the parallel recession of the cross section while maintaining a constant slope was accurately predicted in the calculation. Along transect $X = 9$ m, the development of a berm of 5 cm height after 1 hr, as shown in Fig. 15(b), was observed in both the experiment and simulation, but the location of the berm was slightly seaward in the simulation. After 8 hr, however, the berm location had moved landward and a stable barrier island had formed. These experimental and calculated changes are in good agreement. Furthermore, no beach changes occurred on the flat shallow seabed because the elongation of the sand spit was too rapid to permit wave intrusion into the flat seabed. Along transect $X = 12$ m, the elongation of the sand spit was limited until 1 hr, as shown in Fig. 15(c), and there was little development of the berm. However, a substantial berm had developed after 8 hr. The experimental and calculated results are in good agreement regarding these points. Along transect $X = 14$ m near the left boundary, although a sand bar did not develop until 1 hr, a large amount of sand had been deposited after 8 hr, forming a barrier island with a 1.2 m width, as shown in Fig. 15(d). In this case, the seabed slope gradually steepened because of the continuous deposition of sand along the offshore steep slope, resulting in the deposition of sand up to a depth of -23 cm, which is approximately two fold larger than the depth of closure of -12 cm.

5.2. Formation of cuspate foreland on steep coast

5.2.1. Bathymetric changes

Figures 16(a)-16(f) show the results for the calculation of the development of a cuspate foreland on a steep coast $t = 0$, 0.5, 1, 2, 4, and 8 hr after the start of wave generation given the same conditions as those in the experiment. The contour lines that extended parallel to

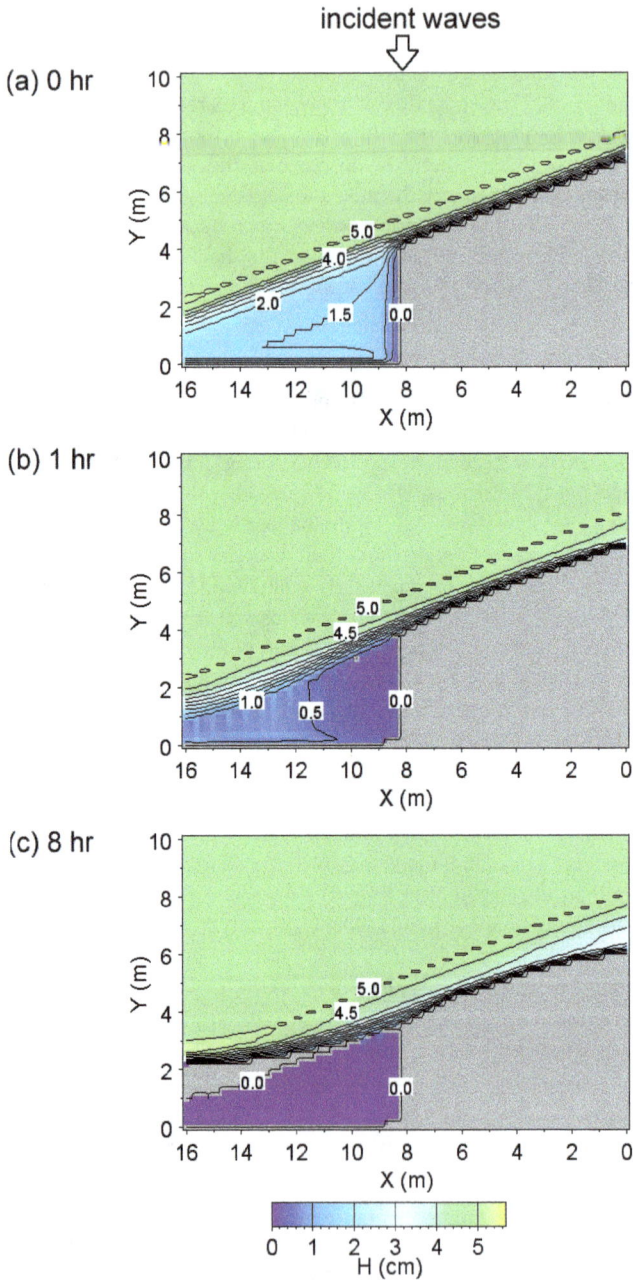

Figure 13. Changes in wave height (Case 1: flat shallow seabed).

BG Model Based on Bagnold's Concept and Its Application to Analysis of Elongation of
Sand Spit and Shore – Normal Sand Bar

21

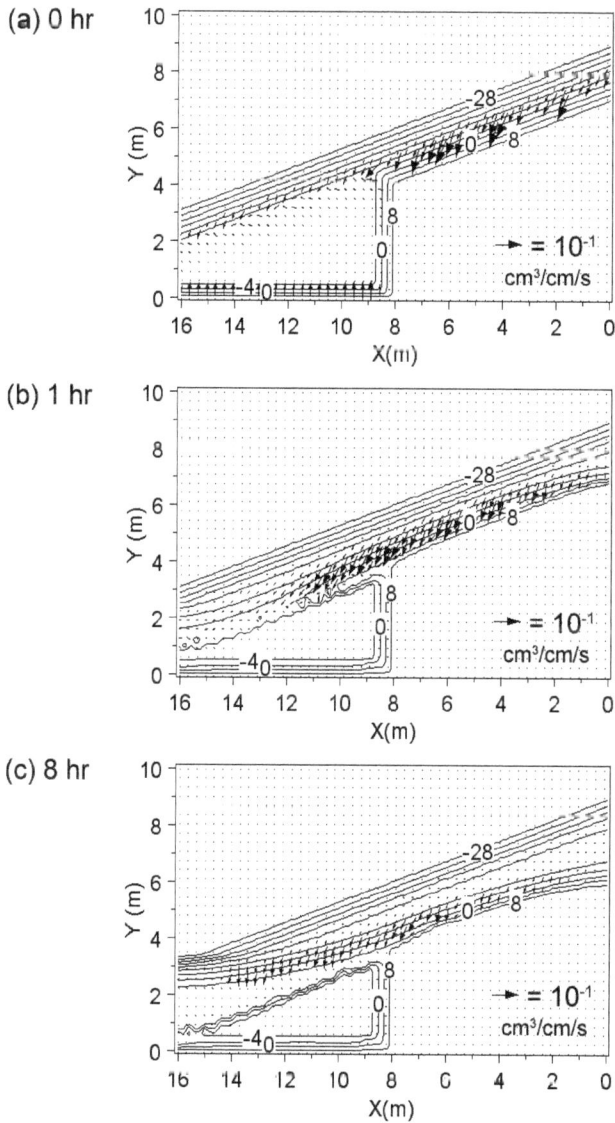

Figure 14. Sand transport flux (Case 1: flat shallow seabed).

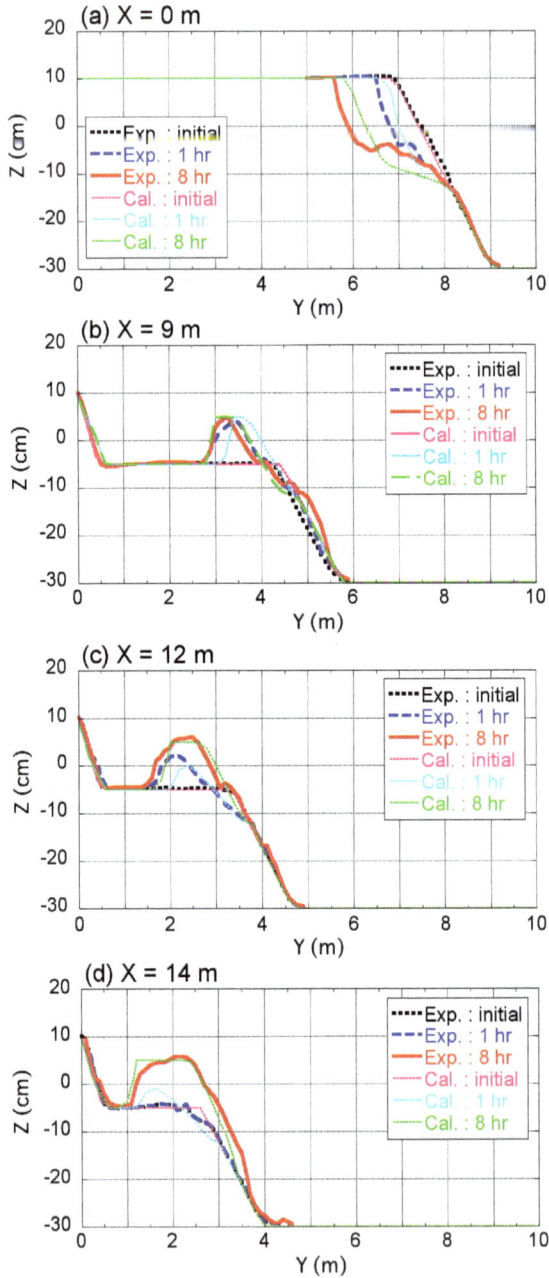

Figure 15. Changes in longitudinal profiles (Case 1: flat shallow seabed).

each other at the initial stage around the location with an abrupt change in the coastline orientation rapidly changed over time, causing sand deposition in the deep zone and the formation of a cuspate foreland after 0.5 hr. After 1 hr, the shoreline protruded further and a neck in the contour had formed downcoast of the sand deposition zone of the cuspate foreland. The size of this neck increased with time, and a sand spit that enclosed a shallow sea inside the neck had formed after 2 hr. This morphology is very similar to that of Miho Peninsula, formed by the extension of a sand spit, and Miho Bay in Shizuoka Prefecture (Uda & Yamamoto, 1994). Because of the continuous sand supply from upcoast, the sand spit elongated and the tip became connected to the downcoast shoreline. As a result, the shallow sea located inside the neck had formed a pond after 4 hr. After 8 hr, a steep slope had formed along the shoreline of the cuspate foreland owing to the continuous sand deposition in the deeper zone, whereas a wave base with a gentle slope had formed on the coast from which sand was supplied.

Figures 17(a)-17(f) show bird's-eye view of the development of a cuspate foreland on a steep coast. Although sand transported from upcoast was deposited on the steep seabed, the deposition of sand started near the location with an abrupt change in the coastline orientation, forming a neck in the contours, as shown in Figs. 17(a) and 17(b). This neck moved landward with time, and finally a sandy beach with a flat surface and a steep slope offshore of the shoreline had formed after 8 hr.

5.2.2. Changes in wave field

Figures 18(a)-18(c) show the distributions of the calculated wave height after wave generation for 0, 1 and 8 hr. At the initial stage, the longshore change in the wave height is large near the location with an abrupt change in the coastline orientation. Although a semicircular cuspate foreland developed until 1 hr, a marked decay in the wave height occurred over the short distance along the protruding shoreline, because the wave height was extremely low behind the protruded shoreline. The change in longshore sand transport due to this decay in wave height was successfully taken into account by the additional term given by Ozasa & Brampton (1980). The spatial change in longshore sand transport is large in this area because of the abrupt decrease in wave height, meaning that sand was rapidly deposited. However, after 8 hr, the area where the wave height abruptly decreased had disappeared and the wave height was smoothly distributed alongshore.

5.2.3. Sand transport flux

Figure 19 shows the sand transport flux after wave generation for 0, 1 and 8 hr. Although the sand transport flux was initially large to the right of $X = 8$ m, which is the sand supply area, as in Case 1, after 1 hr the area with a large sand transport flux had moved leftward with the elongation of the sand spit. Moreover, the sand transport flux was reduced because the angle between the direction normal to the contour lines and the direction of incident waves decreased near the right boundary. After 8 hr, the sand spit had reached the left boundary, and the absolute value of sand transport flux decreased because the angle between the direction normal to the contour lines and the wave incidence direction had been reduced near both boundaries.

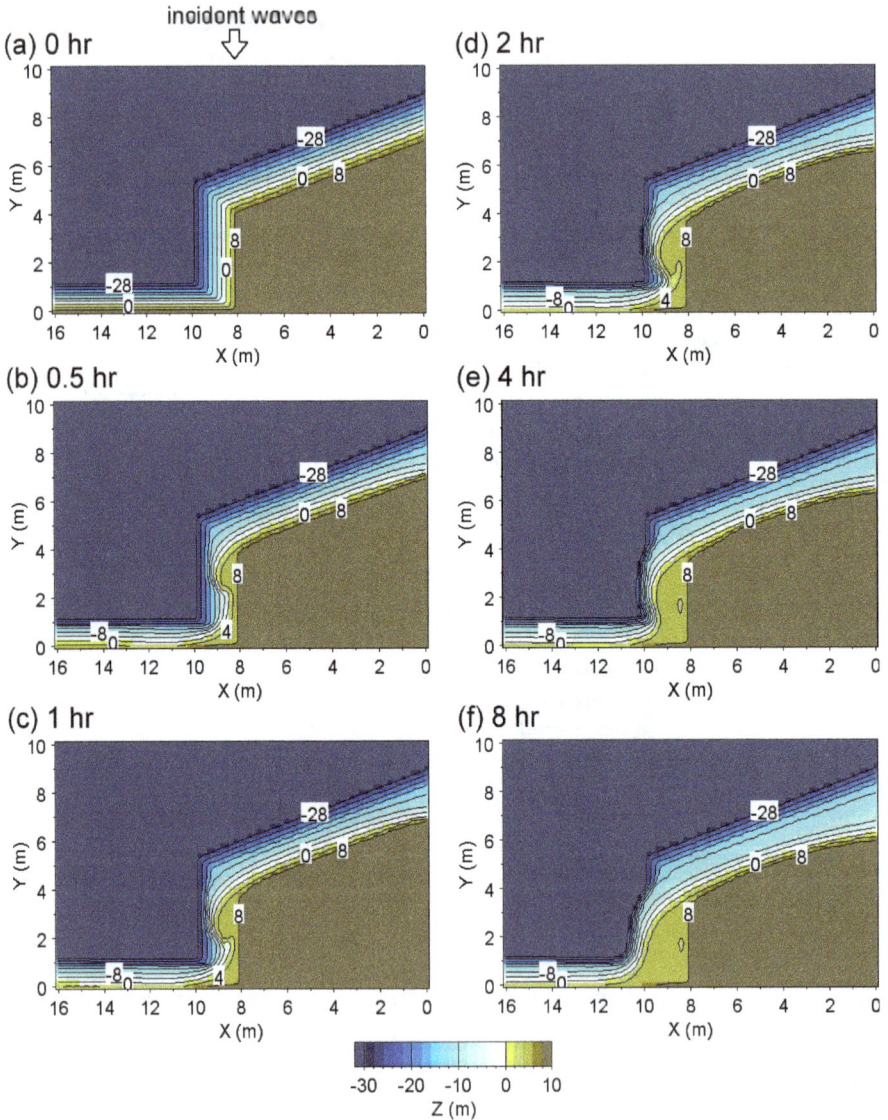

Figure 16. Predicted results for development of cuspate foreland on a coast with abrupt change in coastline orientation (Case 2: steep slope and deep seabed).

BG Model Based on Bagnold's Concept and Its Application to Analysis of Elongation of
Sand Spit and Shore – Normal Sand Bar

25

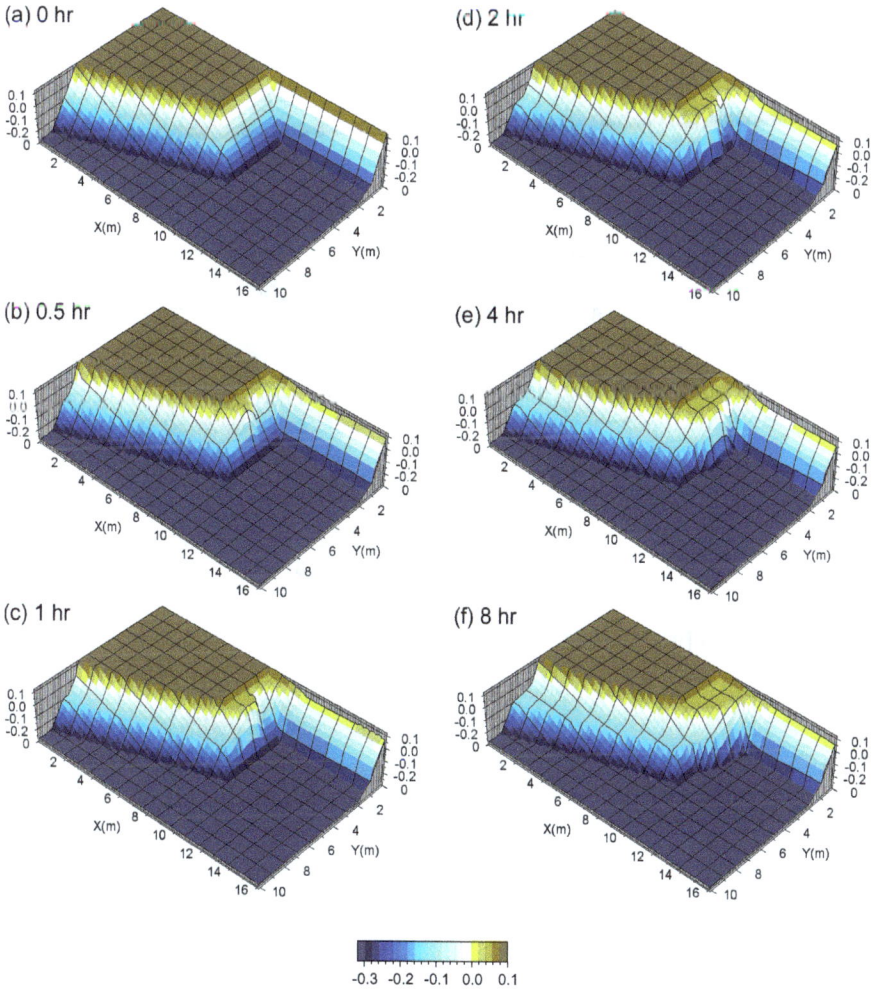

Figure 17. Bird's-eye view of topographic changes (Case 2: steep slope and deep seabed).

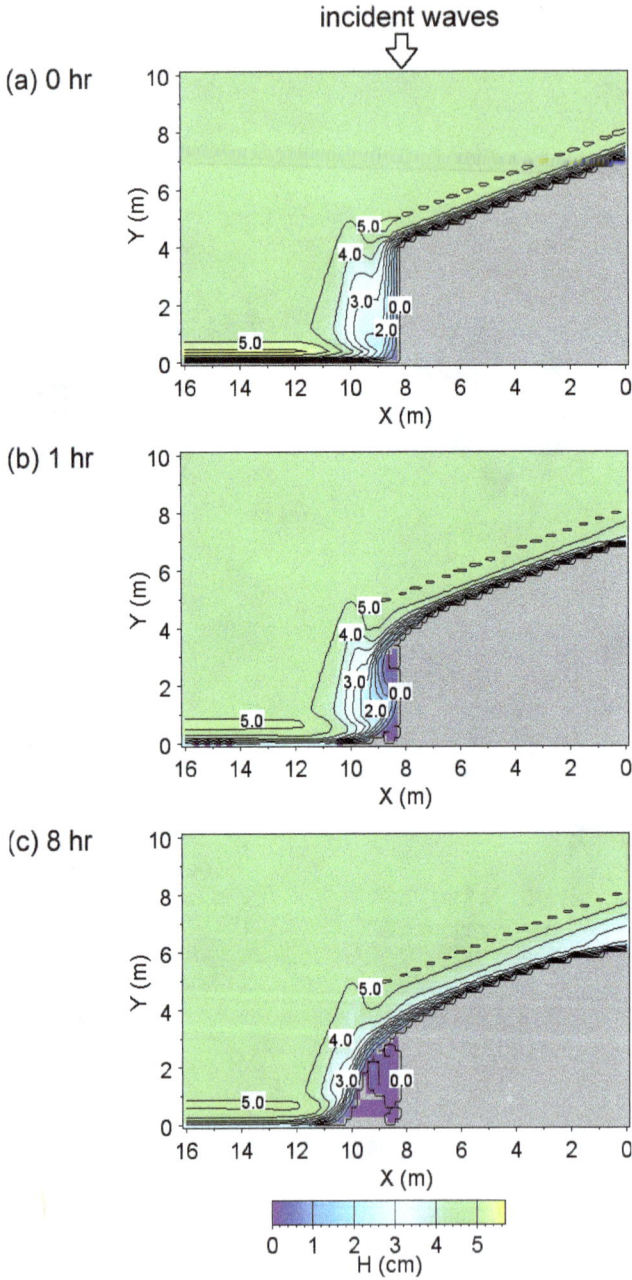

Figure 18. Changes in wave height (Case 2: steep slope and deep seabed).

BG Model Based on Bagnold's Concept and Its Application to Analysis of Elongation of
Sand Spit and Shore – Normal Sand Bar

27

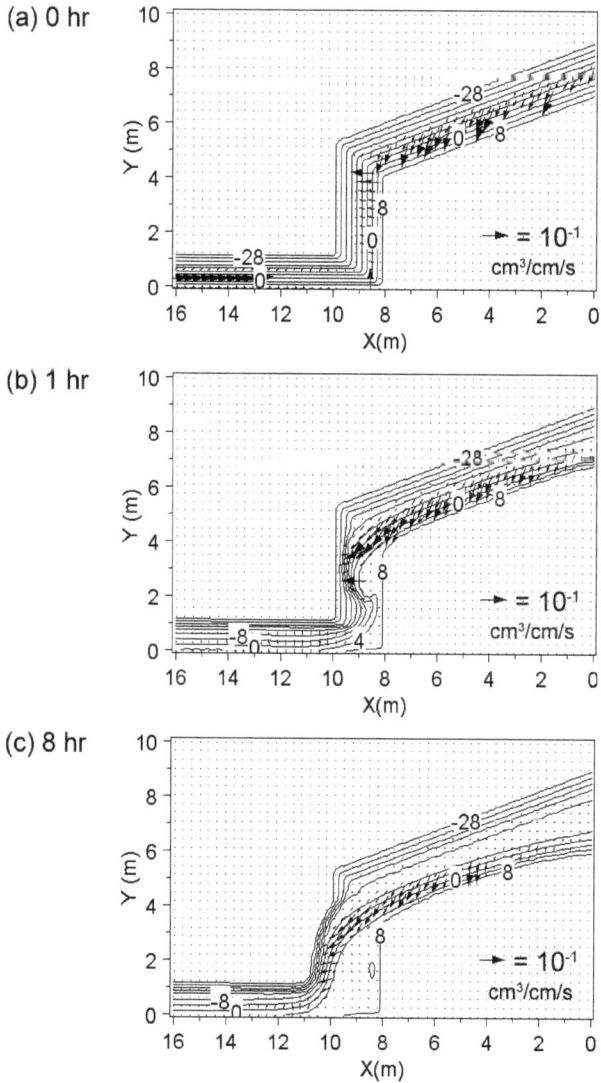

Figure 19. Sand transport flux (Case 2: steep slope and deep seabed).

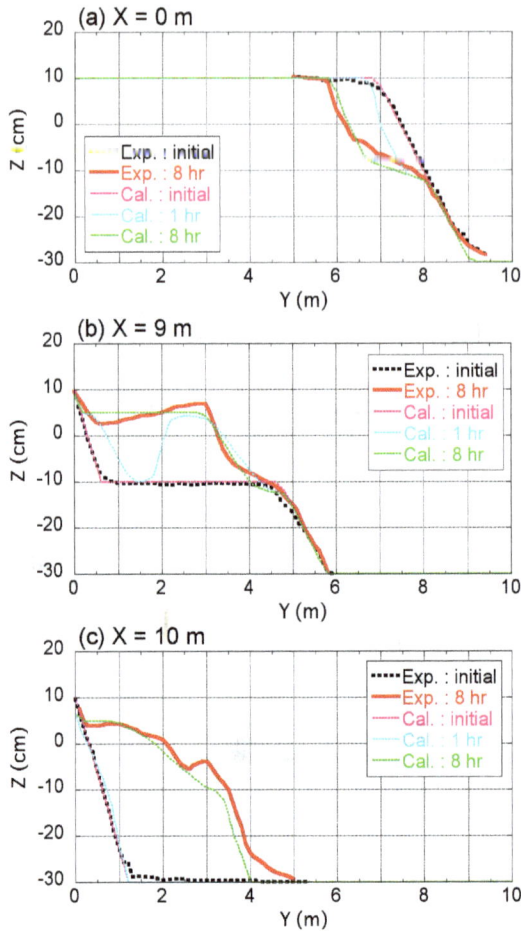

Figure 20. Changes in longitudinal profiles (Case 2: steep slope and deep seabed).

5.2.4. Changes in longitudinal profiles

Figures 20(a)-20(c) show the experimental and predicted changes in longitudinal profiles along transect $X = 0$ m located at the right boundary, transect $X = 9$ m near the location where the coastline orientation abruptly changes, and transect $X = 10$ m, respectively. Along transect $X = 0$ m, the parallel recession of the cross section is accurately reproduced in the calculation, as shown in Fig. 20(a). Along transect $X = 9$ m, a slope that slightly inclined landward had formed in the experiment after 8 hr, whereas a flat surface was predicted in the calculation. With the exception of these points, the experimental and predicted results are in good agreement, as shown in Fig. 20(b). The experimental and calculated results are also in good agreement along transect $X = 10$ m, as shown in Fig. 20(c).

6. Simulation of formation of slender sand bar on Kutsuo coast

To model the accumulation of sand on the tidal flat due to wave action and the formation of a slender sand bar, a point source in a single mesh (5 m×5 m) was assumed. The intensity of the point source was determined by trial and error, and it was set to 3.75×10^4 m³/yr at the point (x, y) = (200 m, 0 m). In addition, an upper limit of 0.5 m was assumed as the elevation of the sandy island. When the elevation of the island reached this height during the calculation, the wave energy was set to 0.

Given the simplified initial topography and the conditions of the annual energy-mean waves of the Kutsuo coast (significant wave height H_i of 0.4 m and wave period T of 4 s), and assuming wave incidence from the direction normal to the coastline, the beach changes were predicted. The depth of closure was given by h_c = 2.5H (H: wave height at a local point). The water depth of the initial bottom of the tidal flat was assumed to be 2 m. This initial flat bottom was assumed to be a solid bed, and a sandy beach with a slope of 1/10 was set at the landward end of the flat bottom. The berm height was assumed to be 0.5 m, and angles of the equilibrium slope and repose slope were set as 1/10 and 1/2, respectively. The calculation domain was divided by $\Delta x = \Delta y$ = 5 m intervals in the cross-shore and longshore directions, respectively, and a calculation for up to 5000 hr (5×10^4 steps) was carried out using time intervals of Δt = 0.1 hr. Table 2 shows the calculation conditions.

6.1. Bathymetric changes

Figure 21 shows the calculation results obtained after every 10^4 steps. Initially, only a flat bottom extended offshore of the sandy beach with a straight shoreline. The solid circle in Fig. 21(a) shows the location of the sand source, and a sandy beach with a slope of 1/10 extended along the marginal line between the tidal flat and the land, as observed on the Kutsuo coast. Owing to the wave action under these conditions, a slender submerged sand bar started to form after 10^4 steps, as shown in Fig. 21(b), as a result of the deposition of sand supplied from the sand source. The landward end of the slender sand bar was sharp and similar to a comet tail formed on the lee of an island. On the other side of the slender sand bar, longshore sand transport toward the lee of the slender sand bar was induced from the nearby coast, resulting in the formation of a cuspate foreland because of the wave-sheltering effect of the sand bar. After 2×10^4 steps, the submerged sand bar had become a sandy island because of its continuous development (Fig. 21(c)). After 2×10^4 steps, the cuspate foreland behind the sandy island was more developed than that after 10^4 steps. The beach width in the zone between x = 50 and 100 m was very small and a neck was formed. After 3×10^4 steps, the widths of the sandy island and the neck behind the island had increased, and sand that had originally been supplied from the offshore point source had reached the beach, resulting in the connection of the sandy island to the beach (Fig. 21(d)). Finally, the island developed a spoonlike shape with the shoreline of a tombolo connected to the slender island.

The development of the sandy island continued up to 5×10^4 steps, and the widths of the sandy island and the neck between the island and the tombolo continued to increase (Figs. 21(e) and 21(f)). The numerical results for the development of a slender sand bar and the resultant sandy island successfully explain the results observed on the Kutsuo coast, as shown in Fig. 3, and the fact that the slender sand bar has a neck near the landward end, as shown in Figs. 8-10.

Figures 22(a)-22(f) show bird's-eye view of the development of a slender sand bar developed on a flat seabed. Sand supplied from a sand source at $(x, y) = (200 \text{ m}, 0 \text{ m})$ was deposited to form an island with the gradual shoreward movement of sand. After 2×10^4 steps, a slender island connected to the land extended with the formation of a tombolo because of the wave-sheltering effect of the island itself. Because of the continuous supply of sand, the width of the island increased and the scale of the tombolo increased with increasing number of steps. The final configuration of the slender bar was very similar to that observed on the Kutsuo coast, as shown in Fig. 9.

Wave conditions	Incident waves: $H_I = 0.4$ m, $T = 3$ s, wave direction $\theta = 0°$, normal to initial shoreline
Tide condition	H.W.L. = +2.0 m
Berm height	$h_R = 0.5$ m
Depth of closure	$h_c = 2.5H$ (H: wave height)
Equilibrium slope	$\tan\beta_c = 1/10$
Angle of repose slope	$\tan\beta_g = 1/2$
Coefficients of sand transport	Coefficient of longshore sand transport $K_s = 0.05$ Coefficient of Ozasa & Brampton (1980) term $K_2 = 1.62 \, K_s$ Coefficient of cross-shore sand transport $K_n = 0.2 \, K_s$
Mesh size	$\Delta x = \Delta y = 5$ m
Time intervals	$\Delta t = 0.1$ hr
Duration of calculation	5000 hr (5×10^4 steps)
Boundary conditions	Shoreward and landward ends: $q_x = 0$, left and right boundaries: $q_y = 0$
Calculation of wave field	Energy balance equation (Mase, 2001) • Term of wave dissipation due to wave breaking: Dally et al. (1984) model • Wave spectrum of incident waves: directional wave spectrum density obtained by Goda (1985) • Total number of frequency components $N_F = 1$ and number of directional subdivisions $N_\theta = 8$ • Directional spreading parameter $S_{max} = 10$ • Coefficient of wave breaking $K = 0.17$ and $\Gamma = 0.3$ • Imaginary depth between minimum depth h_0 and berm height $h_R:h_0 = 0.5$ m • Wave energy = 0 where $Z \geq h_R$ • Lower limit of h in term of wave decay due to breaking: 0.2 m
Remark	Point source in single mesh (3.75×10^4 m³/yr)

Table 2. Calculation conditions.

BG Model Based on Bagnold's Concept and Its Application to Analysis of Elongation of
Sand Spit and Shore – Normal Sand Bar

31

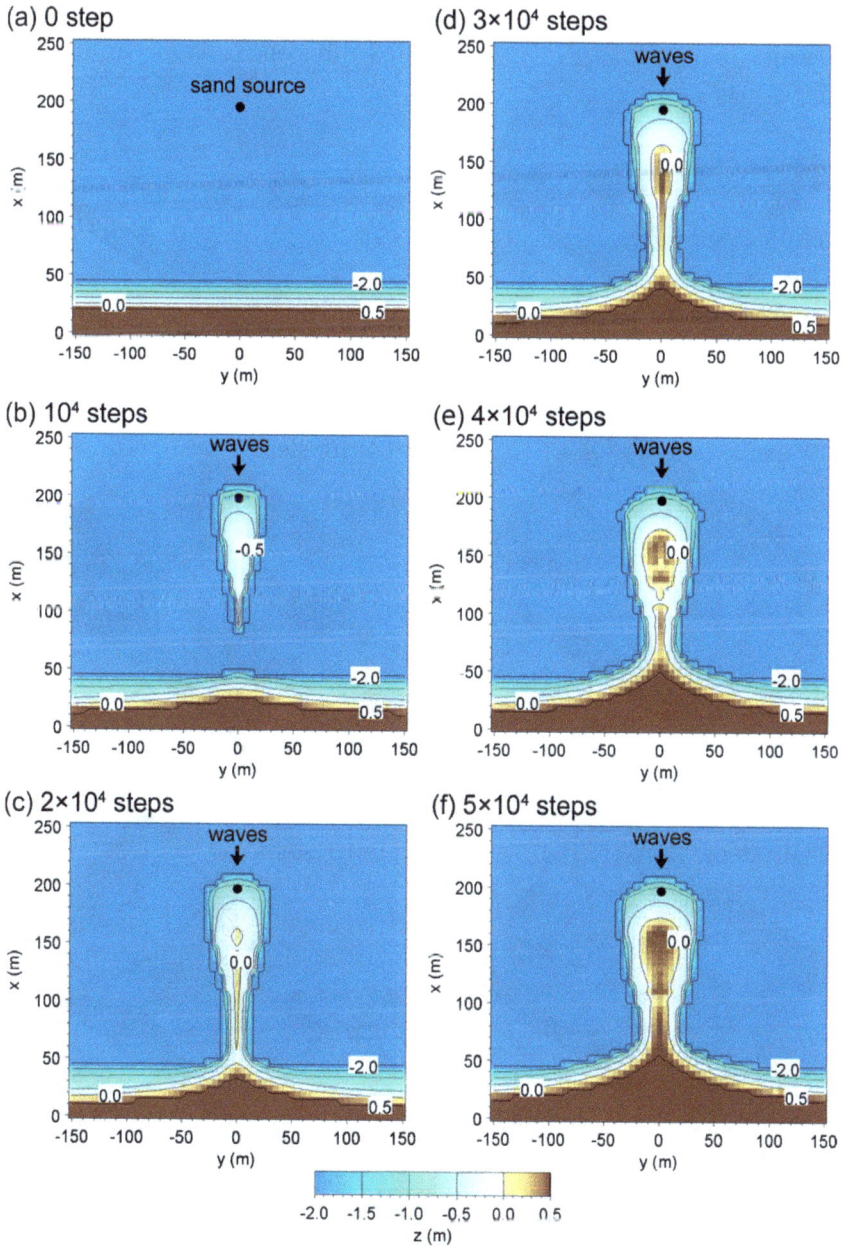

Figure 21. Calculation results observed after every 10^4 steps.

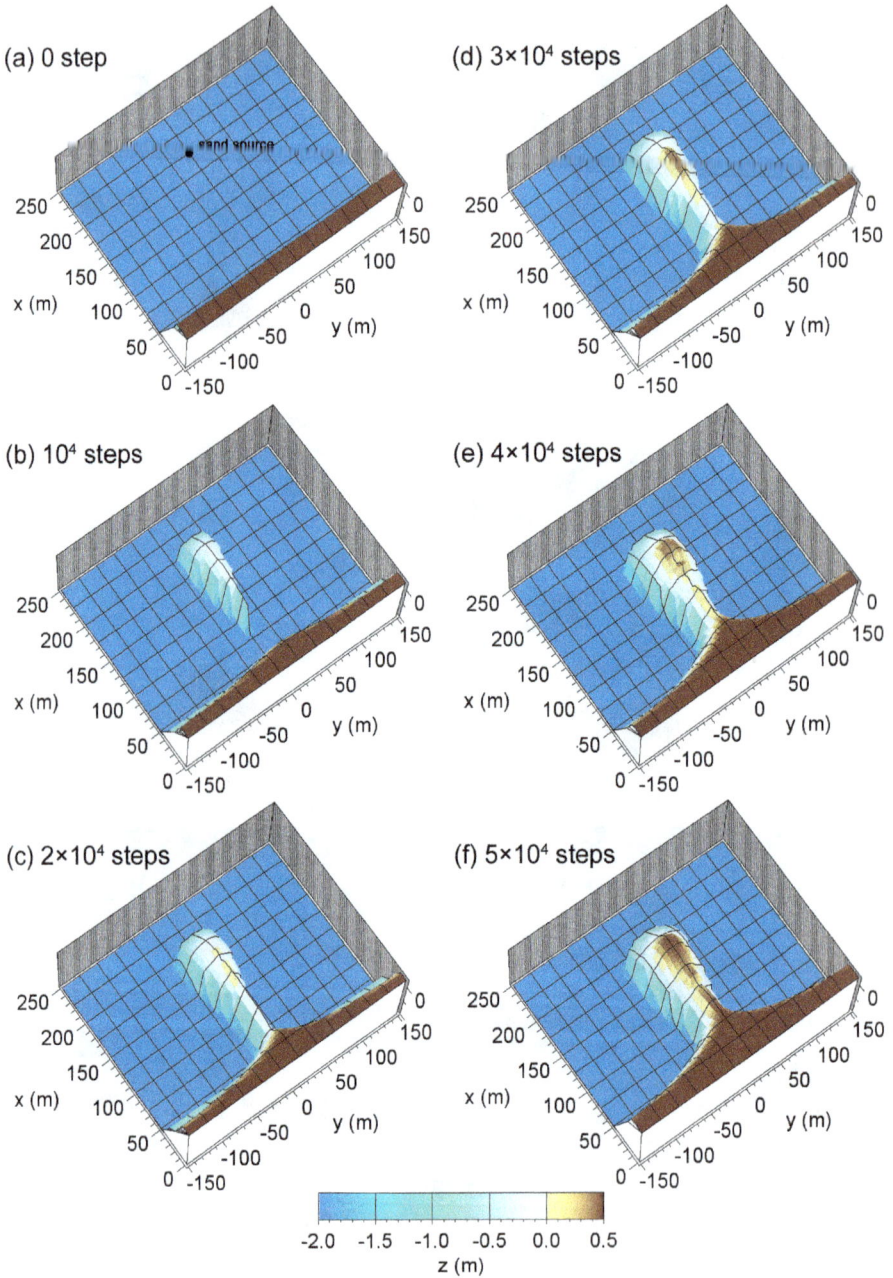

Figure 22. Bird's-eye view of topographic changes.

BG Model Based on Bagnold's Concept and Its Application to Analysis of Elongation of
Sand Spit and Shore – Normal Sand Bar

33

6.2. Changes in longitudinal and transverse profiles

Figure 23 shows the changes in the longitudinal profile along transect $y = 0$ m, which passes through the center of the slender sandy island, as shown in Fig. 21(f). Because the development of the sandy island along this transect was very rapid, the results obtained after not only 1×10^4 steps but also 5×10^3 and 1.5×10^4 steps are also shown. Although the sand bar gradually extended landward, the wave height along the side slopes of the sand bar (or the resultant sandy island) was reduced owing to the wave-sheltering effect induced by the sand bar itself, resulting in a decrease in the depth of closure along the side slope. For example, a steep slope with an angle of repose of 1/2 had formed in the zone deeper than $Z = -0.3$ m after 1×10^4 steps. The elevation of the beach connecting the land and the sandy island increased to reach the berm height of $h_R = 0.5$ m.

Figure 24 shows the changes in the cross section along transect $x = 100$ m, which passes through the neck of the slender sandy island, and transect $x = 50$ m near the boundary between the sandy beach and the tidal flat under the initial conditions, as shown in Fig. 21(f). After 10^4 steps, an island with a sharp top and a side slope with a steep angle of repose on both sides of the island had formed along transect $x = 100$ m. The increase in the width of the sandy beach over time was slow. In contrast, because of the increase in the wave-sheltering effect owing to the development of the sandy island, the cuspate foreland was more developed along transect $x = 50$ m than along transect $x = 100$ m, and its width increased over time.

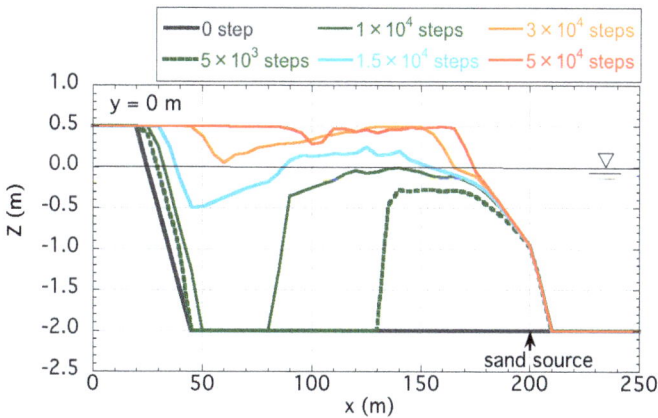

Figure 23. Changes in longitudinal profile along transect y = 0 m passing through center of slender sandy island.

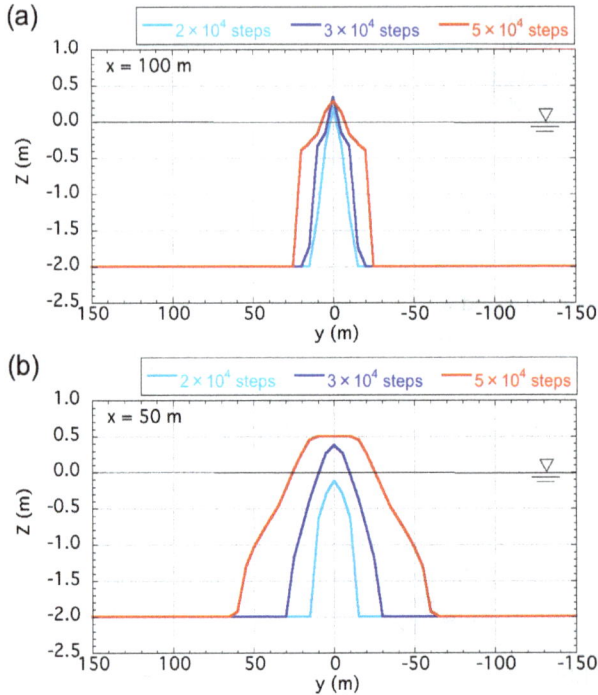

Figure 24. Changes in cross section along transect x = 100 m passing through neck of slender sandy island and transect x = 50 m near boundary between sandy beach and tidal flat.

6.3. Wave height and direction

The wave fields corresponding to the beach changes are shown in Fig. 25. The wave height on the tidal flat was initially uniform (0 step) and uniform wave decay due to wave breaking occurred on the sandy beach extending between the land and the tidal flat. After 1×10^4 steps, a submerged sand bar had been formed by the accumulation of sand supplied from a point source. This sand bar induced a change in the wave field; the wave height was reduced along the side slope of the slender sand bar, and oblique wave incidence occurred on both sides of the sand bar owing to wave diffraction by the sand bar itself. This oblique

BG Model Based on Bagnold's Concept and Its Application to Analysis of Elongation of
Sand Spit and Shore – Normal Sand Bar

35

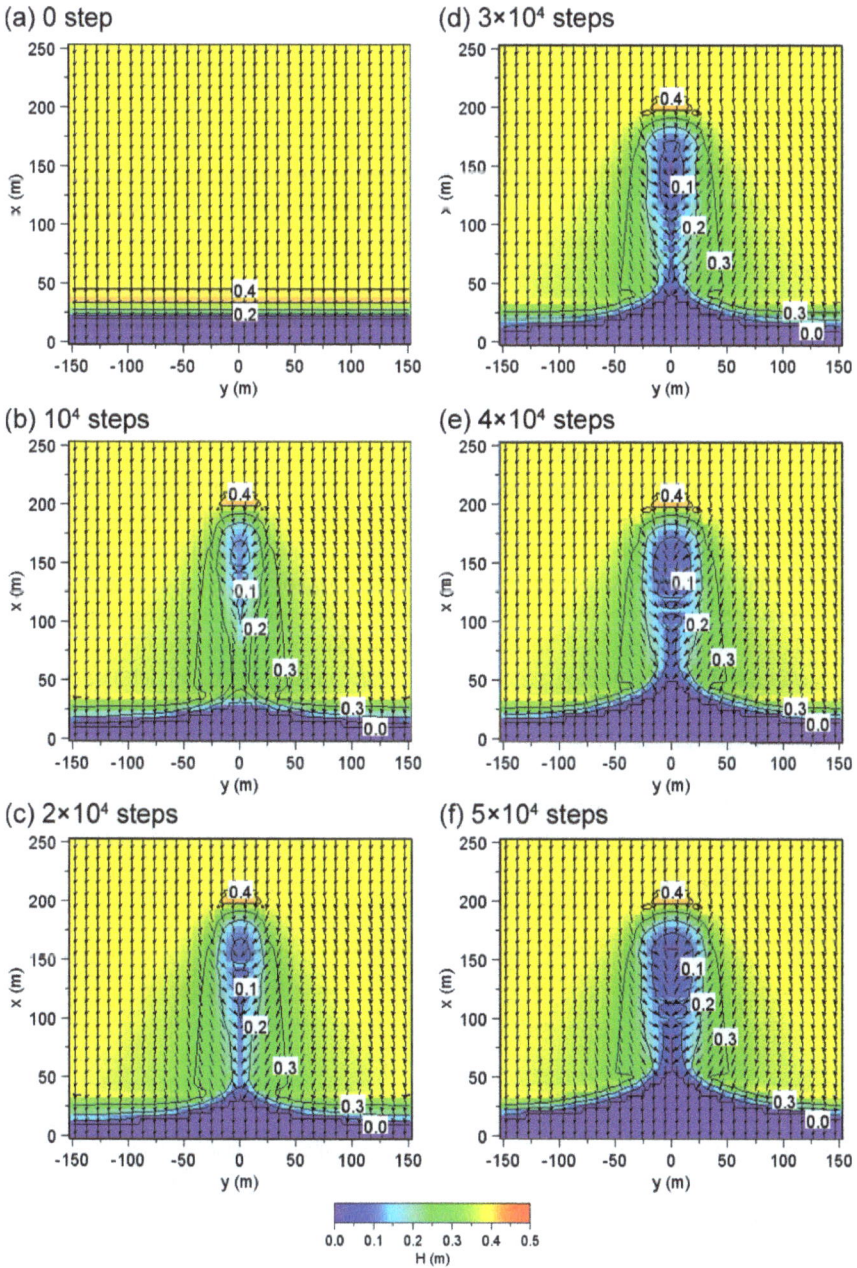

Figure 25. Wave fields corresponding to beach changes.

wave incidence caused shoreward sand transport flux along the contours of the sand bar. On the other hand, on the lee of the submerged sand bar, the wave height was reduced, similarly to in the case of a detached breakwater, inducing longshore sand transport toward the lee of the sand bar from outside the wave-shelter zone. After 2×10^4 steps, as a result of the landward extension of the slender sand bar and the decrease in water depth, wave breaking occurred there, causing a marked reduction in wave height. At the same time, the oblique wave incidence continued in this area and landward sand transport from the island to the land continuously occurred. After 4×10^4 and 5×10^4 steps, the same beach changes as those until 2×10^4 steps continued.

7. Conclusions

The BG model was applied to predict the development of a sand spit on a flat shallow seabed and the formation of a cuspate foreland on coasts with an abrupt change in the shoreline orientation. The results of the model were validated by comparison with the experimental results obtained by Uda & Yamamoto (1992). The predicted and measured results were in good agreement. As another type of beach change due to waves on a coast with a shallow flat seabed, the shoreward transport of sand originally supplied from the offshore zone of a tidal flat, forming a slender sand bar, and the landward deposition of such sand were observed on the Kutsuo coast, which has a very wide tidal flat and faces the Suo-nada Sea, part of the Seto Inland Sea. We investigated these phenomena by field observations and then performed a numerical simulation using the BG model. The observed results were successfully explained by the results of the numerical simulation. Although the BG model has been used to predict the development of river mouth bars, a single sand spit and a bay barrier (Serizawa et al., 2009b, 2009a; Uda & Serizawa, 2011), another application was demonstrated in the present study.

Author details

Takaaki Uda
Public Works Research Center, Taito, Tokyo

Masumi Serizawa and Shiho Miyahara
Coastal Engineering Laboratory Co., Ltd., Shinjuku, Tokyo

8. References

Ashton, A.; Murray, A. B. & Arnault, O. (2001). Formation of coastline features by large-scale instabilities induced by high angle waves, *Nature*, Vol. 14, No. 6861, (November 2001), pp. 296-300, ISSN 0028-0836

Bagnold, R. A. (1963). Mechanics of Marine Sedimentation, In: *The Sea*, Hill, M. N. (editor), Vol. 3, pp. 507-528, Wiley, ISBN 978-0674017306, New York

Bailard, J. A. & Inman, D. L. (1981). An energetics bedload model for a plane sloping beach: Local transport, *J. Geophys. Res.*, Vol. 86, C3, pp. 2035-2043

Dally, W. R.; Dean, R. G. & Dalrymple, R. A. (1984). A model for breaker decay on beaches, *Proc. 19th ICCE*, pp. 82-97,Houston, Texas, USA, September 1984

Goda, Y. (1985). *Random Seas and Design of Maritime Structures*, University of Tokyo Press, ISBN 978-9810232566, Tokyo

Inman, D. L. & Bagnold, R. A. (1963). Littoral Processes, In. *The Sea*, M. N. Hill (editor), Vol. 3, pp. 529-533, Wiley, ISBN 978-0674017306, New York

Mase, H. (2001). Multidirectional random wave transformation model based on energy balance equation, *Coastal Eng. J.*, JSCE, Vol. 43, No. 4, (December 2001),pp. 317-337.

Ozasa, H. & Brampton, A. H. (1980). Model for predicting the shoreline evolution of beaches backed by seawalls, *Coastal Eng.*, Vol. 4, pp. 47-64

Serizawa, M.; Uda, T.; San-nami, T. & Furuike, K. (2006). Three-dimensional model for predicting beach changes based on Bagnold's concept, *Proc. 30th ICCE*, pp. 3155-3167, ISBN 978-981-270-636-2, San Diego, California, USA, September 2006

Serizawa, M.; Uda, T.; San-nami, T.; Furuike, K. & Ishikawa, T. (2009a). Prediction of topographic changes of sand spit using BG model, *J. Coastal Res.*, SI 56, pp. 1060-1064

Serizawa, M.; Uda, T.; San-nami, T.; Furuike, K. & Ishikawa, T. (2009b). Model for predicting recovery of a river mouth bar after flood using BG model, *Asian and Pacific Coasts 2009, Proc. 5th International Conf.*, Vol. 3, pp. 96-102, Singapore, October 2009

Serizawa, M. & Uda, T. (2011). Prediction of formation of sand spit on coast with sudden change in coastline using improved BG model, *Coastal Sediments '11*, pp. 1907-1919, Miami, Florida, USA, May 2011

Serizawa, M.; Uda, T. & Miyahara, S. (2011). Model for predicting formation of slender sand bar due to shoreward sand transport on shallow tidal flat, *Asian and Pacific Coasts 2011, Proc. 6th International Conf.*, pp. 1477-1484, ISBN 978-981-4366-48-9, Hong Kong, December 2011

Uda, T. & Kawano, S. (1996). Development of a predictive model of contour line change due to waves, *Proc. JSCE*, No. 539/ II -35, pp. 121-139. (in Japanese)

Shimizu, T.; Kumagai, T. & Watanabe, A. (1996). Improved 3-D beach evolution model coupled with the shoreline model (3D-SHORE), *Proc. 25th ICCE*, pp. 2843-2856, Orlando, Florida, USA, September 1996

Uda, T. & Yamamoto, K. (1992). On relationship between seabed topography and formation of spit, *Trans. Jpn. Geomor. Union*, Vol. 13, pp. 141-157 (in Japanese)

Uda, T. & Yamamoto, K. (1994). Beach erosion around a sand spit-an example of Mihono-Matsubara sand spit-, *Proc. 24th ICCE*, pp. 2726-2740, Kobe, Japan, October 1994

Uda, T. & Serizawa, M. (2010). Model for predicting topographic changes on coast composed of sand of mixed grain size and its applications (Chap. 16), In. *Numerical simulations-examples and applications in computational fluid dynamics*,Angermann, L. (Ed.), pp. 327-358, INTECH,http://www.intechopen.com/articles/show/title/model-for-

predicting-opographic-changes-on-coast-composed-of-sand-of-mixed-grain-size-and-
its-appli

Uda, T. & Serizawa, M. (2011). Model for predicting formation of bay barrier in flat shallow
sea, *Coastal Sediments '11*, pp. 1176-1189, Miami, Florida, USA, May 2011

Watanabe, S., Serizawa, M. & Uda, T. (2004). Predictive model of formation of a sand spit,
Proc. 29th ICCE, pp. 2061-2073, ISBN 981-256-298-2, Lisbon, Portugal, September 2004

Numerical Simulations of Water Waves' Modulational Instability Under the Action of Wind and Dissipation

Julien Touboul and Christian Kharif

Additional information is available at the end of the chapter

1. Introduction

The seek of uniform, propagative wave train solutions of the fully nonlinear potential equations has been a major topic for centuries. [5] was the first to propose an expression of such waves, the so called Stokes' waves. However, pioneer works of [6] emphasized that such waves might be unstable, providing a geometric condition for this stability problem. Later on, [1] showed analytically that Stokes' waves of moderate amplitude are unstable to long wave perturbations of small amplitude travelling in the same direction. This instability is named the Benjamin-Feir instability (or modulational instability). This result was derived independently by [7] in an averaged Lagrangian approach, and by [8] who used an Hamiltonian formulation of the water wave problem. Using this approach, the latter author derived the nonlinear Schrödinger equation (NLS), and confirmed the previous stability results.

Within the last fifty years, the study of this instability became central for fundamental and applied research. The modulation instability is one of the most important mechanisms for the formation of rogue waves [9]. A complete review on the various phenomena yielding to rogue waves can be found in the book of [10]. In the absence of forcing and damping, Stokes' waves of specific initial steepness are submitted to this instability, when they encounter perturbations of specific wave numbers [11, 12]. In this case, they encounter a nonlinear quasi-recursive evolution, the so called Fermi-Pasta-Ulam recurrence phenomenon ([13]). This phenomenon corresponds to a series of modulation - demodulation cycles, during which initially uniform wave trains become modulated, leading possibly to the formation of a huge wave. Modulation is due to an energy transfer from the wave carrier to the unstable sidebands. In the wave number space, these unstable sidebands are located in a finite narrow band centered around the carrier wave number. During the demodulation, the energy returns to the fundamental component of the original wave train. Using the Zakharov equation, [14] questions the relevance of the Benjamin-Feir index to indicate the intensity of modulational instability. Indeed, this index is often used to quantify the intensity of interactions between a carrier wave and the finite amplitude sidebands. However, [14] emphasized that nonlinear interactions occur also for sidebands located beyond the Benjamin-Feir instability domain.

A damped nonlinear Schrödinger equation (dNLS) was derived by [15] who revisited the Benjamin-Feir instability in the presence of dissipation. They studied numerically the evolution of narrow bandwidth waves of moderate amplitude. More recently [2] investigated theoretically the modulational instability within the framework of the dNLS equation and demonstrated that any amount of dissipation stabilizes the modulational instability in the sense of Lyapunov. Namely, they showed that the zone of unstable region, in the wavenumber space, shrinks as time increases. As a result, any initially unstable mode of perturbation will finally become stable. [2] have confirmed their theoretical predictions by laboratory experiments for waves of small to moderate amplitude. Later, [3] developed fully nonlinear numerical simulations which agreed with the theory and experiments of [2].

From the latter study we could conclude that dissipation may prevent the development of the Benjamin-Feir instability. This effect questions the occurrence of modulational instability of water wave trains in the field. [16] speculated about the effect of dissipation on the early development of rogue waves and raised the question whether or not the Benjamin-Feir instability was able to spawn a rogue wave.

Nevertheless, these authors did not take the effect of wind into account. When considering the occurrence of modulational instability in the field, the role of wind upon this instability in the presence of dissipation needs to be addressed. Based on this assumption, [4] derived a forced and damped nonlinear Schrödinger equation (fdNLS), and extended the analysis of [2] when wind input is introduced. The influence of wind was introduced through a pressure term acting at the interface, in phase with the wave slope, accordingly to Miles' theory [17]. This quasi-laminar theory of wind wave amplification is based on the Miles' shear flow instability. This mechanism of wave amplification is a resonant interaction between water waves and a plane shear flow in air which occurs at the critical height where the wind velocity matches the phase velocity of the surface waves. Stokes waves propagating in the presence of such a forcing, when not submitted to modulational instability, encounter an exponential growth. They demonstrated, within the framework of fdNLS equation, that Stokes' waves were unstable to modulational instability as soon as the friction velocity is larger than a threshold value. Conversely, for a given friction velocity it was found that only carrier waves presenting frequencies (or wavenumbers) lower than a threshold value are subject to Benjamin-Feir instability. Otherwise, due to dissipation, modulational instability restabilizes in the sense of Lyapunov.

As it was mentioned, this physical result is based on the solution of an approached model, the fdNLS equation. Thus, a proper verification is required. However, the phenomenon at hand is based on the long-time behavior of the modulated wave train when propagating in the presence of wind and dissipation. This remark explains the difficulty to provide an experimental verification of the theory. This physical problem is then especially well adapted for a numerical verification. This verification was performed in a first time by [18], who investigated the development of the modulational instability under wind action and viscous dissipation within the framework of fully nonlinear potential equations. This work is an extension of that of [3] when wind input is considered. Later on, [19] emphasized that the equations empirically introduced by [3] were not completely representative of the dispersion relation in the presence of damping, and corrected the equations to overcome this problem, in accordance with the demonstration of [20] and [21].

This work aims to emphasize how numerical simulations can provide useful information to validate long term results based on weakly nonlinear theory. Furthermore, the numerical approach presented here constitutes an extension of the results of [4] to higher orders of nonlinearity and larger band spectra, too. The long time evolution of modulated wave trains can be investigated in a way not allowed by fdNLS equation. The numerical simulations enable to produce results concerning the long time behavior of the modulated wave train. Especially, the phenomenon of permanent frequency downshift will be investigated.

In section 2, the governing equations of the problem are presented. Section 3 presents the weakly nonlinear model obtained by [4], and summarizes their results. The numerical model used to investigate the long time behavior of the modulated wave train is developed in section 4. The initial conditions used to support the numerical strategy for validating the theory introduced by [4] is presented in section 5. Finally, the results obtained are described in section 6.

2. Governing equations of the problem

The approach used in this study is based on the potential flow theory. The fluid is assumed to be incompressible, inviscid, and animated by an irrotational motion. Thus, the fluid velocity derives from a potential ϕ. However, non-potential effects due to wind and viscosity can be taken into account through a modification of the boundary conditions at the surface.

The wind has already been introduced in the dynamic boundary condition through a pressure term acting at the free surface in several numerical potential models. Among them, one may cite [22], [23] and [24] who introduced and discussed this approach for BIEM methods and [25], [26], and [27] who extended it to HOS methods. The pressure term used here is based on the Miles' theory [17], accordingly to the approach of [4]. The viscosity was introduced heuristically by [3] who used the HOS method to address the question raised in [2] on the restabilisation of the Benjamin-Feir instability of a Stokes wave train in the presence of dissipation. The introduction was made through the addition of a damping term in the dynamic boundary condition. However, a proper derivation of the kinematic and dynamic boundary condition in the presence of viscosity was made by [20], and later on by [21]. A modification of both kinematic and dynamic condition was found, resulting in a slight difference in the dispersion relation, as it was discussed by [19]. Finally, the system of equations corresponding to the potential theory, in the presence of wind and viscous damping reads

$$\phi_{xx} + \phi_{zz} = 0 \text{ for } -\infty < z < \eta(x, t) \tag{1}$$

$$\nabla\phi \to 0 \text{ for } z \to -\infty \tag{2}$$

$$\eta_t + \phi_x \eta_x - \phi_z - 2\nu \eta_{xx} = 0 \quad \text{for} \quad z = \eta(x,t) \tag{3}$$

$$\phi_t + \frac{1}{2}\left[(\phi_x)^2 + (\phi_z)^2\right] + g\eta = -\frac{P_a}{\rho} - 2\nu\phi_{zz} \quad \text{for} \quad z = \eta(x,t), \tag{4}$$

where $\phi(x,z,t)$ refers to the velocity potential, $\eta(x,t)$ is the free surface elevation, $P_a(x,t)$ is the atmospheric pressure due to the wind action, applied at the free surface, and where g, ρ, and ν are respectively the gravity, the water density and the water kinematic viscosity. In this system of equations, the influence of wind has to be specified. In the absence of wind, the term P_a/ρ is equal to zero. Otherwise, the wind action is modeled through the term initially introduced by [17], which reads

$$P_a(x,t) = \frac{\rho_{air}\beta u_*^2}{\kappa^2}\frac{\partial \eta}{\partial x}(x,t), \tag{5}$$

where ρ_{air} is the air density, u_* the friction velocity, κ is the von Karman constant, and β a parameter depending on the friction velocity u_* and the wave carrier velocity c_0.

3. Weakly nonlinear approach: The nonlinear Schrödinger equation

The Nonlinear Schrödinger equation can be obtained from the fully nonlinear potential theory by using the multi-scale method. The equations are expanded in Taylor series, around a small parameter, ε, the wave steepness. In the presence of forcing and dissipation, this work was performed initially by [15], who obtained a forced and damped version of this equation. The equation obtained is an approximation of the system of equations (1 - 4), correct to the third order in ε. Recently, [4] used the forced and damped nonlinear Schrödinger equation (fdNLS),

$$i(\psi_t + c_g\psi_x) - \frac{\omega_0}{8k_0^2}\psi_{xx} - 2\omega_0 k_0^2|\psi|^2\psi = i\frac{W\omega_0 k_0}{2g\rho}\psi - 2i\nu k_0^2\psi \tag{6}$$

to investigate both damping and amplification effects on the Benjamin-Feir instability. Herein, $W = \rho_{air}\beta u_*^2/\kappa^2$ represents the wind effect, $c_g = \omega_0/2k_0$ is the group velocity of the carrier wave, and where all the parameters ν, ρ, ρ_{air}, g, u_*, and κ are the parameters defined in previous section. Equation (6) describes the spatial and temporal evolution of the envelope of the surface elevation, ψ, for weakly nonlinear and dispersive gravity waves on deep water when dissipation, due to viscosity, and amplification, due to wind, are considered. If considering the right hand side of this equation, it can be rewritten as

$$i\left(\frac{W\omega_0 k_0}{2g\rho} - 2\nu k_0^2\right)\psi = i\mathcal{K}\psi. \tag{7}$$

[4] found that the stability of the envelope depends on the sign of the constant $\mathcal{K}$. For values of $\mathcal{K} < 0$, solutions are found to be stable, while for values of $\mathcal{K} \geq 0$, solutions are unstable. Physically, they interpreted this result in terms of frequency of the carrier wave ω_0 and friction velocity u_* of the wind over the waves. They plotted the critical curve separating stable envelopes from unstable envelopes. Namely, they showed that for a given friction velocity u_*, only carrier wave of frequency ω_0 which satisfies the following condition are unstable to

modulational perturbations

$$\frac{4\nu\kappa^2\omega_0}{\beta(\kappa c_0/u_*)su_*^2} < 1 \qquad (8)$$

This condition can be rewritten as follows

$$\frac{\mathcal{A}^2\Omega}{\beta(\mathcal{A})} < 1 \qquad (9)$$

where $\mathcal{A} = \kappa c_0/u_*$ is associated to wind whereas $\Omega = \omega_0/(sg^2/4\nu)^{1/3}$ is associated to dissipation. Note that $\mathcal{A}^2\Omega/\beta(\mathcal{A})$ is equal to the ratio between the rate of damping and the rate of amplification and illustrates the competition between dissipative effects and wind input. The non dimensional numbers $\mathcal{A}$ and Ω correspond to the wave age and non dimensional carrier wave frequency. The modulational instability was found to be sustained as soon as the friction velocity is larger than a threshold value. Conversely, for a given friction velocity, it was found that only carrier waves presenting frequencies (or wavenumbers) lower than a threshold value are subject to Benjamin-Feir instability. Otherwise, due to dissipation, modulational instability restabilizes in the sense of Lyapunov.

4. Fully nonlinear approach: The High Order Spectral method

Within the framework of two-dimensional flows, a High-Order Spectral Method is used to solve numerically the basic partial differential equations corresponding to equations (1 - 4). The lateral conditions correspond here to space-periodic conditions. The horizontal bottom condition corresponds to infinite depth. The velocity potential is expanded in a series of eigenfunctions fulfilling both these lateral and bottom conditions. A spectral treatment is well adapted to investigate numerically the long time behavior of periodic water waves encountering the modulational instability.

4.1. Mathematical formulation

We first introduce the following dimensionless variables into equations (1), (2), (3) and (4):

$$\tilde{x} = k_0 x, \ \tilde{z} = k_0 z, \ \tilde{\eta} = k_0\eta, \ \tilde{t} = \sqrt{gk_0}t, \ \tilde{\phi} = \phi/\sqrt{g/k_0^3} \text{ and } \tilde{p} = p/(\rho g/k_0), \qquad (10)$$

where x, z, η, t, ϕ and p are dimensional variables, and where k_0 is a reference wave number. Hence, the kinematic and dynamic boundary conditions become

$$\frac{\partial\tilde{\eta}}{\partial\tilde{t}} + \frac{\partial\tilde{\phi}}{\partial\tilde{x}}\frac{\partial\tilde{\eta}}{\partial\tilde{x}} - \frac{\partial\tilde{\phi}}{\partial\tilde{z}} - 2\frac{\nu k_0}{c_0}\frac{\partial^2\tilde{\eta}}{\partial\tilde{x}^2} = 0 \ \text{ on } \ \tilde{z} = \tilde{\eta}(\tilde{x},\tilde{t}), \qquad (11)$$

$$\frac{\partial\tilde{\phi}}{\partial\tilde{t}} + \frac{\nabla\tilde{\phi}^2}{2} + \tilde{\eta} + \tilde{P}_a + 2\frac{\nu k_0}{c_0}\frac{\partial^2\tilde{\phi}}{\partial\tilde{z}^2} = 0 \ \text{ on } \ \tilde{z} = \tilde{\eta}(\tilde{x},\tilde{t}). \qquad (12)$$

Following [8], we introduce the velocity potential at the free surface $\tilde{\phi}_s(\tilde{x},\tilde{t}) = \tilde{\phi}(\tilde{x}, \tilde{z} = \tilde{\eta}(\tilde{x},\tilde{t}),\tilde{t})$ into these equations, and it comes, after dropping the tilde for sake of readability,

$$\frac{\partial \eta}{\partial t} = -\phi_{s_x} \eta_x + w \left(1 + \eta_x^2\right) + 2 \frac{\nu k_0}{c_0} \eta_{xx} \quad \text{on } z = \eta(x,t) \tag{13}$$

$$\frac{\partial \phi^s}{\partial t} = -\eta - \frac{1}{2} \phi_{s_x}^2 + \frac{1}{2} w^2 \left(1 + \eta_x^2\right) + 2 \frac{\nu k_0}{c_0} \frac{\phi_{s_{xx}} - w \eta_{xx}}{1 + \eta_x^2} - P_a \quad \text{on } z = \eta(x,t) \tag{14}$$

with

$$w = \frac{\partial \phi}{\partial z}(x, z = \eta(x,t), t) \tag{15}$$

The main difficulty in this approach is the computation of the vertical velocity at the free surface, w. Following [28], the potential $\phi(x,z,t)$ is written in a finite perturbation series up to a given order M,

$$\phi(x,z,t) = \sum_{m=1}^{M} \phi^{(m)}(x,z,t). \tag{16}$$

The term $\phi^{(m)}$ is of order $\mathcal{O}(\varepsilon^m)$, where ε, a small parameter, is a measure of the wave steepness. Then expanding each $\phi^{(m)}$ evaluated on $z = \eta$ in a Taylor series about $z = 0$, we obtain

$$\phi_s(x,t) = \sum_{m=1}^{M} \sum_{l=0}^{M-m} \frac{\eta^l}{l!} \frac{\partial^{(l)}}{\partial z^l} \left(\phi^{(m)}(x, z = 0, t) \right). \tag{17}$$

At a given instant of time, ϕ_s and η are known, and we can estimate $\phi^{(m)}$ at each order ε^m:

$$\phi^{(1)}(x, z = 0, t) = \phi_s(x,t), \quad m = 1, \tag{18}$$

$$\vdots$$

$$\phi^{(m)}(x, z = 0, t) = -\sum_{l=1}^{m-1} \frac{\eta^l}{l!} \frac{\partial^l}{\partial z^l} \phi^{(m-l)}(x, z = 0, t), \quad m \geq 2. \tag{19}$$

The boundary conditions, together with the Laplace equations $\nabla^2 \phi^{(m)} = 0$ define a series of Dirichlet problems for $\phi^{(m)}$. For $2\pi-$periodic conditions in x, say, $\phi^{(m)}$ can be written as follows in deep water

$$\phi^{(m)}(x,z,t) = \sum_{j=1}^{\infty} \phi_j^{(m)} e^{-jz} e^{ijx}. \tag{20}$$

Note that $\phi^{(m)}(x,z,t)$ satisfies automatically the Laplace equation and the boundary condition $\nabla \phi^{(m)} \to 0$ when $z \to -\infty$.

4.2. Computation of the vertical velocity

Substitution of equation (20) into the set of equations (18 - 19) provides an expression of the modes $\phi_j^{(m)}$. The vertical velocity at the free surface is then

$$w(x,t) = \sum_{m=1}^{M} \sum_{l=0}^{M-m} \frac{\eta^l}{l!} \frac{\partial^{l+1}}{\partial z^{l+1}} \phi^{(m)}(x, z = 0, t) \tag{21}$$

This expression might be substituted into the kinematic and dynamic boundary conditions (13) and (14), yielding to the evolution equations for ϕ_s and η. Another version of HOSM developed by [29] can be used. The difference between both methods lies in the way of computing w from ϕ_s. [29] assume a power series for w as

$$w(x, l) = \sum_{m=1}^{M} w^{(m)}, \tag{22}$$

where

$$w^{(m)} = \sum_{l=0}^{m-1} \frac{\eta^l}{l!} \frac{\partial^{l+1}}{\partial z^{l+1}} \phi^{(m-l)}(x, z = 0, t). \tag{23}$$

In fact, the version of [29] differs from the version of [28] not only in the expression of the approximated vertical velocity at the surface, but also in its subsequent treatment in the free surface equations. According to [29], the surface equations must be truncated at consistent nonlinear order if they are to simulate a conservative Hamiltonian system. This requires to treat carefully all nonlinear terms containing w in the prognostic equations. In contrast to the series used by [28], those used by [29] are naturally ordered with respect to the nonlinear parameter . The [28] formulation is not consistent, after truncation, with the underlying Hamiltonian structure of the canonical pair of free-surface equations. Thus, the formulation of [29] preserves the Hamiltonian structure of the prognostic equations.

5. Initial conditions for the numerical simulations

From a numerical point of view, one part of the initial condition is obtained by considering a Stokes wavetrain $(\bar{\eta}, \bar{\phi})$ which is computed using the approach first introduced by [30]. A very high-order Stokes wave of amplitude a_0 and wavenumber k_0 is calculated iteratively. In the absence of wind and dissipation, the infinitesimal perturbation components (η', ϕ') calculated through a perturbative approach developed by [31] correspond to a Benjamin-Feir instability of wavenumber δk. The perturbed Stokes wave is obtained by adding the infinitesimal perturbations at the sidebands $k_0 \pm \delta k$ of the fundamental and its harmonics. For fixed values of $(\mathcal{A}, \Omega)$ two kinds of initial conditions are used when wind and dissipation are considered. The first kind (unseeded case) corresponds to the unperturbed Stokes' wave $(\eta, \phi) = (\bar{\eta}, \bar{\phi})$, whereas the second kind (seeded case) corresponds to the perturbed Stokes' wave $(\eta, \phi) = (\bar{\eta}, \bar{\phi}) + \varepsilon(\eta', \phi' + \bar{\phi}_z \eta')$, with $\varepsilon = 10^{-3}$. In both cases, we consider a Stokes wavetrain such as $a_0 k_0 = 0.11$ and $k_0 = 5$. The wavenumber of the modulational instability is $\delta k = 1$. This choice of the perturbation wave number corresponds to the closest approximation of the most unstable wave number that can be fitted in the computational domain. The order of nonlinearity was taken equal to $M = 6$. In other words, nonlinear terms have been retained up to sixth-order. The highest wavenumber taken into account in the simulations is $k_{max} = 50$, corresponding to the ninth harmonic of the fundamental wavenumber. The number of mesh points was taken equal to $N = 750$, satisfying the stability criterion $N > (M + 1) \times k_{max}$. In the absence of wind and damping, the unperturbed initial condition leads to the steady evolution of the Stokes' wavetrain, whereas the perturbed initial condition leads to the well known Fermi-Pasta-Ulam recurrence. We propagate these initial wavetrains under various conditions of wind and dissipation, to analyze the behavior of the modulational instability of the Stokes wavetrain.

6. Results and comparisons

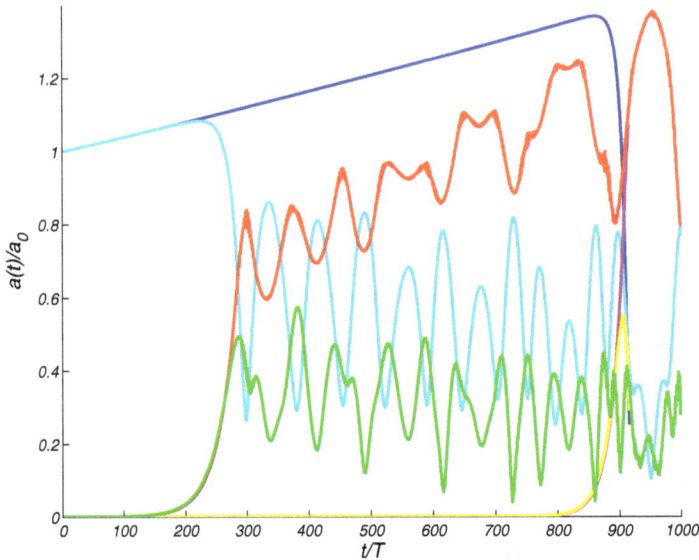

Figure 1. Time evolution of the normalized amplitudes of the fundamental mode ($k = 5$), subharmonic mode ($k = 4$) and superharmonic mode ($k = 6$) for $(\mathcal{A}, \Omega) = (4, 0.59)$. Fundamental mode amplitude (—), subharmonic mode amplitude (—) and superharmonic mode amplitude (—) for an initially unperturbed Stokes' wave (unseeded case). Fundamental mode amplitude (—), subharmonic mode amplitude (—) and superharmonic mode amplitude (—) for an initially perturbed Stokes' wave (seeded case). T is the fundamental wave period.

One of the difficulties involved in this study is to define clearly the stability. Indeed, since Stokes' waves are propagating under the action of wind and viscosity, this flow cannot be considered stationary nor periodic. Discussing of the combined influence of wind forcing and damping on the modulational instability, however, implies to define a reference flow. In order to do so, we first consider the evolution of the unperturbed Stokes' waves in the presence of forcing and dissipation (unseeded case). It is checked that the instability does not develop spontaneously in the laps of time considered. Afterwards, we consider the evolution of the initially perturbed Stokes' wave train under the same conditions of wind forcing and damping (seeded case). The nonlinear evolution of the Stokes' wavetrain perturbed by the modulational instability in the presence of wind and dissipation is then compared to that of the reference flow. In that way, the deviation from the reference flow can be interpreted in terms of modulational instability, and the influence of wind forcing and dissipation can be analyzed. Following our previous works [18, 19], the evolution of the energy of the perturbation is thus obtained.

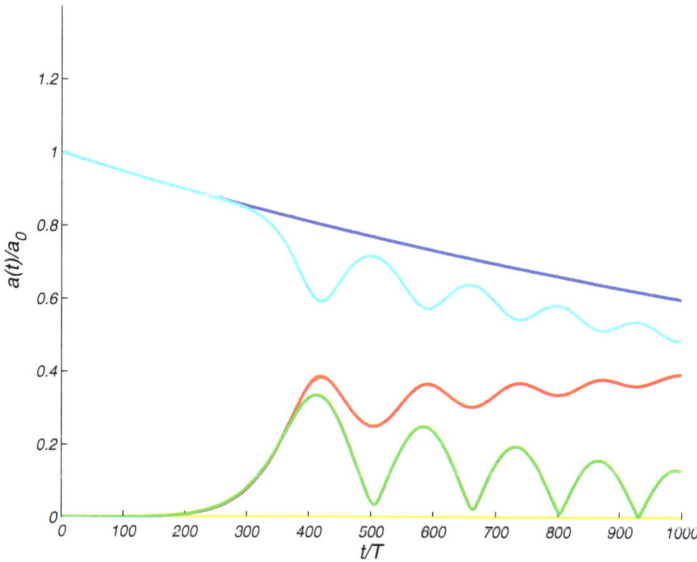

Figure 2. Time evolution of the normalized amplitudes of the fundamental mode ($k = 5$), subharmonic mode ($k = 4$) and superharmonic mode ($k = 6$) for $(\mathcal{A}, \Omega) = (4, 0.61)$. Fundamental mode amplitude (—), subharmonic mode amplitude (—) and superharmonic mode amplitude (—) for an initially unperturbed Stokes' wave (unseeded case). Fundamental mode amplitude (—), subharmonic mode amplitude (—) and superharmonic mode amplitude (—) for an initially perturbed Stokes' wave (seeded case). T is the fundamental wave period.

Figures 1 and 2 present the time evolution of the amplitudes of three components of the water waves' spectrum. The mode $k = 5$ is the fundamental mode, while modes $k = 4$ and $k = 6$ are sidebands, respectively the subharmonic and the superharmonic. Each of these figures present to two kinds of initial conditions, namely the unseeded and the seeded cases. The two figures correspond to two different conditions of wind forcing and damping.

Figure 1 shows the time evolution of the normalized amplitudes $a(t)/a_0$ of the fundamental mode $k = 5$, subharmonic mode $k = 4$ and superharmonic mode $k = 6$ with and without perturbations for the modulational instability. For both cases, the simulations correspond to a wind parameter $\mathcal{A} = 4$ and to a viscosity parameter $\Omega = 0.59$. Within the framework of the NLS equation, [4] showed that the wave train is unstable to modulational instability for these values of $\mathcal{A}$ and Ω. From this figure, it appears that both wavetrains (unseeded and seeded cases) present a similar evolution during the first hundred periods of propagation, T being the fundamental wave period. Then, the behavior of the wavetrain is strongly affected by the development of the modulational instability. For the unperturbed case (unseeded case), the fundamental component increases, since no occurrence of the modulational instability is expected. However, due to the accumulation of numerical errors, the spontaneous occurrence of the modulational instability cannot be avoided, but not before $t = 900T$. On figure 3 one can observe the persistence of the modulational instability through the evolution of the free surface. Indeed, it is observed that the wave packet is alternatively modulated, leading to the formation of a large wave, and demodulated, corresponding to a state which is closer from the origin. This is an expression of the Fermi-Pasta-Ulam quasi-recurrence.

Figure 2 corresponds to $(\mathcal{A}, \Omega)=(4, 0.61)$. Wind condition is similar to the previous numerical simulation, but the dissipative effect considered is stronger. This case correspond to a linearly stable case of the modulational instability, as obtained by [4] in the framework of the NLS equation. From this figure, one can see that wind energy goes to the subharmonic mode whereas dissipation reduces the fundamental and superharmonic components, as previously observed. However, modulation of modes decrease, and they present a monotonic behavior. For unseeded case, as expected, we observe an exponential decay of the fundamental mode. Note that there is no natural occurrence of the subharmonic mode of the modulational instability as it was found in figure 1. For seeded case, the first maximum of modulation that occurs at $t = 410T$ is followed by partial damped modulation/demodulation cycles. Figure 4 illustrates the disappearance of the modulational instability through the evolution of the free surface. In this case dissipation prevails over amplification due to wind and [2] have obtained linear and nonlinear stability of modulational perturbations within the framework of the dissipative NLS equation. More specifically they showed that dissipation reduces the set of unstable wavenumbers as time increases. Consequently every mode becomes stable. The result of this numerical simulation agrees with that of [2] and [3] who considered only dissipation. In their approach, a solution is said to be stable if every solution that starts close to this solution at $t = 0$ remains close to it for all $t > 0$, otherwise the solution is unstable. To include nonlinear stability analysis they introduced a norm and considered stability in the sense of Lyapunov.

In our previous work [18], we assumed that the dominant mode describes the main behavior of a wave train, and we introduced a norm measuring the distance between the fundamental modes of the unperturbed and perturbed Stokes wave corresponding to unseeded case and seeded case respectively. However, it is more consistent to consider the energy of the perturbation, as it was stated in [19]. Thus, another norm can be introduced as

$$E_N(t) = \frac{\int_{-\infty}^{\infty} (a_{k_s}(t) - a_{k_{us}}(t))^2 dk}{\int_{-\infty}^{\infty} a_{k_{us}}^2(0) dk}, \tag{24}$$

where $a_{k_{us}}(t)$ is the amplitude of the component of water elevation η of wave number k, for the initially unperturbed wave train (unseeded case), and $a_{k_s}(t)$ is the amplitude of the component of water elevation η of wave number k, for the initially perturbed wave train (seeded case). This norm corresponds to the potential energy of the perturbation. Its value characterizes the deviation of the perturbed solution from the unperturbed solution. Figure 5 shows the time evolution of this norm for two sets of parameters $(\mathcal{A}, \Omega) = (4, 0.59)$ and $(\mathcal{A}, \Omega) = (4, 0.61)$. For the two cases we can observe two regimes. The first regime corresponds to the development of the modulational instability and shows that it is the nonlinear interaction between the fundamental mode and its sidebands which dominates with a weak effect of the wind forcing and the dissipation. The second regime corresponding to the oscillatory evolution of the norm is dominated by the competition between wind forcing and dissipation. The nonlinear interaction between the fundamental mode and the sidebands is

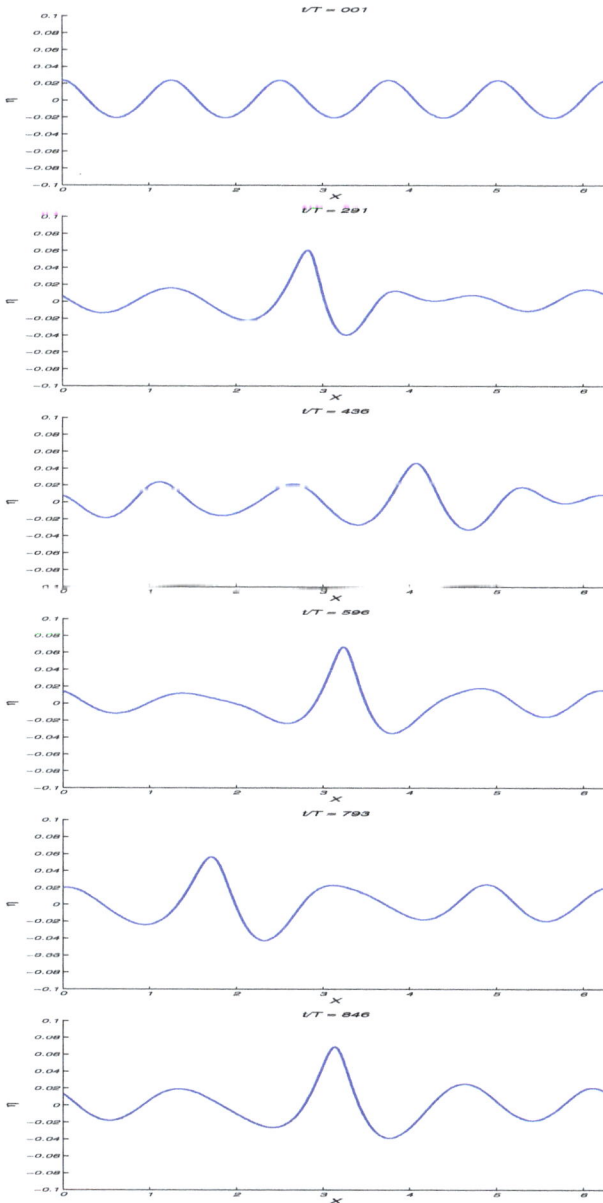

Figure 3. Surface wave profiles at different times, obtained while propagating initial condition corresponding to seeded case with $(\mathcal{A}, \Omega) = (4, 0.59)$. From top to bottom $t/T = 1, 291, 436, 596, 793, 846$.

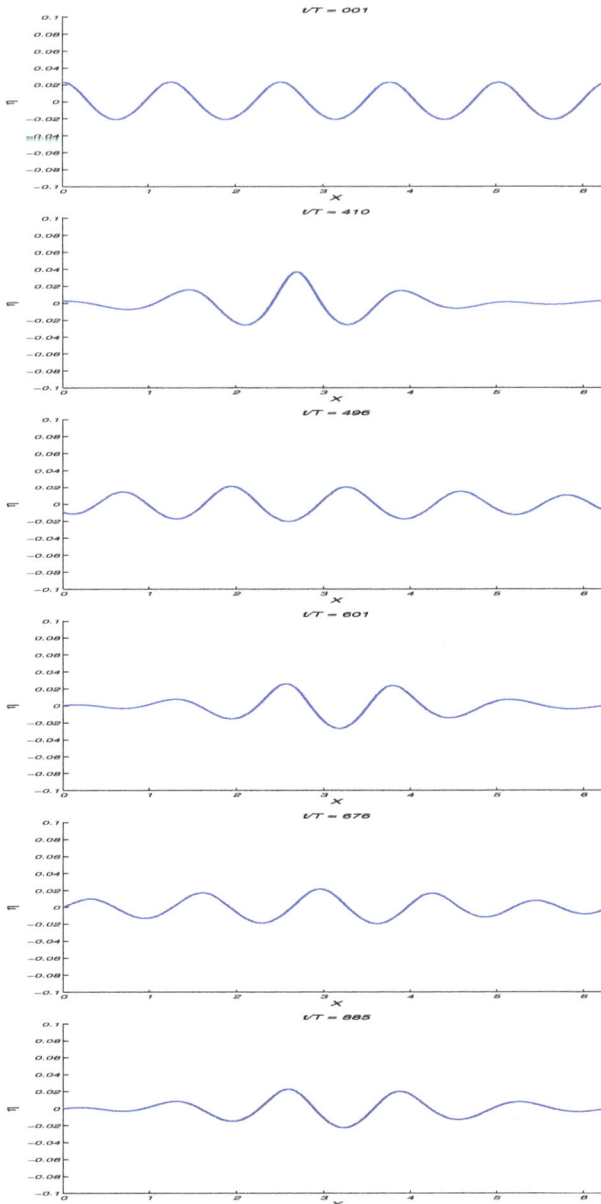

Figure 4. Surface wave profiles at different times, obtained while propagating initial condition corresponding to seeded case with $(\mathcal{A}, \Omega) = (4, 0.61)$. From top to bottom $t/T = 1, 410, 496, 601, 676, 885$.

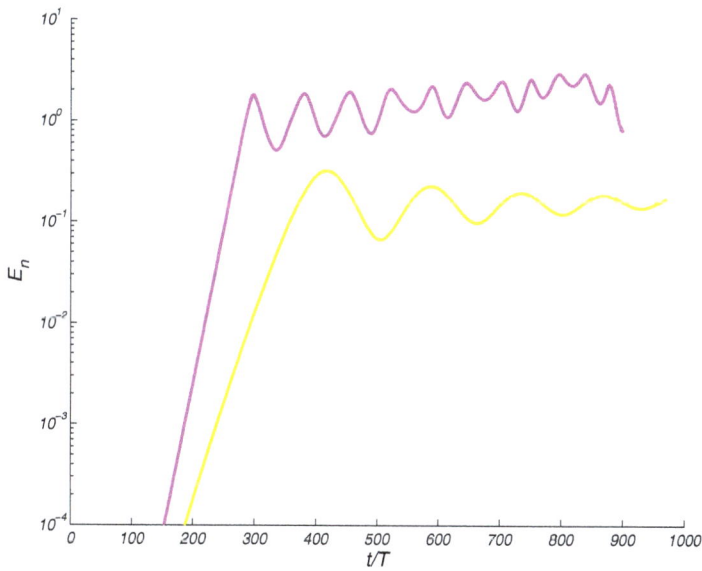

Figure 5. Time evolution of the norm E_n for $(\mathcal{A}, \Omega) = (4, 0.59)$ (—) and $(\mathcal{A}, \Omega) = (4, 0.61)$ (—).

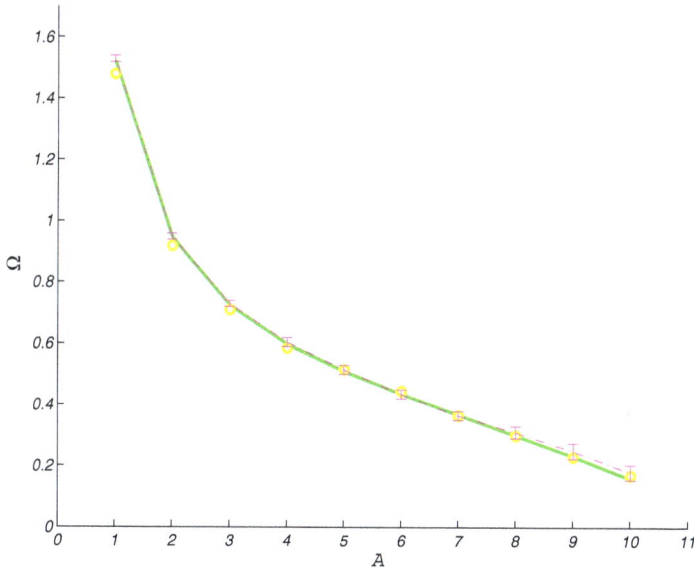

Figure 6. Theoretical (—) and numerical (- · -) marginal stability contour lines. ( ) correspond to numerical results obtained in the framework of the equations suggested by [3]. The theoretical curve corresponds to the figure 1 of [4].

not the dominant mechanism. The magenta curve exhibits oscillations around an averaged value growing exponentially, whereas the yellow curve exhibits the same oscillations around a constant value. We can claim that the norm, E_N, presents globally exponential growth or asymptotical saturation corresponding to instability and stability respectively. Herein, the stability can be interpreted in terms of asymptotic stability. The first case is said to be unstable whereas the second case corresponds to a stable solution. In the latter case we expect that the solution will remain close to the unperturbed solution. In other words, nonlinear interactions are affected by the non conservative effects that are wind and dissipation, leading to a long time disappearance of these interactions.

Many numerical simulations have been run for various values of the parameters $\mathcal{A}$ and Ω. Figure 6 shows a stability diagram which presents comparison between the present numerical results and those of [4] obtained theoretically. The marginal curve corresponding to the fully nonlinear equations is very close to the theoretical marginal curve obtained within the framework the NLS equation. The region above the critical curve corresponds to stable cases, whereas the region beneath corresponds to unstable cases. Bars in figure 6 correspond to uncertainty on stability or instability. Numerical results obtained in our previous work [18] within the framework of equations suggested by [3] are plotted for the sake of reference ( ). The way of introducing damping effect into the kinematic boundary condition has little influence on the results, especially for young waves. The present numerical simulations demonstrate that the results derived by [4] within the framework of the NLS equation are correct in the context of the fully nonlinear equations.

This result provides a validation of the weakly nonlinear theory obtained in the framework of nonlinear Schrödinger equation. However, the numerical approach allows to investigate the long time evolution of the wave train, taking into account the strongly nonlinear behavior of water waves. One phenomenon especially illustrates this nonlinear behavior: the permanent frequency downshift. This phenomenon was discussed by [32] and [33] within the framework of gravity waves. These authors considered that dissipation due to breaking wave was responsible for this permanent downshift. [34] modeled the phenomenon in the presence of wind and eddy viscosity, and latter on [35] in the presence of only molecular viscosity. All these works are based on equations valid up to fourth order in nonlinearity, or higher. Indeed, it is well known that the frequency downshift cannot be observed in the framework of nonlinear Schrödinger equation, which preserves the symmetry between subharmonic and superharmonic components. In fact, [36] concluded that in the absence of wind and dissipation, it was not possible to observe the phenomenon even with higher order equations.

If going back to figure 1, for the initially perturbed case (seeded case), the development of the modulational instability is responsible for the frequency downshift observed at around $t = 500T$. Indeed, one can see that the subharmonic component increases continuously whereas the fundamental and superharmonic component decrease. The superharmonic component decreases faster than the fundamental component. Hence, wind energy goes to the subharmonic mode whereas dissipation reduces the fundamental and superharmonic components. During the modulation process, a broadening of the spectrum is observed, even if not presented here for the sake of clarity. Beyond $t = 500T$, the subharmonic mode $k = 4$ is dominant in the spectrum. To investigate the effect of wind and damping, another series of simulations is performed. Namely, the values of $\mathcal{A}$ and Ω where chosen to fulfill the condition

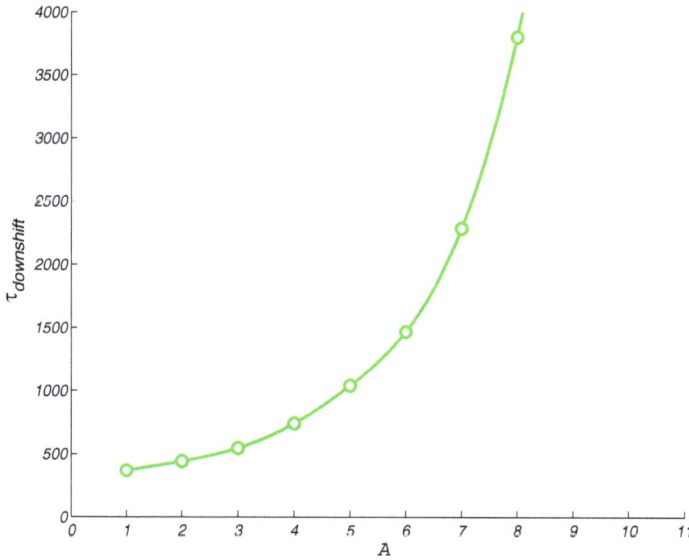

Figure 7. Nondimensional time $\tau_{downshift} = t/T$ for which the permanent frequency downshift is observed, plotted as a function of the wave age parameter $\mathcal{A}$. Ω is chosen here to be on the marginal stability curve presented on figure 6.

$\mathcal{K} = 0$ (see equation (7) for the definition of $\mathcal{K}$). This might be interpreted in terms of balance between wind forcing and damping, corresponding to an equilibrium state. Furthermore, in these conditions, the forced and damped nonlinear Schrödinger equation reduces to the canonical NLS equation. It becomes obvious that the weakly nonlinear theory does not predict any variation in behavior, since the NLS equation remains unchanged for each values of $\mathcal{A}$ and Ω. However, the results of the numerical simulations are different. Indeed, figure 7 presents the nondimensional time $\tau_{downshift}$ after which the permanent frequency downshift is observed, as a function of $\mathcal{A}$. From this figure, it seems obvious that this time is strongly dependant on the wind speed. The youngest the waves are, the fastest is the downshift.

7. Conclusion

In this work, it was evidence how numerical simulations can provide a good demonstration of a weakly nonlinear theory that cannot be achieved by means of experimental demonstration. In this study, an extension of the work of [4] to the fully nonlinear case was suggested. Within the framework of the NLS equation the latter authors considered the modulational instability of Stokes wave trains suffering both effects of wind and dissipation. The results they obtained show that the modulational instability depends on both frequency of the carrier wave and strength of the wind velocity. They plotted the curve corresponding to marginal stability in the $(\mathcal{A}, \Omega)$-plane. Here, a numerical verification is performed, by means of a fully nonlinear approach. The long term behavior of a wave group propagating under both actions of wind and dissipation was obtained thanks to this method. To distinguish stable solutions from unstable solutions, a norm based on the potential energy of the perturbations was introduced. A nonlinear stability diagram resulting from the numerical simulations of the fully nonlinear equation has been given in the $(\mathcal{A}, \Omega)$-plane which coincides with the linear stability analysis of [4]. In the presence of wind, dissipation and modulational instability it is found that wind energy goes to the subharmonic sideband whereas dissipation lowers the amplitude of the fundamental mode of the wave train yielding to a permanent frequency-downshifting. This permanent frequency downshift is strongly influenced by the wind and dissipation parameter. If the wave group is at equilibrium in energy input and dissipation, the fdNLS equation reduces to the classical NLS equation, and predict no influence of the wind. However, by considering the asymmetry between wave components, induced by strong nonlinearity (higher than fourth order), a strong influence of the wind and dissipation is observed.

Author details

Julien Touboul
Mediterranean Institute of Oceanography (MIO), Aix-Marseille Univ., Université du Sud Toulon-Var, CNRS/INSU, UMR 7294, IRD, UMR235, France

Christian Kharif
Institut de Recherche sur les phénomènes hors équilibre (IRPHE), Aix-Marseille Univ., Ecole Centrale Marseille, CNRS/INSIS UMR 7342, France

8. References

[1] T. B. Benjamin and J. E. Feir. The disintegration of wave trains on deep water. *J. Fluid Mech.*, 27:417–430, 1967.

[2] H. Segur, D. Henderson, J. Carter, J. Hammack, C. M. Li, D. Pheiff, and K. Socha. Stabilizing the benfamin-feir instability. *J. Fluid Mech.*, 539:229–271, 2005.

[3] G. Wu, Y. Liu, and D. K. P. Yue. A note on stabilizing the benjamin-feir instability. *J. Fluid Mech.*, 556:45–54, 2006.

[4] C. Kharif, R. Kraenkel, M. Manna, and R. Thomas. The modulational instability in deep water under the action of wind and dissipation. *J. Fluid Mech.*, 664:417–430, 2010.

[5] G. G. Stokes. On the theory of oscillatory waves. *Trans. Camb. Phil. Soc.*, 8:441–455, 1847.

[6] M. J. Lighthill. Contributions to the theory of waves in non-linear dispersive systems. *J. Inst. Math. Appl.*, 1:269–306, 1965.

[7] G. B. Whitham. Linear and nonlinear waves. Wiley-Interscience (New-York), 1974.

[8] V. E. Zakharov. Satbility of periodic waves of finite amplitude on the surface of a deep fluid. *J. Appl. Tech. Phys.*, 9:190–194, 1968.

[9] C. Kharif and E. Pelinovsky. Physical mechanisms of the rogue wave phenomenon. *Eur. J. Mech. B Fluids*, 22 (6):603–633, 2003.

[10] C. Kharif, E. Pelinovsky, and A. Slunyaev. Rogue waves in the ocean. In *Mathematical Aspects of Vortex Dynamics*. Springer (ISBN: 978-3-540-88418-7), 2009.

[11] J. W. Dold and D. H. Peregrine. Water-wave modulation. In *Proc. 20th Intl. Conf. Coastal Eng. (Taipei)*, volume 1, pages 163–175. 1986.

[12] M. L. Banner and X. Tian. On the determination of the onset of breaking for modulating surface gravity water waves. *J. Fluid Mech.*, 367:107–137, 1998.

[13] E. Fermi, J. Pasta, and S. Ulam. Studies of non linear problems. *Amer. Math. Month.*, 74 (1), 1967.

[14] L. Shemer. On benjamin feir instability and evolution of a nonlinear wave with finite amplitude sidebands. *Nat. Hazards Earth Syst. Sci.*, 10:1–7, 2010.

[15] S. W. Joo, A. F. Messiter, and W. W. Schultz. Evolution of weakly nonlinear water waves in the presence of viscosity and surfactant. *J. Fluid Mech.*, 229:135–158, 1991.

[16] H. Segur, D. Henderson, and J. Hammack. Can the benjamin-feir instability instability spawn a rogue wave? In *Proc. 14th 'Aha Huliko' a Hawaiian winter worshop*, pages 43–57. 2005.

[17] J. W. Miles. On the generation of surface waves by shear flow. *J. Fluid Mech.*, 3:185–204, 1957.

[18] C. Kharif and J. Touboul. Under which conditions the benjamin-feir instabilty may spawn a rogue wave: a fully nonlinear approach. *Eur. Phys. J. Special Topics*, 185:159–168, 2010.

[19] J. Touboul and C. Kharif. Nonlinear evolution of the modulational instability under weak forcing and damping. *Nat. Hazards Earth Syst. Sci.*, 10 (12):2589–2597, 2010.

[20] T. S. Lundgren. A free surface vortex method with weak viscous effects. In R. E. Calfish, editor, *Mathematical Aspects of Vortex Dynamics*, pages 68–79. SIAM Proceedings, 1989.

[21] F. Dias, A. I. Dyachenko, and V. E. Zakharov. Theory of weakly damped free-surface flows: A new formulation based on potential flow solutions. *Phys. Lett. A*, 372:1297–1302, 2008.

[22] M.I. Banner and J. Song. On determining the onset and strength of breaking for deep water waves. part ii: Influence of wind forcing and surface shear. *J. Phys. Oceanogr.*, 32 (9):2559–2570, 2002.

[23] J. Touboul, J.-P. Giovanangeli, C. Kharif, and E. Pelinbovsky. Freak waves under the action of wind: Experiments and simulations. *Eur. J. Mech. B. Fluid.*, 25:662–676, 2006.

[24] J. Touboul, C. Kharif, E. Pelinbovsky, and J.-P. Giovanangeli. On the interaction of wind and steep gravity wave groups using miles' and jeffreys' mechanisms. *Nonlin. Processes Geophys.*, 15:1023–1031, 2008.

[25] J. Touboul and C. Kharif. On the interaction of wind and extreme gravity waves due to modulational instability. *Phys. Fluids*, 18(108103), 2006.

[26] J. Touboul. On the influence of wind on extreme wave events. *Nat. Hazards Earth Syst. Sci.*, 7:123–128, 2007.

[27] C. Kharif, J.-P. Giovanangeli, J. Touboul, L. Grare, and E. Pelinovsky. Influence of wind on extreme wave events: experimental and numerical approaches. *J. Fluid Mech.*, 594:209–247, 2008.

[28] D. G. Dommermuth and D. K. P. Yue. A high-order spectral method for the study of nonlinear gravity waves. *J. Fluid Mech.*, 184:267–288, 1987.

[29] B. J. West, K. A. Brueckner, R. S. Janda, M. Milder, and R. L. Milton. A new numerical method for surface hydrodynamics. *J. Geophys Res.*, 92:11803–11824, 1987.

[30] M. S. Longuet-Higgins. Bifurcation in gravity waves. *J. Fluid Mech.*, 151:457–475, 1985.

[31] C. Kharif and A. Ramamonjiarisoa. Deep water gravity wave instabilities at large steepness. *Phys. Fluids*, 31:1286–1288, 1988.

[32] K. Trulsen and K. B. Dysthe. Frequency down-shift through self modulation. In A. Torum and O. T. Gud-Mestad, editors, *Proc. Water Wave Kinematics*, volume 178, pages 561–572. Kluwer Academic Publishers, 1990.

[33] K. Trulsen and K. B. Dysthe. Action of windstress and breaking on the evolution of a wavetrain. In M. L. Banner and R. H. J. Grimshaw, editors, *Breaking Waves*, pages 243–249. Springer Verlag, 1992.

[34] T. Hara and C. C. Mei. Frequency downshift in narrowbanded surface waves under the influence of wind. *J. Fluid Mech.*, 230:429–477, 1991.

[35] C. Skandrani, C. Kharif, and J. Poitevin. On benjamin feir instability and evolution of a nonlinear wave with finite amplitude sidebands. *Cont. Math.*, 200:157–171, 1996.

[36] E. Lo and C. C. Mei. A numerical study of water wave modulation based on a higher order nonlinear schrödinger equation. *J. Fluid Mech.*, 150:395–416, 1985.

Numerical Simulation of Droplet Dynamics in Membrane Emulsification Systems

Manabendra Pathak

Additional information is available at the end of the chapter

1. Introduction

An emulsion is a two-phase liquid system of two immiscible liquids, where the liquid with lower mass fraction is dispersed in form of small droplets in other surrounding liquid of higher mass fraction. Emulsions are widely used to produce sol–gel, drugs, synthetic materials, and food products. Based on the size of droplet, emulsions can be classified as micro and macro emulsion. Karbstein and Schubert, (1995) have made a limiting droplet size of 0.1 µm, below which the emulsion is termed as micro emulsion and above that size the emulsion is termed as macro emulsion. Size and size distribution of droplets play important roles in the stability of emulsion. There are also other factors such as sedimentation, skimming, droplet aggregation and coalescence, which may affect the stability of the droplets. Thus for making a stable emulsion it is necessary to convert the dispersed phase into tiny droplets and stabilize them against coalescence. Some amount of energy is required in the process to break the dispersed phase into droplets. The amount of energy put in the dispersing phase also controls the resulting droplet size. The stability of newly formed droplets depends on how fast the used emulsifiers are able to occupy the newly created interfaces and how well they stabilize them. The common devices used to produce emulsions are rotor-stator-systems, stirrers and high-pressure homogenizers. During last two decades, new technologies of making emulsion have been developed. Compared to conventional method of emulsification such as rotor-stator method, these new techniques of emulsification have several advantages such as low energy consumption, controllable droplet size with proper distribution and easy scalability. These new methods are based on the microdroplet formation in micrometer sized channels. Three such new methods are T-junction emulsification, flow focusing emulsification, and membrane emulsification. In all these methods, controllable droplet formations are achieved by properly maintaining the combination of continuous and dispersed phase flow rate.

In membrane emulsification process, micro or macro porous membranes are used to generate droplets by pressing the dispersed phase through the porous matrix of the membrane towards the continuous phase. At the interface, the dispersed phase forms droplets near the region of pores openings and detached by the cross-flowing continuous phase. Sometimes surfactants are used to stabilize the droplets. Compared to the conventional method, membrane emulsification process requires a lower energy input (10^5 – 10^6 J/m^3) to generate micro sized droplets (Schubert and Behrend, 2003). Since small droplets are directly formed at the micro-pores of a membrane, rather than by disruption in zones of high energy density, smaller amount of stress is required in the process compared to the conventional method. The main disadvantage of the process is the requirement of longer production time compared to the conventional processes because of the slow rate at which the dispersed phase flows through the membrane (Joscelyne and Trägårdh, 1991, 2000). The longer production time can be reduced by increasing the flow rate of dispersed phase fluid. With the increase in dispersed phase flow rate, the droplet diameter increases first than decreases and the process shifts towards jetting phenomenon (Pathak, 2011). Thus there should be an optimum dispersed phase flux for optimum production time and droplet size in a membrane emulsification system.

The schematic diagram of a cross-flow membrane emulsification has been shown in Fig. 1. Some commonly used membranes are tubular micro-porous glass (MPG) and shirasu porous glass (SPG) membrane. Some metallic oxides such as ceramic α-Al$_2$O$_3$ or α-Al$_2$O$_3$ coated with titainia oxide or zirconia oxide are also used as membrane. These membranes contain cylindrical, interconnected, uniform micro-pores having pore sizes, typically ranging about 0.05–14 μm. In membrane emulsification system, the time of droplets formation, size and stability of droplets are three important parameters which control the emulsification system. Thus understanding the droplet dynamics in detail may enable to explore the possibilities and limits of membrane emulsification for various applications.

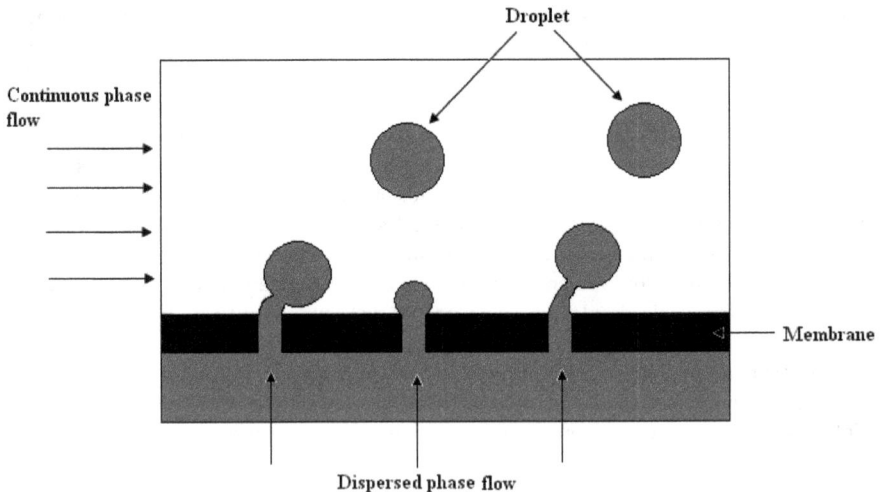

Figure 1. Schematic diagram of a cross-flow membrane emulsification process

In membrane emulsification process, the size distribution of pores and their relative spatial distribution in the membrane surface control the production of mono-disperse emulsions. The growth and detachment of the droplet i.e. droplet dynamics in membrane emulsification depends on several parameters. Luca et al., (2004) has classified them into three broad categories.

i. **Operating parameters:** cross-flow velocity, transmembrane pressure and disperse phase flux,
ii. **Membrane parameters:** pore size, active pores, distance between the pores, membrane hydrophobicity/hydrophilicity;
iii. **Phase parameters:** interfacial tension, viscosity and density of the processed phases.

Based on these parameters, the size of the droplet in membrane emulsification system depends upon the pore diameter and the dependence of the droplet diameter on pore diameter can be expressed as:

$$D_p = x\, D_0 \tag{1}$$

where D_p is the droplet diameter and D_0 is the pore diameter. Katoh et al., (1996) experimentally observed the value of x in the range of 2 to 12. The major factors affecting the value of x are: (i) the shear rates of continuous cross-flow fluid (ii) the dynamic interfacial tension, γ; and (iii) the disperse phase flux (J_d). Other parameters those implicitly control the value of x are: the average velocity of the continuous phase flow $\langle u_c \rangle$, the viscosities of the disperse phase and the continuous phases (μ_d and μ_c) the density of the continuous phase (ρ_c) the thermodynamic temperature, (T), the transmembrane pressure (P_m). The viscosity and dynamic surface tension depend upon temperature and that way temperature can influence the emulsification process.

Cross-flow velocity of continuous phase imparts drag force on the growing droplet for which the droplet detaches at the pore. With the increase in cross-flow velocity, the droplet diameter decreases. The dispersed phase flow rate or velocity influences the droplet dynamics via the inertial force competing with other forces such as drag and surface tension force. The difference between the pressure of dispersed phase in the dispersed phase channel and the average pressure of the continuous phase in the main channel is termed as transmembrane pressure.

$$\Delta p_m = p_d - \overline{p_c} \tag{2}$$

The average pressure of the continuous phase is defined as:

$$\overline{p_c} = (p_{c,in} + p_{c,out})/2 \tag{3}$$

where $p_{c,in}$ and $p_{c,out}$ are the pressure of continuous phase at the inlet and outlet of the main channel. Total transmembrane pressure consists of two parts: one is the capillary pressure (p_γ) and other is the effective transmembrane pressure (p_{eff}) or drag pressure inside the pore. Due to curvature of the droplet and dynamic interfacial tension, a small amount of pressure is required to inflate the droplet. This pressure is the capillary pressure. The capillary pressure is maximum at the starting of formation of the droplet at the pore and it decrease as the droplet grows. The capillary pressure becomes the minimum at the detachment time of the droplet. The effective or drag pressure part is responsible for the flow rate of dispersed phase. The effective pressure determines the throughput and thus the productivity of the membrane emulsification system. The transmembrane pressure controls the size of the droplet formed in membrane emulsification process. With the increase in transmembrane pressure, several researchers (Katoh et al, 1996; Peng and Williams, 1998; Schröder and Schubert, 1999) have observed the increase in droplet diameter while others (Abrahamse et al., 2002; Vladisavljevic and Schubert, 2003; Vladisavljevic et al., 2004) have observed the decrease in droplet diameter. The variation in droplet diameter with transmembrane pressure results the wide range of droplet size distribution (Abrahamse et al., 2002; Vladisavljevic et al., 2004). The wide range of droplet size also is affected by the steric hindrance between droplets or by the membrane being wetted by the dispersed phase, causing coalescence on the membrane.

The design and pore distribution of the membrane are important factors controlling the droplet dynamics in membrane emulsification. Due to presence of multi-pore and multi-droplet formation, there is a change in hydrodynamic effects caused by neighboring droplet and interactions between the droplets. The separation distance between the pores controls those hydrodynamic effects. If the separation distance of pores in the flow direction is small, the continuous phase velocity decreases and the boundary layer thickness increases as the flow approaches consecutive rows after crossing the first row. These would lead to an increase in the size of the droplets. With the increase in droplet size, there would be a caution of stability loss and coalescence of the droplets. For high efficiency of the emulsification process, narrow droplet size distribution and higher dispersed phase velocity is required. However, with the increase in dispersed phase flow rate, the droplet formation phenomenon shifts towards jetting (Pathak, 2011) and this requires a greater distance between the pores in the direction of the cross-flowing continuous phase in order to prevent drops from colliding and coalescing. In several experimental studies (Sugiura et al., 2002; Kobayashi et al, 2003, 2006) the droplet size distribution has been observed narrow up to a specific velocity of the dispersed phase, above which the diameter of the droplet distribution has been increased. Timgren et al. (2009) have investigated the effects of pore size distribution on hydrodynamic effects of droplet size and distribution. They observed that for small pore separation distance and with a low dispersed phase velocity the drop formation process was uniform, resulting an emulsion with a narrow drop size distribution. For shortest pore separation distance, with the increase in dispersed phase velocity, they observed the formation of poly dispersed emulsion, whereas pore separations of 15 and 20 times the pore diameter gave nearly mono dispersed emulsions.

The wetting behavior of membrane surface also controls the droplet growth. The wetting behavior of a membrane is represented by the static contact angle between the two liquid phases and the solid boundary. The static contact angle between the two phases and walls controls the evolution of the dispersed phase inside the micro-pore and in the continuous phase flow channel. If the angle is less than 90° the wall is said to be wetting and if it is greater than 90°, the wall is called the non-wetting.

Among different properties of the phase, surface tension controls the droplet dynamics in a greater way than any other properties. Surface tension force holds the droplet and offers the resistance against any deformation. The viscosities of both the phases have also effect on the droplet deformation. The drag force imparted by the continuous phase on the droplet depends upon the viscosity ratio of dispersed phase and continuous phase. For fixed flow rate of continuous phase the drag force increases with the increase in viscosity ratio up to some extent. After that value, the drag force becomes independent of the viscosity ratio. The densities of both phases enter into the droplet dynamics through the buoyancy or gravity force. In micro- or nano-fluidics flow the value of gravity force is very small and it can be neglected without loss of much accuracy.

1.1. Different forces acting in membrane emulsification.

All the parameters discussed in above control the droplet dynamics in membrane emulsification process with different magnitudes and output of the process can be analyzed on the basis of these operating parameters. Besides the individual effects, many of these parameters exhibit coupling effects. Different types of hydrodynamic forces act in the emulsification process. The droplet growth and deformation in membrane emulsification can be explained from the action of these and the final droplet size is a result of the interaction of these forces.

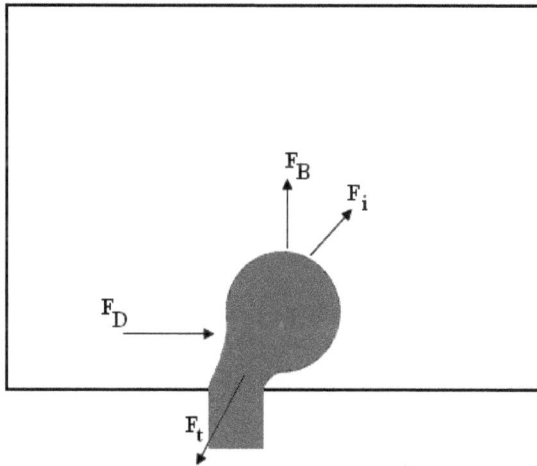

Figure 2. Different forces acting on the emulsification system

The major forces that act in the process are: drag force imparted by the flowing continuous phase, the interfacial tension force, the inertial force of the dispersed phase and the buoyancy or gravitational force. Different forces acting in the droplet formation process are shown in Fig. 2. Among these forces, interfacial tension force is the attaching force and other are detaching force. The droplet is detached from the pore when the detaching forces overcome the attaching force.

These four forces can be approximated as follows:

Drag force:

$$F_D = \frac{1}{2}C_d\rho_{cp}(v^* - v_{dn})^2(\pi D_P^2 / 4) \tag{4}$$

Surface tension force:

$$F_t = \pi\sigma D_p \tag{5}$$

Inertial force of dispersed phase:

$$F_i = \frac{\pi\rho_{dp}D_0^2 v_{dp}^2}{4} \tag{6}$$

Buoyancy force:

$$F_B = \rho_c g V_{dr} - p_c A_n \tag{7}$$

In above v_{dp} is the velocity of dispersed phase ($v_{dp} = v_o$), v^* is the local continuous phase velocity at the centre of the drop, D_p is the diameter of the drop, D_0 is the pore diameter and V_{dr} is the droplet volume. The local continuous phase velocity is given by:

$$v^* = 2u_{cp}\left[1 - \left(\frac{D_h - D_p}{D_h}\right)^2\right] \tag{8}$$

where u_{cp} is the average velocity of the continuous phase in the channel. D_h is the hydraulic diameter given by:

$$D_h = 2bh / (b + h) \tag{9}$$

where b and h are the width and height of the continuous phase channel respectively. The drag coefficient C_d depends upon the Reynolds number of the droplet (Re_p) and the viscosity ratio λ (μ_{dp} / μ_{cp}). The drop Reynolds number is defined as

$$Re_p = \frac{\rho_{dp}\left|v^* - v_{dp}\right|D_p}{\mu_{dp}} \tag{10}$$

Out of these forces, the only attaching force is the surface tension force and remaining forces, drag and inertial and buoyancy force are detaching forces. Neglecting the buoyancy or gravity force, the balance of forces at the moment of droplet detachment can be written as

$$\frac{1}{2}C_d\rho_{cp}(v^* - v_{dp})^2(\pi D_P^2 / 4) + \frac{\pi\rho_{dp}D_0^2 v_{dp}^2}{4} = \pi\sigma D_p \tag{11}$$

For low value of *We* number, the value of inertial force (F_i) is very low. In that situation the diameter of the drop can be approximated as

$$D_p \propto \frac{\sigma}{(C_d \rho_{cp})(v^* - v_{dp})^2} \tag{12}$$

Thus analytically, droplet diameter increases with the increase in surface tension value.

1.2. Numerical simulation of droplet dynamics

Droplet formation and deformation have been studied for a long time due to complexity with the problem and practical utilities of the phenomenon. Droplet formation in a two-phase flow system possesses a rich dynamics with the involvement of several parameters such as average velocity of the liquids, their viscosities, densities, surface tension, surface chemistry and the flow geometry. Droplets formation results the creation of new surfaces which enhance the heat and mass transfer between the phases. Due to enhanced heat and mass transfer, the process has been used for wide ranges of phase-contact applications. Particularly the droplet formation in micro or nano size has received significant attention during last several years. Due to miniature size, the fabrication of experimental facility is expensive and reliable experimentation of microfluidic is very intricate. Hence the viable alternate is the numerical tools for investigating the problems. With the development of high speed computer and advanced algorithm, numerical modeling and simulation have become an essential part in the design and development of numerous engineering systems. Numerical simulations of droplet dynamics i.e. the investigations of two-phase flow in micro scale have been extensively undertaken during last several years. Various types of numerical techniques have been developed to solve the governing equations of the two-phase flow.

1.3. Different numerical methods

In numerical simulation of droplet dynamics or as a whole in the simulation of two-phase flow, there are several challenges which need to be carefully tackled to obtain reliable results. The main challenge is capturing the moving interface of the two phases accurately, which is not known priory. The accurate tracking of the interface and investigation of two-phase flow topology should be the essentiality of a good numerical method. There are several numerical methods based on interface kinematics to track the interface in free surface flows. Among them are: volume of fluid methods, front tracking methods, level set methods, phase field formulations, continuum advection schemes, boundary integral methods, particle-based methods, and moving mesh methods.

Volume of fluid method (VOF), earlier known as the volume tracking method, were originally developed by Nichols and Hirt (1975), Noh and Woodward (1976) and further extended by Hirt and Nichols (1981). Since then, the method has been extensively used and significantly improved over the years (Rudman,1997; Rider and Kothe,1998). The VOF method is based on the conservation of the volume fraction function F with respect to time and space, expressed as

$$\frac{\partial F}{\partial t} + (v.\nabla)F = 0 \qquad (13)$$

In VOF method, the computational grid is kept fixed and the interface between the two fluids is tracked within each cell through which it passes. In a computational grid cell, the interface can be effectively represented by line of the slope. To reconstruct the interface, the piecewise linear interpolation calculation (PLIC) method developed by Youngs (1982) is used in the computation. The interfacial surface forces are incorporated as body forces per unit volume in the Navier-Stokes equations; hence no extra boundary condition is required across the interface.

The basic working principle of front tracking method is based on the marker and cell (MAC) formulation (Harlow and Welch, 1966; Daly, 1967). The interface is represented discretely by Lagrangian markers connected to form a front which lies within and moves through a stationary Eulerian mesh. As the front moves and deforms, interface points are added, deleted, and reconnected as necessary. Further details of the method may be found in (Glimm et al., 1985; Churn et al., 1986; Tryggvason et al., 1998).

Level set methods have been developed by Osher and Sethian (1988). This method can compute the geometrical properties of highly complicated interface without explicitly tracking the interface. The basic principle of the level set method is to embed the propagating interface $\Gamma(t)$ as the zero level set of a higher dimensional function ϕ, defined as $\phi(x, t=0) = \pm d$, where d is the distance from x to $\Gamma(t=0)$. The function is chosen to be positive (negative) if x is outside (inside) the initial position of the interface $\Gamma(t=0) = \phi(x, t=0)=0$. Afterwards a dynamical equation for $\phi(x, t=0)$ that contains the embedded motion for $\Gamma(t)$ as the level set $\phi=0$ can be derived similarly as in the volume of fluid conservation equation (13)

In phase field method, interfacial forces are modeled as continuum forces by smoothing interface discontinuities and forces over thin but numerically resolvable layers. This smoothing allows conventional numerical approximations of interface kinematics on fixed grids. The method has been used for investigating the problems governed by Navier-Stokes equations (Antanovskii, 1995; Jacqumin, 1996).

Boundary integral methods are designed to track the interface explicitly, as in front tracking methods, although the flow solution in the entire domain is deduced solely from information possessed by discrete points along the interface. The advantage of these methods is the reduction of the flow problem by one dimension involving quantities of the interface only.

Particle-based methods use discrete "particles" to represent macroscopic fluid parcels. Here, Lagrangian coordinates are used to solve the Navier-Stokes equations on "particles" having properties such as mass, momentum, and energy. The nonlinear convection term is modeled simply as particle motion and by knowing the identity and position of each particle, material interfaces are automatically tracked. By using particle motion to approximate the convection terms, numerical diffusion across interfaces (where particles change identity) is virtually zero; hence interface widths are well defined.

In moving mesh methods, the position history of discrete point's x_i lying on the interface is tracked for all time by integrating the evolution equation, forward in time.

$$\frac{dx_i}{dt} = v_i \qquad (14)$$

A moving mesh is Lagrangian if every point is moved, and mixed (Lagrangian-Eulerian) if grid points in a subset of the domain are moved. Mixed methods are used for mold filling simulations, where the mold computational domain can be held stationary and the molten liquid is followed with a Lagrangian mesh.

Besides these individual methods, there are some combined method such as coupled level set and volume of fluid method. This method has been developed to tackle the inefficiency of VOF method in calculating complex geometrical properties and problem of mass imbalance of level set method. In the coupled method, the LS function is used only to compute the geometric properties (normal and curvature) at the interface while the void fraction is calculated using the VOF approach. Ohta et al. (2007) and Sussman et al. (2007) developed a novel coupled LS-VOF method to determine the sharp interface for incompressible, immiscible two-phase flows for large value of density ratio.

1.4. Issues of numerical simulation

Numerical investigation of active droplet formation is reasonably complex as it is potentially a multi-physics problem governed by a large number of partial differential equations (PDEs). Numerical simulation of droplet dynamics in membrane emulsion process requires accurate capturing of the evolving liquid–liquid interfaces. This gives sufficient challenge since the boundary between the two phases is not known priori and it is a part of the solution. In addition, the solution technique has to deal with different properties of both the phases such as density, viscosity, and velocity ratios. Numerical investigation of membrane emulsification process requires two main issues to be dealt with: one is permeation of the disperse phase through the membrane pores; the other is the mechanism of droplet detachment. Several earlier investigations have treated these issues separately; however both of them simultaneously contribute to droplet evolution and should be considered in the same framework. Both these issues have been considered in the present work. In the present work, numerical simulation of droplet dynamics in a membrane emulsification has been carried out considering multipore membrane. The hydrodynamic effects due to multipores and the effect of different operating parameters on the droplet dynamics have been investigated.

2. Problem description

The flow configuration of membrane emulsification considered in the present work has been shown in Fig. 3. Here the membrane emulsification process with two pores has been considered. A uniform pore arrangement in the membrane surface has been considered and the simulation has been mode for a row of pores consisting of two pores. The dispersed phase liquid has been injected through cylindrical pores of 10 μm diameter and a length of 100 μm. The distance between the pores in cross-flow direction has been considered as 100 μm. The height of the rectangular channel through which the continuous phase flows has been considered as 150 μm (y-direction). The width of the computational domain has been considered as 150 μm (z-direction) and a length of 500 μm (x-direction) has been considered. Since the present computational domain is a small element of whole membrane emulsification process, symmetrical boundary conditions have been considered in both the sides of computational domain in z-direction. The fluid properties used in the present simulation are within the range of properties of o/w emulsion system. Simulations have been made for different values of the non-dimensional numbers and other flow properties as shown in Table 1.

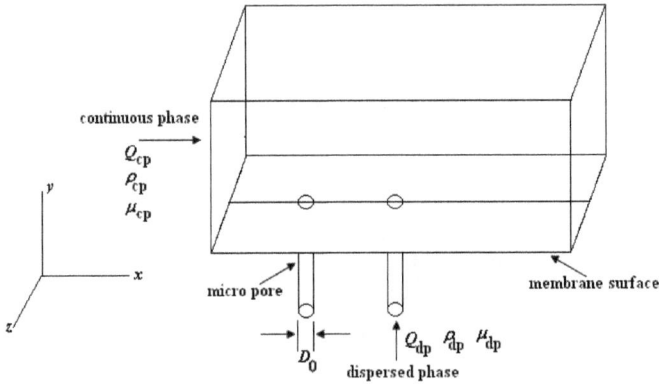

Figure 3. Schematic of computational domain

2.1. Governing equations

In the simulation of membrane emulsification system, both the phases have been considered incompressible, isothermal and laminar flow. The phase properties and the surface tension have been assumed to be constant throughout the flow domain. The conservation of mass on the whole domain (both the fluid phases and interface) leading to continuity equation is written as:

$$\frac{\partial \rho}{\partial t} + \nabla.(\rho V) = 0 \tag{15}$$

The momentum equation or unsteady Navier-Stokes equation is written as:

$$\rho\left(\frac{\partial V}{\partial t} + V.\nabla V\right) = -\nabla p + \rho g + \nabla.\left[\mu(\nabla V + \nabla V^T)\right] + f_s \tag{16}$$

In above f_S is the surface force, which includes surface tension force of the interface of the two fluids.

Parameters	Ranges
Diameter of pore	10 μm
Flow rate of continuous phase	0.27 and 0.54 liter/h
Flow rate of dispersed phase	0.0014 to 0.007 liter/h
Viscosity of continuous phase	0.001 Pa.s
Viscosity of dispersed phase	0.0036 to 0.014 Pa.s
Density of continuous phase	1000 kg/m³
Density of dispersed phase	827 kg/m³
Surface tension	0.0008 to 0.0024 N/m
Weber no (We)	0.0021 to 0.215
Capillary no (Ca)	0.0208 to 0.0625
Froude no (Fr)	6.37 to 637

Table 1. Values of physical properties used in the simulation

2.2. Interface tracking

Volume of fluid method (VOF) has been used for tracking the interface. In this method, the distribution of volume fraction is solved from its transport equation.

$$\frac{\partial F}{\partial t} + V.\nabla F = 0 \tag{17}$$

With the inclusion of volume fraction in the calculation, the volume averaged density and viscosity (ρ and μ) are defined as:

$$\rho = \rho_1 + F(\rho_2 - \rho_1) \tag{18}$$

$$\mu = \mu_1 + F(\mu_2 - \mu_1), \tag{19}$$

where subscript 1 and 2 denote the continuous and dispersed phase respectively. The surface force f_S appearing in momentum equation has been calculated using continuous surface force (CSF) model proposed by Brackbill et al. (1992) . It is defined as:

$$f_S = -\sigma\kappa\nabla F, \tag{20}$$

where σ and κ are the surface tension and surface curvature respectively. Surface curvature is calculated from:

$$\kappa = \nabla.\frac{\nabla F}{|\nabla F|}. \tag{21}$$

2.3. Boundary and initial conditions

Various types of boundary conditions have been used in the simulation of the membrane emulsification. A fully developed laminar duct flow has been considered at the inlet to the continuous phase channel and the flow has been assumed to be dominant along the stream direction thus: $u = u_{cp}$, $v = 0$, $w = 0$. At the inlet of continuous phase, the value of volume fraction has been set to zero ($F = 0$). At the inlets to the micro-pores, a velocity inlet type boundary condition has been used: $v = v_0$. The other components of the velocity have been put as $u = 0$, $w = 0$ and the value of volume fraction has been set to $F = 1$. Outflow boundary condition has been put at the outlet of the flow channel. Symmetry boundary conditions have been put at both sides of the computational domain. The top of the channel has been treated as walls where no-slip and impermeability boundary conditions have been set. At the starting of the computation, the whole flow domain has been assumed to be filled up with the continuous phase fluid. The static contact angle between the two phases and the wall influences the droplet growth in the dispersed and continuous phase channel. It has been observed that the effect of contact angle in the droplet formation is negligible when its value is greater than 165° (Sang et al., 2009). The value of static angle used in the present simulation has been set to 170°.

To reduce the number of dependable variables, the governing equations have been expressed in dimensionless form. The diameter of the pore (D_0) has been used as length scale and the average velocity (v_0) of the dispersed phase has been used as velocity scale for making non-dimensional form of the equation. The non-dimensional equations are:

$$\frac{\partial \rho^*}{\partial t^*} + \nabla.(\rho^* V^*) = 0 \tag{22}$$

$$\left(\frac{\partial V^*}{\partial t^*} + V^*.\nabla V^*\right) = -\nabla p^* + \frac{\rho^*}{Fr} + \frac{1}{Re}\nabla.\left[(\nabla V^* + \nabla V^{*T})\right] + \frac{1}{We}.\nabla.\frac{\nabla F}{|\nabla F|} \tag{23}$$

In above starred quantities are non-dimensional parameters. The three non-dimensional numbers appeared in the problem are: Reynolds number (Re), Weber number (We) and Froude number (Fr). They are defined as:

$$Re = \frac{\rho_{dp} v_0 D_0}{\mu_{dp}}; \ We = \frac{\rho_{dp} v_0^2 D_0}{\sigma}; \ Fr = \frac{v_0^2}{g D_0};$$

The Reynolds numbers (Re) is defined as the ratio of viscous force to the inertia force, Weber number (We) is defined as the ratio of inertia force relative to the surface tension force and Froude number (Fr) is defined as the ratio of inertia to gravity force. In the present work to incorporate the effect of the continuous phase fluid, the Capillary number has been introduced and is defined as:

$$Ca = \frac{\mu_{cp} u_{cp}}{\sigma} = \frac{RWe}{\lambda \, \text{Re}}$$

where λ is the viscosity ratio ($\lambda = \mu_{dp} / \mu_{cp}$) and R is the velocity ratio ($R = u_{cp} / v_0$).

2.4. Numerical method

Commercial code Ansys Fluent (V12) based on finite volume method has been used in the simulation. The momentum and volume fraction transport equation have been discretized with 2nd order upwind scheme. The PISO (pressure implicit with splitting of operators) algorithm has been used for pressure correction. The VOF/CSF techniques have been used to track the fluid interface between the two immiscible fluids. A geometry reconstruction scheme has been used in the simulation to avoid the diffusion at the interface. The interface was reconstructed by the piecewise-linear interface calculation (PLIC) technique (Youngs, 1982). The unsteady term was treated with first-order implicit time stepping. Simulations were made with very small time steps ($\sim 10^{-7}$ s). The solutions have been assumed to be converged and therefore iterations have been terminated when the normalized sum of residual mass was less than 10^{-4} and variation of other variables in successive iteration was less than 10^{-2}. A non-uniform grid was used in the simulation where grids were clustered near the walls and the injection portion of the dispersed phase. The channel was decomposed into 12×10^5 numbers of control volumes and the pore with 12×10^3 after a grid independent study.

3. Results and discussions

In the present work, the dynamics of droplet formation in two pores of membrane emulsification has been investigated for different flow rates (velocities) of dispersed phase, continuous phase, surface tension and viscosity of the two phases. It is to be noted that In case of membrane emulsification process, the dispersed phase fluid gets more space to interact with the continuous phase fluid compared to the confined geometries in case of T-junction emulsification. Due to this the evolution of dispersed phase becomes different than the case of T-junction emulsion and the dependence of the process on different properties of both the phases also changes.

3.1. Growth and detachment of the droplets

In membrane emulsification system, the flow rate of dispersed phase controls the droplet dynamics via its inertial force competing with the drag force imparted by the continuous phase and interfacial tension force. In order to investigate the effect of dispersed phase flow rate, simulations have been made for different values of We number by changing the dispersed phase velocity i.e. inertial force and keeping the surface tension force fixed. Before discussing the effects of We, the growth and detachment of the droplet for a constant values of We (0.0086) and Ca (0.028) at different time levels have been shown in Fig. 4. It has been observed that both the droplet grow at their respective micro-pore and detached by the continuous phase at the same location. As the first droplet is being detached and carried away by continuous phase fluid, the second droplet starts to grow at the pore and the repetition of droplets detachment takes place periodically with constant volume of droplet. Simulation has also revealed that the growth rate of the droplet at the downstream micro pore is different than the upstream micro pore and the droplet at that pore requires more detachment time. The presence of upstream droplet changes the hydrodynamic effects and reduces the viscous drag force of continuous phase on the downstream droplet. Due to reduction in drag force, the surface tension force can hold the droplet for longer time and detachment takes place lately. Due to this the droplet size is greater than the droplet formed at the upstream pore. Moreover due to low drag force of the continuous phase, the inertia force of dispersed phase has some effect in forming the necking of the droplet in the downstream pore. The diameter of the droplet after detachment has been found as 41.3 μm, which is about 4.13 times the pore diameter. This ratio is within the range of 1-12 observed by Katoh et al. (1996).

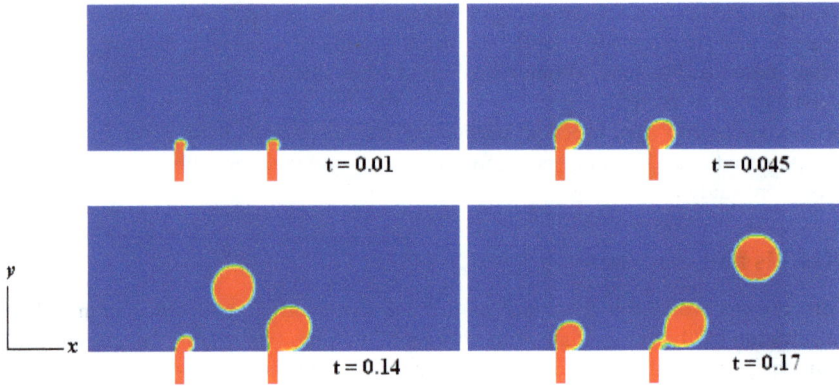

Figure 4. Growth and detachment of droplet at different time level, $We = 0.0086$, $Ca = 0.0208$

Thus the distance between the two pores should be at least five times the pore diameter for simultaneous growing of two spherical droplets without any hindrance from the neighboring pore. On the other hand the droplet deforms in the direction of flow of the continuous phase as shown in Fig. 4. Considering the deformation and growth of the droplet, it can be concluded that a distance of almost 10 times the pore diameter between the two pores in x-direction i.e. in cross-flow direction is needed to avoid contact, and thus avoid coalescence at two neighboring pores for this particular flow rate of dispersed phase and surface tension.

The investigation of droplet growth in transverse direction i.e. z-direction is important to design the pore distance in transverse direction. The growth of the droplet in a horizontal plane (x-z) at at $y/D_0 = 25$ μm has been shown in Fig. 5 at different time levels for $We = 0.0086$ and $Ca = 0.0208$. As the time progresses, the droplet at the first pore grows and deforms along the cross-flow fluid direction. Thus the distance between the two forming droplets decreases. It has been observed that in the transverse (z-) direction the droplet remains almost spherical throughout the formation process; therefore, the distance between the pores in that direction can be fixed at seven times the pore diameter.

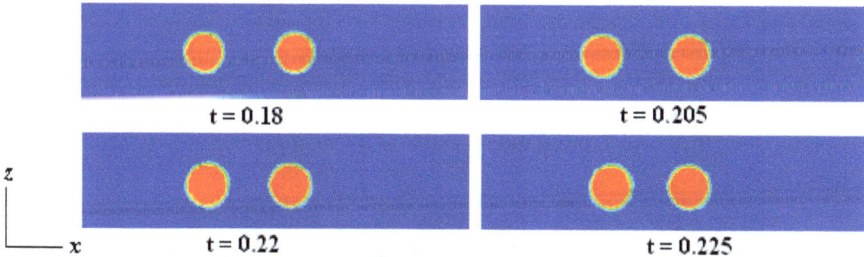

Figure 5. Growth of droplets at horizontal x-z plane at different time level, $We = 0.0086$, $Ca = 0.0208$

With the fixed distance along cross-flow and transverse direction, the maximum porosity of the membrane can be calculated. The porosity is defined as the ratio of the total pore cross-sectional area and the total membrane surface area.

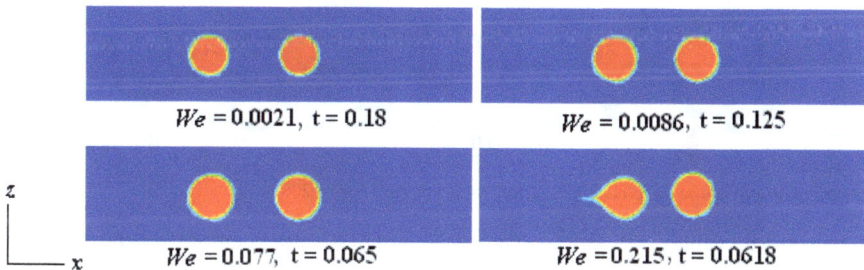

Figure 6. Growth of droplets at horizontal x-z plane for different values of We, $Ca = 0.0208$

To check the effect of We on the droplet growth along transverse direction, the simulated results of droplet growth at horizontal plane for different values of We and at different time levels have been shown in Fig 6. With the increase in We the droplet size increases and the distance between the two droplets decreases along the continuous phase fluid flow direction. At high value of We the necking phenomenon has been observed which makes the insufficient distance between the pore for avoiding coalescence. On the other hand at high value of We the growth of the droplet in transverse direction is same as the case in lower value of We. Thus the spacing of pores in transverse direction can be fixed based on droplet diameter and irrespective of the dripping or jetting mode.

3.2. Effect of dispersed phase flow rate

In order to show the effect of dispersed phase flow rate in droplet dynamics, the droplet growth before the detachment has been shown in Fig.7 for different values of We (0.0021 to 0.215) number. The qualitative difference in droplet growth for different values of We can be seen in the figure. Some key phenomena such as dripping at low We number, necking and jetting at higher We number have been observed. During the droplet formation in membrane emulsification process, the inertial force of the dispersed flow acts as detaching force, and acts against the attaching force of surface tension. When inertia force of the dispersed is less than the drag force or interfacial tension force, it cannot influence the droplet dynamics and the droplet growth and detachment are controlled by the drag force competing with the surface tension force. At the low value of We (We =0.0021) number, the droplet forms and breakups at the micro-pore which is termed as the dripping mode.

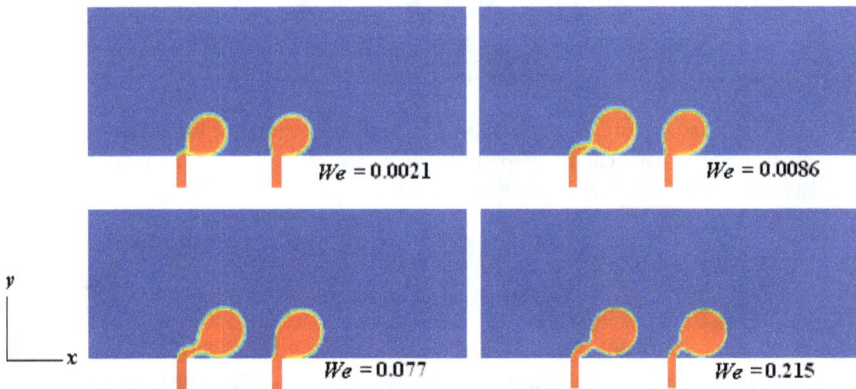

Figure 7. Growth of the droplets for different values of We, $Ca = 0.0208$

At intermediate values of Weber number (0.0086 and 0.077), the point of droplet detachment has been moved away from the micro-pore, and formation of dispersed phase thread and necking have been observed. With the increase in inertial force of the dispersed phase, the effective pressure overcomes the capillary pressure inside the liquid thread leading to a stretched filament and also distends the droplet neck noticeably. With the increasing in dispersed phase flow rate further (0.215), two nodes form in the liquid filament and extension of the droplet neck occurs. The detachment point of the droplets moves further downstream from the pore. Thus jetting occurs and the droplet forms at the tip of the droplet. A decrease in the resultant droplet size can be observed. Thus at higher value of We number, due to formation of the jetting and the droplet formation from the tip, the distance of 10 pore diameter between the pores is not sufficient to avoid the contact and coalescence of two neighboring droplets.

3.3. Velocity field during droplet growth

The droplet dynamics in membrane emulsification process is controlled by the evolving velocity field outside and inside the dispersed phase since the drag force is correlated to velocity. The velocity fields for $We = 0.0086$ in the central vertical plane (x-y) at the time level

t = 0.125 have been shown in Fig. 8. Recirculating flows have been observed inside the both droplets and the center of recirculation is different in both the droplets.

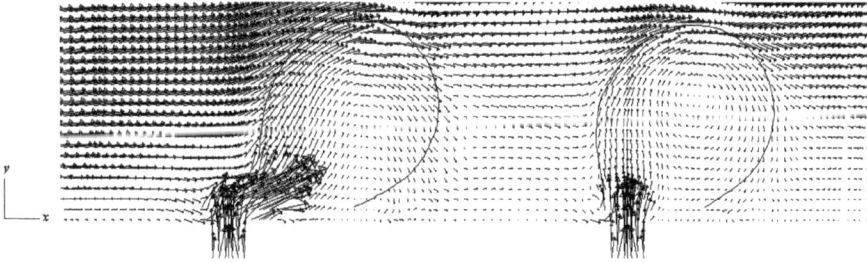

Figure 8. Velocity vector at the central plane, We = 0.0086, Ca = 0.0208

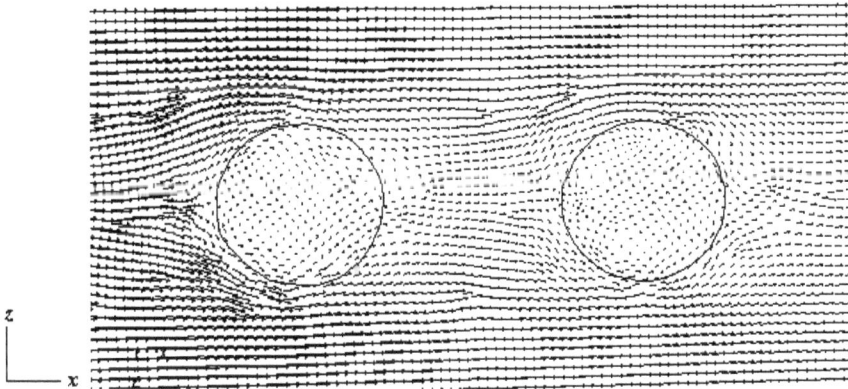

Figure 9. Velocity vector at the horizontal plane, We = 0.0086, Ca = 0.0208

As the continuous liquid phase interacts the dispersed phase, it imparts a viscous drag force on the evolving interface between the two phases. The viscous drag force produces shear stress along the interface that faces the continuous phase fluid. This initiates the recirculation inside the both droplets. The acceleration of the dispersed phase out of the pore also affects the motion inside the forming drop, especially at an early stage of drop formation. The centre point of the rotational flow inside the drop is at the top of the drop, which is controlled by the above two factors. The dispersed phase inside the interface front finally flows along the continuous liquid phase and is accelerated by the viscous drag. From the velocity diagram it can be seen that the upstream droplet has disturbed the approaching velocity field for the downstream growing droplet. Thus the droplet that grows in the "shade" of another droplet experiences a different velocity profile.

The velocity field inside and outside the dispersed phase in horizontal plane (x-z) at the time level (0.125) for We = 0.0086 has been shown in Fig. 9. The wake formed by the first droplet is visible from the velocity field, where velocity field is weak.

3.4. Effect of continuous phase velocity

To show the effect of continuous phase velocity, the simulation has been made for same range of Weber number (0.00 to 0.215) but with higher continuous phase flow rate (Q_{cp} = 0.54 l/h). The comparison of droplet diameter for two values of continuous phase velocities i.e. flow rates has been shown in Fig. 10. It has been observed that for the higher continuous phase flow rate, the diameter of the droplets is smaller compared to lower continuous phase flow rate. At a particular value of continuous phase flow rate, the droplet diameter increases with the increase in droplet diameter, than decreases with the increase in droplet diameter due to jetting phenomenon. It can be seen that with the increase in continuous phase velocity, the droplet diameter decreases. Due to increase in velocity, the continuous phase fluid imparts higher drag force and droplet detaches within short interval. It has been also observed that the reduction of droplet diameter during transition from dripping to jetting is less in case of higher value of continuous phase velocity compared to its lower value.

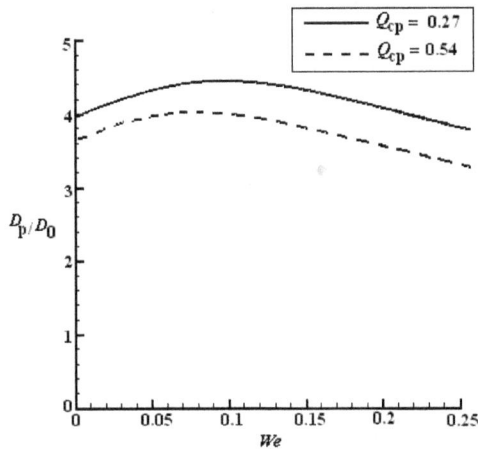

Figure 10. Effect of continuous phase velocity on droplet diameter

The detachment time of droplets for two values of flow rates of continuous phase have been shown in Fig. 11. With lower value of continuous phase flow rate, the detachment time decreases exponentially with the increase in We number up to some value of We number, after that the decrease rate reduces. At a particular We number, the detachment time has been observed less for higher continuous phase flow rate compared to lower continuous phase flow rate. For lower flow rate of continuous phase, the detachment time decreases with the increase of We number, but it does not follow the same trend as for the lower flow rate.

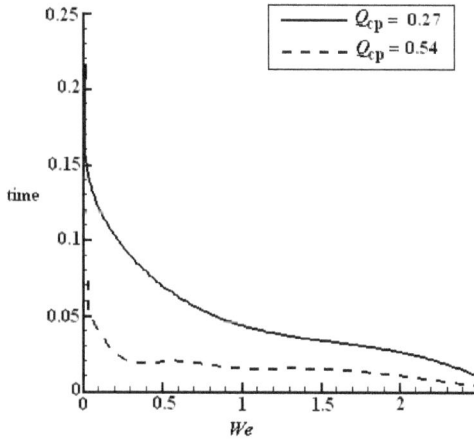

Figure 11. Detachment time of the droplet for different *We* number

3.5. Effect of surface tension

To show the effect of surface tension on the droplet formation, the simulation has been made for different values of surface tension (*Ca* = 0.0208 to 0.0625) at constant value of inertial force. The droplet diameters for different values of *Ca* and two values of continuous phase flow rate have been shown in Fig. 12. With the decrease in surface tension force i.e. increase of *Ca*, the droplet diameter decreases. At low value of surface tension, the attaching surface tension force cannot hold the droplet for longer time against the other detaching forces leading to formation of droplet with smaller size. At very low value of surface tension (0.0008 N/m) even at low value of *We* (0.0021) the phenomenon shows the jetting behavior as shown in Fig. 13. Due to jetting phenomenon, the spacing between the micro-pore is not sufficient to avoid the coalescence of the droplets at two neighbor pores.

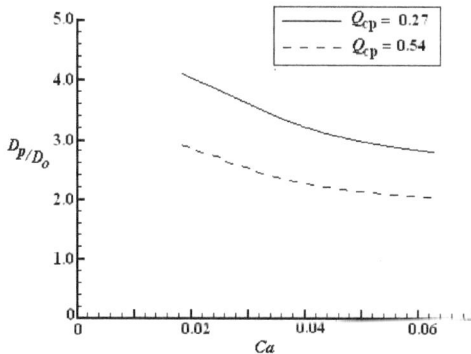

Figure 12. Effect of surface tension on droplet diameter

From the present investigation, it can be seen that dripping to jetting transition can be possible in two ways: one at constant surface tension while varying the inertial force (varying the dispersed phase flow rate) and another at constant inertial force while varying the surface tension force. A qualitative difference in flow pattern in both the transitions has been observed. During dispersed phase controlled transition, the diameter of the drop increases first then decreases rapidly while in surface tension controlled transition, the size of the droplet continuously decreases (Fig. 10 and Fig. 12).

Surface tension force is dominant over the inertial force in dripping mode, hence the droplet size increases and in jetting mode inertial force overcomes the surface tension force for which drop size decreases. In surface controlled breakup, the drag force and interfacial force take part in the droplet formation.

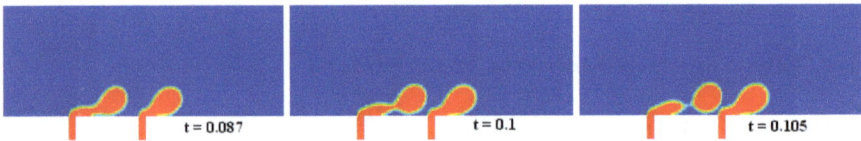

Figure 13. Jetting behavior at low surface tension value

3.6. Effect of viscosities

The viscosities of both the phases also effect the droplet growth and deformation as discussed in introduction section. The drag force which detaches the droplet depends upon the viscosity ratio of dispersed phase and continuous phase. At a constant value of continuous phase flow rate, the drag force increases with the increase in viscosity ratio up to some extent. After that, the drag force becomes independent of the viscosity ratio. Pathak (2011) has observed negligible effect of dispersed phase viscosity in the membrane emulsification system. Moreover it has been also observed that the effect viscosity on droplet dynamics is not very significant if the ratio of dispersed phase and continuous phase viscosity is high (van Dijke et al., 2010). Since the viscosity ratio of dispersed phase and continuous phase fluid is above 3, the effects of viscosities have not been investigated in the present work.

4. Conclusions

Droplet formation in a two-pore membrane emulsification has been numerically investigated in this chapter. The dynamics of droplet formation has been investigated by solving the two-phase governing equations using VOF method. The effects of various parameters, viz., dispersed and continuous phase flow rate, surface tension and viscosities on the droplet dynamics have been investigated. The dynamics of evolution of dispersed phase and droplets formation show the dripping and jetting behavior depending upon the operating conditions and properties of two-phase liquids in the emulsification system. At constant continuous phase flow rate, the dripping phenomenon occurs at low dispersed phase velocity i.e. at low We number and transits towards jetting with the increase in dispersed phase flow rate. At constant continuous phase flow rate, with the increase in dispersed phase flow rate, the droplet size increases initially but decreases as the system transits towards jetting. At constant dispersed phase flow rate, with the increase in continuous phase flow rate, the droplet size decreases and also detachment time. Two ways of dripping to jetting transition have been observed, one with the increasing dispersed phase flow rate at constant continuous phase flow rate and other way is reducing the surface tension at constant dispersed phase flow rate. Both the transitions show different physical structures. The effect of inertia force has been observed the negligible for high value of surface tension and significant for lower surface tension value. The distance between the pore in continuous flow direction depends upon the operating parameters leading to dripping to jetting mode but the pore distance in transverse direction is not affected by the dripping or jetting behavior. Thus at higher value of We number, due to formation of the jetting and the droplet formation from the tip, the distance of 10 pore diameter between the pores is not sufficient to avoid the contact and coalescence of two neighboring droplets. The droplet size in the process scales with four main forces: drag forces imparted by the continuous phase, inertia force imparted by dispersed phase, surface tension force and the gravity force. In dripping mode inertial force of dispersed phase has negligible effect as the surface tension and drag force are dominant whereas in jetting mode inertial force of dispersed phase and surface tension force take part in the droplet formation. The evolving vortices are observed in the initial stage of dripping mode but it disappears in later stage. Three important factors must be considered in order to obtain a high production rate in membrane emulsification. (i) A proper combination of continuous phase, dispersed phase flow rate and surface tension so that droplet formation is made just before the starting of surface instability in jetting region. (ii) A proper distribution of pores so that coalescence of droplets does not occur during the droplet growth. (iii) The crossflow velocity must be high enough to provide a sufficient wall shear stress at the membrane surface to transport the drops away from the pore opening and, thus avoid the static hindrance and drop coalescence.

Author details

Manabendra Pathak
Department of Mechanical Engineering, Indian Institute of Technology Patna, India

Nomenclature

A_n	Area of the droplet neck ((μm)^2)
b	Width of the continuous phase channel (μm)
C_a	Capillary number

C_d	Drag coefficient
D_h	Hydraulic diameter of the continuous phase channel (μm)
D_0	Diameter of micro-pore (μm)
D_p	Diameter of the droplet (μm)
f_s	Surface force (N)
F	Volume fraction of dispersed phase fluid
g	Acceleration due to gravity (m/s^2)
h	Height of the continuous phase channel (μm)
k_1	Constant
k_2	Constant
p	Pressure of the flow (Pa)
p_c	Pressure of the continuous phase fluid flow (Pa)
p_d	Pressure of the dispersed phase fluid flow (Pa)
p_γ	Capillary pressure (Pa)
Q_{cp}	Flow rate of continuous phase fluid (m^3/s)
Q_{dp}	Flow rate of dispersed phase fluid (m^3/s)
R	Viscosity ratio of continuous phase to dispersed phase fluid
Re	Reynolds number of the flow
R_p	Reynolds number of the drop
u	Velocity in x- direction (m/s)
u_{cp}	Velocity of continuous phase at the inlet to the channel (m/s)
v	Velocity in y- direction (m/s)
v^*	Local velocity of continuous phase fluid in the channel (m/s)
v_0	Velocity of dispersed phase at the inlet to the micro-pore (m/s)
V	Velocity vector
V_{dr}	Droplet volume((μm)^2)
w	Velocity in z- direction (m/s)
We	Webber number

Greek letters

r	Position vector of interface of the two phase
ϕ	Level set function
κ	Surface curvature (m^{-1})
λ	Viscosity ratio of dispersed phase to continuous phase fluid
μ_{cn}	Viscosity of continuous phase fluid (m^2/s)
μ_{dn}	Viscosity of dispersed phase fluid (m^2/s)
ρ_{cn}	Density of continuous phase fluid (kg/ m^3)
ρ_{dn}	Density of dispersed phase fluid (kg/ m^3)
σ	Surface tension coefficient (N/m)

5. References

Abrahamse, A. J., van der Padt, A. & Boom, R. M. (2002). Analysis of droplet formation and interactions during cross-flow membrane emulsification, J. Membr. Sci. 204: 125-137.

Antanovskii, L. K. (1995) A Phase Field Model of Capillarity. Phys. Fluids 1: 747-753.

Brackbill, J. U., Kith, D. B. & Zemach, C. (1992). A Continuum method for modeling surface tension. J Comp. Phys 100: 335–354.

Churn, I. L., Glimm, J., McBryan, O., Plohr, B. & Yanic, S. (1986). Front tracking for gas dynamics. J. Comp. Phys. 62: 83-110.

Daly, B. J. (1967). Numerical study of two fluid Rayleigh-Taylor instabilities. Phys. Fluids 10:297-307.

Glimm, J. & McBryan, O. A. (1985). A computational model for interfaces. Adv. Appl. Math. 6: 422-435.

Harlow, F. H. & Welch, J. E. (1966). Numerical study of large-amplitude free-surface motion. Phys. Fluids 9: 842-851.

Hirt, C. W. & Nichols, B. D. (1981). Volume of fluid (VOF) method for the dynamics of free boundaries. J. Comp. Phys. 39: 201-225.

Jacqumin, D. (1996). An energy approach to the continuum surface method. AIAA paper, 96-0858.

Joscelyne, S. M. & Trägårdh, G. (1991). Food emulsions using membrane emulsification: conditions for producing small droplets. J. Food Eng., 39: 59-64.

Joscelyne, S. M. & Trägårdh, G. (2000). Membrane emulsification – a literature review. J. Membr. Sci.169: 107-117.

Karbstein, H., & Schubert, H. (1995). Developments in the continuous mechanical production of oil-in-water macro-emulsions. Chem. Engg. Process. 34: 205-211.

Katoh, R., Asano, Y., Furuya, A., Sotoyama, K. & Tomita, M. (1996). Preparation of food emulsions using membrane emulsification system. J. Membr. Sci. 113: 131-135.

Kobayashi, I., Nakajima, M. & Mukataka, S. (2003). Preparation characteristics of oil-in-water emulsions using differently charged surfactants in straight-through microchannel emulsification. Colloids Surf. A 229: 33–41.

Kobayashi, I., Uemura, K. & Nakajima, M. (2006). CFD study of the effect of a fluid flow in a channel on generation of oil-in-water emulsion droplets in straight-through microchannel emulsification. J. Chem. Eng. Jpn. 39: 855–863.

Luca, G. D., Sindona, A., Giorno, L. & Drioli, E. (2004). Quantitative analysis of coupling effects in cross-flow membrane emulsification. J. Membr. Sci. 229: 199-209.

Nichols, B. D. & Hirt, C. W. (1975). Methods for calculating multi-dimensional, transient free surface flows past bodies. Technical Report LA-UR-75-1932, Los Alamos National Laboratory. NM.

Noh, W. F. & Woodward, P. R. (1976). SLIC (simple line interface methods). In: Voore A I V, P. J. Zandbergen P J, editors. Lecture Notes in Physics 59: 330-340.

Ohta, M., Kikuchi, D., Yoshida, Y. & Sussman, M. (2007). Direct numerical simulation of the slow formation process of single bubbles in a viscous liquid. J. Chem. Eng. Jpn. 40: 939-943.

Osher, S. & Sethian, J. A. (1988). Fronts propagating with curvature-dependent speed: algorithms based on Hamilton-Jacobi formulation. J. Comp. Phys. 79:12-49.

Pathak, M. (2011). Numerical simulation of membrane emulsification: effect of flow properties in the transition from dripping to jetting. J. Membr. Sci. 382: 166-176.

Peng, S., J. & Williams, R. A (1998) Controlled production of emulsions using a crossflow membrane, Part I: droplet formation from a single pore. Chem. Eng. Res. Des., 76: 894-901.

Rider, W. J. & Kothe, D. B. (1998). Reconstructing volume tracking. J. Comp. Phys 141:112-52.

Rudman, M. (1997). Volume tracking methods for interfacial flow calculations. Int. J. Num. Meth. Fluids. 24: 671-691.

Sang, L., Hong, Y. & Wang, F. (2009). Investigation of viscosity effect on droplet formation in T-shaped microchannels by numerical and analytical methods. Microfluid Nanofluid 6: 621–635.

Schröder, V. & Schubert, H. (1999). Production of emulsions using microporous, ceramic membranes. Colloids Surf A 152: 103-109.

Schubert, H., Ax K, & Behrend, O. (2003). Product engineering of dispersed systems. Trends Food Sci & Technol. 14: 9-16.

Sugiura, S., Nakajima, M., Kumazawa, N., Iwamoto, S. & Seki, M. (2002). Characterization of spontaneous transformation-based droplet formation during microchannel emulsification. J. Phys. Chem. B 106: 9405–9409.

Sussman, M., Smith, K. M., Hussaini, M. Y., Ohta, M. & Zhi-Wei, R. (2007). A sharp interface method for incompressible two-phase flows. J. Comp. Phys. 221: 469-505.

Timgren, A., Trägårdh, G. & Trägårdh. C. (2009). Effects of pore spacing on drop size during cross-flow membrane emulsification—A numerical study. J. Membr. Sci. 337: 232-239.

Tryggvason, G., Bunner, B., Ebrat, O. & Tauber, W. (1998). Computations of multi-phase flow by a finite difference/front tracking method. I. Multi-fluid flows. In Lecture Notes for the 29th Computational Fluid Dynamics Lecture Series, Karman Institute for Fluid Mechanics, Belgium.

van Dijke, K., Kobayashi, I., Schroe¨n, K., Uemura, K., Nakajima, M. & Boom. R. (2010). Effect of viscosities of dispersed and continuous phases in microchannel oil-in-water emulsification, Microfluid Nanofluid. 9: 77–85.

Vladisavljevic, G. T. & Schubert, H. (2003). Preparation of emulsions with a narrow particle size distribution using microporous α-alumina membranes. J. Disp. Sci. Technol. 24: 811-819.

Vladisavljevic, G. T., Lambrich, U., Kakajima, M. & Schubert, H. (2004). Production of o/w emulsion using SPG membranes, ceramic α-aluminium oxide membranes, microfluidizer anda silicon microchannel plate - a comparative study. Colloids Surf. A 232: 199-207.

Youngs, D. L. (1982). Time-dependent multi-material flow with large fluid distortion. In: Morton K W, Baines M J, editors, Numerical Methods for Fluid Dynamics, Academic Press 1982.

Chapter 4

Spectral Modeling and Numerical Simulation of Compressible Homogeneous Sheared Turbulence

Mohamed Riahi and Taieb Lili

Additional information is available at the end of the chapter

1. Introduction

This chapter is mainly in the area of the use of Rapid Distortion Theory (RDT) to clarify and to well better increase our understanding of the physics of the compressible turbulent flows. This theory is a computationally viable option for examining linear compressible flow physics in the absence of inertial effects. In this linear limit, the statistical evolution of incompressible homogeneous turbulence can be described completely in terms of closed spectral covariance equations (see Refs. (Hunt, 1990 & Savill, 1987) and references therein). Many papers in literature deal with homogeneous compressible turbulence and RDT solution (Cambon et al., 1993; Coleman & Mansour, 1991; Blaisdell et al., 1993, 1996; Durbin & Zeman, 1992; Jacquin et al., 1993; Livescu & Madnia, 2004; Riahi et al., 2007; Riahi, 2008; Riahi & Lili, 2011; Sarkar, 1995; Simone, 1995; Simone et al., 1997). These studies have yielded very valuable physical insight and closure model suggestions. In all the above works, the fluctuation equations are solved directly to infer turbulence physics. For the case of viscous compressible homogeneous shear flow in the RDT limit no analytical solutions are known. Simone et al. (1997) performed RDT simulations of homogeneous shear flow and showed that the role of the distortion Mach number, M_d, on the time variation of the turbulent kinetic energy is consistent with that found in the direct numerical simulation (DNS) results. In this chapter, numerical solutions to the RDT equations for the special case of mean shear is described completely by finding numerical solutions obtained by solving linear double point correlations equations. Numerical integration of these equations is carried out using a second-order simple and accurate scheme (Riahi & Lili, 2011). Indeed, this numerical method is proved more stable and faster than the previous one which use linear transfer matrix (Riahi et al., 2007 & Riahi, 2008) and allows in particular to obtain accurately the asymptotic behavior of the turbulence parameters (for large values of the non-dimensional times St) characteristic of equilibrium states. To perform this work, RDT code solving linearized equations for compressible homogeneous shear flows is validated by comparing RDT results to those of direct numerical simulation (DNS) of Simone et al. (1997) and Sarkar (1995) for various values of initial gradient Mach number M_{g0} (Riahi & Lili, 2011).

A study of compressibility effects on structure and evolution of a sheared homogeneous turbulent flow is carried out using this theory (RDT). An analysis of the behavior of different terms appearing in the turbulent kinetic energy and the Reynolds stress equations permit to well identify compressibility effects which allow us to analyze performance of the compressible model of Fujiwara and Arakawa concerning the pressure-dilatation correlation (Riahi et al., 2007). The evaluation of this model stays in the field of RDT validity (Riahi & Lili, 2011).

Equilibrium states of homogeneous compressible turbulence subjected to rapid shear can be studied using rapid distortion theory (RDT) for large values of St ($St > 10$) in particular for large values of the initial gradient Mach number M_{g0} describing various regimes of flow. In fact, the study of the behavior of the non-dimensional turbulent kinetic energy $q^2(t)/q^2(0)$ (with $q^2 = \overline{u_i u_i}$) allows to check relevance of an incompressible regime for low values of initial gradient Mach number, of an intermediate regime for moderate values of M_{g0} and of a compressible regime for high values of M_{g0} (Riahi et al., 2007). The pressure-released regime is related to infinite values of M_{g0} (Riahi & Lili, 2011). The gradient Mach number M_g appears naturally when scaling the (linearized) RDT equations for homogeneous compressible turbulence (see equations (A.10), (A.11), (A.12) and (A.13) in appendix (Riahi et al., 2007)). This parameter can be viewed as the ratio of an acoustic time $\tau_a = \dfrac{l}{a}$ for a large eddy to the mean flow time scale $\tau_d = \dfrac{1}{S} : M_g = \dfrac{\tau_a}{\tau_d} = \dfrac{Sl}{a}$, where l is an integral lengthscale and a is the mean sound speed. The lengthscale of energetic structure is expressed by $l = \dfrac{q^3}{\varepsilon}$, where ε is the rate of turbulent kinetic energy dissipation. Sarkar (1995) also used $\dfrac{Sl}{a}$ (which he referred to as 'gradient Mach number' M_g), to quantify compressibility effects for homogeneous shear flow. In the case of shear flow, l is chosen to be the integral lengthscale of the streamwise fluctuating velocity in the shearing direction x_2. Another Mach number relevant to homogeneous shear flow and that characterize the effects of compressibility on turbulence is the turbulent Mach number $M_t = u/a$ based on a characteristic fluctuation velocity u and mean sound speed a.

2. RDT equations for compressible homogeneous turbulence

The flow to be considered is a homogeneous, compressible turbulent shear flow where we retain the same RDT equations adopted by Simone (1995), Simone et al. (1997), Riahi et al. (2007) and Riahi (2008). The linearized equations of continuity, momentum and entropy controlling the fluctuating of velocity u_i and pressure p lead to general RDT equations:

$$\left(\frac{\dot{p}}{\gamma \overline{P}}\right) = -u_{i,i},$$ (1)

$$\dot{u}_i + u_j \frac{\partial \overline{U_i}}{\partial x_j} = -\frac{1}{\overline{\rho}} \frac{\partial p}{\partial x_i} + \frac{\nu}{3} \frac{\partial^2 u_j}{\partial x_i \partial x_j} + \frac{\nu}{3} \frac{\partial^2 u_i}{\partial x_j^2}, ,$$ (2)

where γ is the ratio of specific heats, $\overline{\rho}$ is the mean density and ν is the kinematic viscosity.

The dot superscript denotes a substantial derivative along the mean flow trajectories related to mean field velocity $\overline{U_i}$.

Equation (1) is derived from continuity equation

$$\left(\frac{\dot{\rho}}{\overline{\rho}}\right) = -u_{i,i}$$ (3)

associated to isentropic fluctuations hypothesis $\dot{s} = 0$ which is translated by

$$\left(\frac{\dot{p}}{\overline{P}}\right) = \gamma \left(\frac{\dot{\rho}}{\overline{\rho}}\right)$$ (4)

and leads to equation (1). So, the isentropic hypothesis $\dot{s} = 0$ simplifies governing equations by keeping the variables u_i and p and eliminating ρ. We can obviously discard the hypothesis $\dot{s} = 0$ by writing the (exact) energy equation. The linearized derived equation within the framework of RDT can be written as pressure fluctuation equation which contains some viscous terms (Livescu et al., 2004; see equation (8)). Such an equation is compatible with equation (2) which contains also viscous terms and represents the linearized equation of momentum. Concerning now the validity of isentropic hypothesis $\dot{s} = 0$, Blaisdell et al. (1993) introduced a polytropic coefficient n defined by, $\frac{p}{P} = n \frac{\rho}{\rho}$ $n = \gamma$

corresponding obviously to isentropic fluctuations. These authors performed direct numerical simulations (DNS) related to homogeneous compressible sheared turbulence with various initial conditions. They calculated temporal variation of an average polytropic

coefficient $n = \left(\dfrac{\overline{p^2 / \overline{P}^2}}{\overline{\rho^2 / \overline{\rho}^2}}\right)^{\frac{1}{2}}$. During the period of the turbulence establishment (for early

times), the evolution of n with St depends on initial conditions (and in particular of initial entropy fluctuations). However, for all of their simulations and for large values of St, n tends toward an equilibrium value independent of initial conditions and slightly less than $\gamma = 1.4$. Blaisdell et al. (1993) concluded that thermodynamic variables "follow a nearly isentropic process" for large values of St and for compressible sheared turbulence. As a conclusion, we retain that with isentropic fluctuations hypothesis, we can obtained a good prediction of equilibrium states $(St \rightarrow \infty)$ for compressible homogeneous sheared turbulence.

After these remarks, we consider the Fourier transform (denoted here by the symbol " ^ ") of various terms in equations (1) and (2) which leads to the following equations expressed in the spectral space and in the case of turbulent shear flow:

$$\frac{\hat{\dot{p}}}{\gamma \overline{P}} = -\hat{u}_{i,i}, \tag{5}$$

$$\hat{\dot{u}}_i + \lambda_{ij}\,\hat{u}_j + \frac{\nu}{3}k_i\,k_j\,\hat{u}_j + \frac{\nu}{3}k^2\,\hat{u}_i = -Ik_i\frac{\hat{p}}{\rho}, \tag{6}$$

where λ_{ij} is the mean velocity gradient defined by:

$$\lambda_{ij} = S\delta_{i1}\delta_{j2} \tag{7}$$

and $I^2 = -1$.

In the Fourier space, velocity field is decomposed on solenoidal and dilatational contributions by adopting a local reference mark of Craya (Cambon et al., 1993):

$$\hat{u}_i(\vec{k},t) = \hat{\varphi}^1(\vec{k},t)\,e_i^1(\vec{k}) + \hat{\varphi}^2(\vec{k},t)\,e_i^2(\vec{k}) + \hat{\varphi}^3(\vec{k},t)\,e_i^3(\vec{k}), \tag{8}$$

where $\hat{\varphi}^1(\vec{k},t)$ and $\hat{\varphi}^2(\vec{k},t)$ are the solenoidal contributions and $\hat{\varphi}^3(\vec{k},t)$ is the dilatational contribution.

By separating solenoidal and dilatational contributions, we introduce the spectrums of doubles correlations:

$$\Phi_{ij}(\vec{k},t) = \frac{1}{2}\left[\hat{\varphi}^{i*}(\vec{k},t)\hat{\varphi}^j(\vec{k},t) + \hat{\varphi}^i(\vec{k},t)\hat{\varphi}^{j*}(\vec{k},t)\right] \quad i,j=1..4, \tag{9}$$

where $\hat{\varphi}^4(\vec{k},t)$ is related to the pressure and takes the following form:

$$\hat{\varphi}^4(\vec{k},t) = I\frac{\hat{p}}{\rho a} \tag{10}$$

We then write evolution equations of these doubles correlations:

$$\frac{d\Phi_{11}}{dt} = -2\nu k^2\Phi_{11} - 2\frac{Sk_3}{k}\Phi_{12} + 2\frac{Sk_2k_3}{kk'}\Phi_{13}, \tag{11}$$

$$\frac{d\Phi_{12}}{dt} = (-2vk^2 + \frac{Sk_1k_2}{k^2})\Phi_{12} - \frac{Sk_1}{k'}\Phi_{13} - \frac{Sk_3}{k}\Phi_{22} + \frac{Sk_2k_3}{kk'}\Phi_{23}, \tag{12}$$

$$\frac{d\Phi_{13}}{dt} = 2\frac{Sk_1k'}{k^2}\Phi_{12} - (\frac{7}{3}vk^2 + \frac{Sk_1k_2}{k^2})\Phi_{13} - ak\Phi_{14} - \frac{Sk_3}{k}\Phi_{23} + \frac{Sk_2k_3}{kk'}\Phi_{33}, \tag{13}$$

$$\frac{d\Phi_{14}}{dt} = ak\Phi_{13} - vk^2\Phi_{14} - \frac{Sk_3}{k}\Phi_{24} + \frac{Sk_2k_3}{kk'}\Phi_{34}, \tag{14}$$

$$\frac{d\Phi_{22}}{dt} = (-2vk^2 + 2\frac{Sk_1k_2}{k^2})\Phi_{22} - 2\frac{Sk_1}{k'}\Phi_{23}, \tag{15}$$

$$\frac{d\Phi_{23}}{dt} = 2\frac{Sk_1k'}{k^2}\Phi_{22} - \frac{7}{3}vk^2\Phi_{23} - ak\Phi_{24} - \frac{Sk_1}{k'}\Phi_{33}, \tag{16}$$

$$\frac{d\Phi_{24}}{dt} = ak\Phi_{23} + (-vk^2 + \frac{Sk_1k_2}{k^2})\Phi_{24} - \frac{Sk_1}{k'}\Phi_{34}, \tag{17}$$

$$\frac{d\Phi_{33}}{dt} = 4\frac{Sk_1k'}{k^2}\Phi_{23} - 2(\frac{4}{3}vk^2 + \frac{Sk_1k_2}{k^2})\Phi_{33} - 2ak\Phi_{34}, \tag{18}$$

$$\frac{d\Phi_{34}}{dt} = 2\frac{Sk_1k'}{k^2}\Phi_{24} + ak\Phi_{33} - (\frac{4}{3}vk^2 + \frac{Sk_1k_2}{k^2})\Phi_{34} - ak\Phi_{44}, \tag{19}$$

$$\frac{d\Phi_{44}}{dt} = 2ak\Phi_{34}, \tag{20}$$

where $k' = \sqrt{k_1^2 + k_3^2}$ and k_1, k_2, k_3 are the components of the wave vector $\vec{k}$.

Numerical integration of these equations ((11)-(20)) is carried out using a simple second-order accurate scheme:

$$f(t+\Delta t) = f(t) + \Delta t\, f'(t) + \Delta t^2 f''(t)/2, \tag{21}$$

where the derivatives $f'(t)$ and $f''(t)$ are expressed exactly from evolution equations (11)-(20) and Δt is the time-step size.

3. RDTcode validation

3.1. Introduction

In this section, results are presented and used to verify the validity of the RDT code. For this, tests of this code have been performed by comparing RDT results with those of direct numerical simulation (DNS) of Simone et al. (1997) for various values of initial gradient Mach number M_{g0} which describe different regimes of the flow (see below). Comparisons concern the turbulent kinetic energy $q^2(t)/q^2(0)$ [Figs. 1, 2], the non-dimensional production term $\dfrac{2P}{Sq^2} = -2b_{12}$ [Figs. 3, 4] and the turbulent kinetic energy growth rate $\Lambda = \dfrac{1}{Sq^2}\dfrac{dq^2}{dt}$ [Figs. 5, 6]. In the same way, comparisons deal with the solenoidal and the dilatational b_{12} anisotropy tensor components [Figs. 7, 8] (Riahi & Lili, 2011). We note that $b_{12} = \dfrac{\overline{u_1 u_2}}{q^2}$ is the relevant component of the Reynolds stress anisotropy tensor. Tavoularis & Corrsin (1981) have shown that in the incompressible homogeneous shear flow b_{12} tends toward an equilibrium value which is independent of initial conditions.

Concerning the existence of the evolution zones where shocks develop, we neither observe any anomaly in physic parameters (no problem of realisability), nor any instability in the calculation course. In addition, it is important to specify that our code was not conceived to take explicitly into account the existence of shocks. However, we think that in the applications presented, the flow is not probably contaminated by the appearance of shocks.

3.2. Initial parameters

Intrinsic parameters that characterize the flow include the initial turbulent Mach number $M_{t0} = \dfrac{q_0}{a}$ (recall that $\dfrac{1}{2}q_0^2$ is the initial turbulent kinetic energy and a is the mean speed of sound), the initial gradient Mach number $M_{g0} = M_{t0}\, S\dfrac{q_0^2}{\varepsilon_0}$ (S is the constant mean shear and ε_0 the initial total rate of turbulent kinetic energy dissipation) and the initial turbulent Reynolds number $R_{et0} = \dfrac{q_0^4}{\nu\varepsilon_0}$. The ratio of the gradient Mach number to the turbulent Mach number $r = S\dfrac{q_0^2}{\varepsilon_0}$ characterizes the rapidity of the shear.

Table 1 lists ten simulations cases labelled A₁, A₂ ... A₁₀ corresponding to different values of the initial gradient Mach number M_{g0}. In these simulations, M_{g0} increases respectively in cases A₁ ($M_{g0} = 2.7$) to A₁₀ ($M_{g0} = 66.7$) by keeping the same initial value of the turbulent Mach number M_{t0} and the turbulent Reynolds number R_{et0}.

In these cases, the initial turbulent kinetic energy spectrum is similar to that used by Simone (1995),

$$E(K,t_0) = K^4 \exp\left(-2\frac{K^2}{\kappa_{pic}^2}\right),$$

where κ_{pic} denotes the wave number corresponding to the peak of the power spectrum and K is the initial wave number.

Case	M_{t0}	M_{g0}	r_0	R_{et0}
A_1	0.25	2.7	10.7	296
A_2	0.25	4	16	296
A_3	0.25	8.3	33.1	296
A_4	0.25	12	48	296
A_5	0.25	16.5	66.2	296
A_6	0.25	24	96.1	296
A_7	0.25	32	128.1	296
A_8	0.25	42.7	170.8	296
A_9	0.25	53.4	213.5	296
A_{10}	0.25	66.7	266.9	296

Table 1. Initial parameters.

3.3. Results and discussion

All Figures 1-8 validate RDT approach and numerical code for low values of the non-dimensional times St (until 3.5). Indeed, comparisons carried out between our RDT results (Riahi & Lili, 2011) and DNS results of Simone et al. (1997) show that the linear analysis predicted correctly – as well qualitatively as quantitatively – the behavior of turbulence in particular for small values of St.

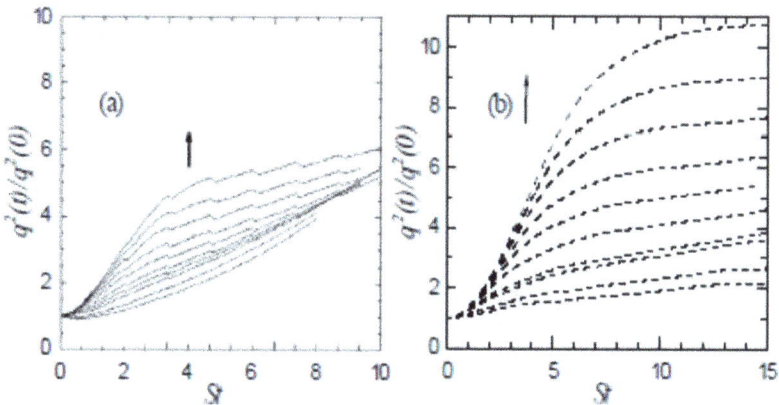

Figure 1. Histories of the turbulent kinetic energy for pure-shear flow: (a) DNS results of Simone et al. (1997) and (b) our RDT results. Initial gradient Mach number M_{g0} ranges from 2.7 to 67 for both DNS and RDT; arrows show trend with increasing M_{g0}.

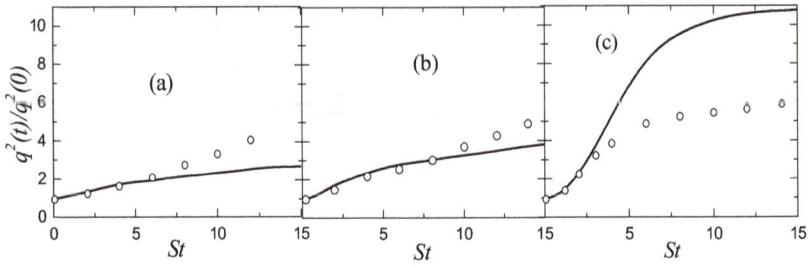

Figure 2. Evolution of the turbulent kinetic energy (a) case A2 ($M_{g0} = 4$), (b) case A4 ($M_{g0} = 12$) and (c) case A10 ($M_{g0} = 66.7$). ——— : RDT results, o : DNS results of Simone et al. (1997).

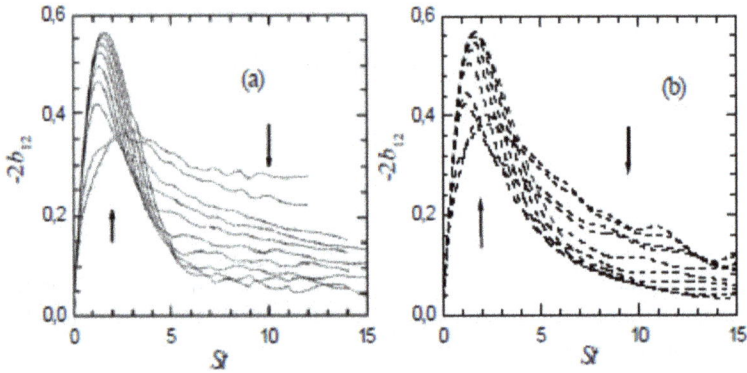

Figure 3. Histories of the non-dimensional production term $-2b_{12}$ for pure-shear flow: (a) DNS results of Simone et al. (1997) and (b) our RDT results. Initial gradient Mach number M_{g0} ranges from 2.7 to 67 for both DNS and RDT; arrows show trend with increasing M_{g0}.

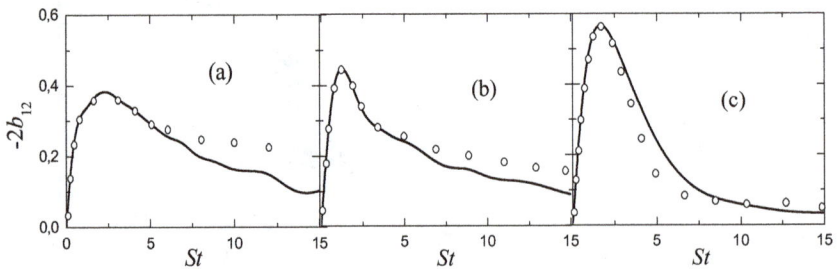

Figure 4. Evolution of the non-dimensional production term (a) case A2 ($M_{g0} = 4$), (b) case A4 ($M_{g0} = 12$) and (c) case A10 ($M_{g0} = 66.7$). ——— : RDT results, o : DNS results of Simone et al. (1997).

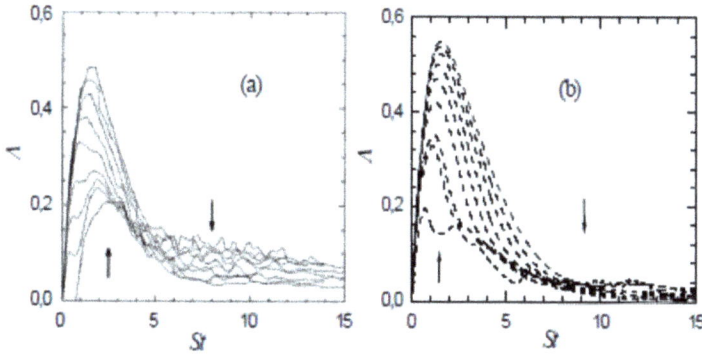

Figure 5. Histories of the temporal energy growth rate for pure-shear flow: (a) DNS results of Simone et al. (1997) and (b) our RDT results. Initial gradient Mach number M_{g0} ranges from 2.7 to 67 for both DNS and RDT; arrows show trend with increasing M_{g0}.

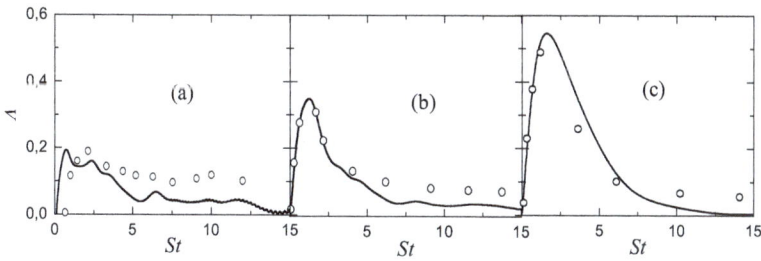

Figure 6. Evolution of the turbulent kinetic energy growth rate (a) case A_2 ($M_{g0} = 4$), (b) case A_4 ($M_{g0} = 12$) and (c) case A_{10} ($M_{g0} = 66.7$). ——— : RDT results, o : DNS results of Simone et al. (1997).

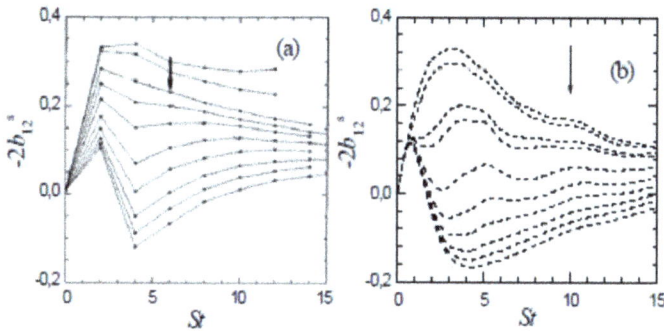

Figure 7. Histories of the solenoidal b_{12} component anisotropy tensor for pure-shear: (a) DNS results of Simone et al. (1997) and (b) our RDT results. Initial gradient Mach number M_{g0} ranges from 2.7 to 67 for both DNS and RDT; arrows show trend with increasing M_{g0}.

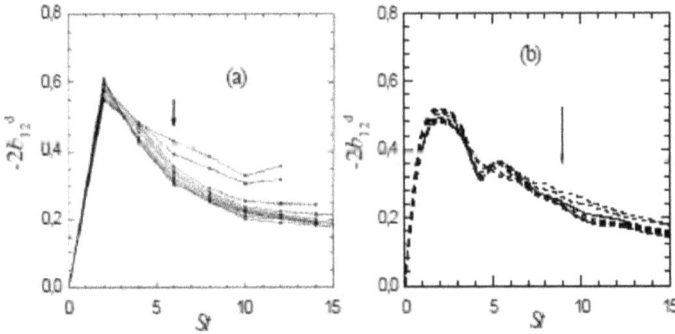

Figure 8. Histories of the dilatational b_{12} component anisotropy tensor for pure-shear flow: (a) DNS results of Simone et al. (1997) and (b) our RDT results. Initial gradient Mach number M_{g0} ranges from 2.7 to 67 for both DNS and RDT; arrows show trend with increasing M_{g0}.

In the compressible regime (M_{t0} = 0.25 and M_{g0} = 66.7), RDT and DNS equilibrium values of the non-dimensional production $-2b_{12}$ of Simone et al. (1997) are very close [Fig. 4(c)]. As one can remark from this figure that for large values of St ($St > 10$), there is a small difference between RDT and DNS results of Simone et al. (1997). Thus, in this compressible regime, RDT predicts correctly equilibrium behavior of the turbulence (asymptotic behavior $(St \to \infty)$) relatively to the non-dimensional production $-2b_{12}$ (strongly compressible regime M_{g0} = 66.7 >> 1 and r_0 >> 1) which is a rapid distortion regime compatible with the framework of RDT. The intermediate regime (case A4, Mg0 = 12) and especially the incompressible regime, for which non-linear effects are preponderant, are poorly estimated by RDT for large values of St. Moreover, Simone et al. (1997) indicate, as general conclusion that compressibility effects on homogeneous sheared turbulence, concerning particularly b_{12} and the non-dimensional production, which are generally associated with non linear phenomena, can be explained in terms of RDT. So, non linear and dissipative effects (through the non linear cascade) are not significant for predicting equilibrium states for strongly compressible regime (Simone et al., 1997). Consequently, we can justify the resort to RDT in order to determine equilibrium states in the compressible regime.

In Figures 9-11, we present evolutions of b_{12} (which is linked to production), b_{11} and b_{22} components anisotropy tensor in the incompressible (M_{g0} = 2.2) and intermediate regimes (M_{g0} = 13.2). In the incompressible regime, as illustrated in Figures 9(a), 10(a), 11(a), we confirm that RDT validity is limited for small values of St; considerable differences appear between RDT and DNS results of Sarkar (1995) beyond. With RDT, it is not possible to predict evolution of sheared turbulence for large values of St. Consequently, we can't predict, by means of RDT, equilibrium states of sheared turbulence in the incompressible regime. We note that in the intermediate regime [Figs. 9(b), 10(b), 11(b)], the behavior of DNS results of Sarkar (1995) concerning b_{12}, b_{11} and b_{22} seems to be compatible with RDT results for large values of St. In the intermediate zone of St (between small and large values of St) differences are appreciable.

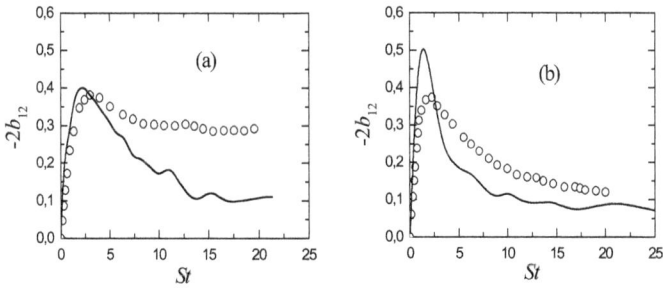

Figure 9. Evolution of the non-dimensional production term with (a): $M_{g0} = 2.2$ and (b): $M_{g0} = 13.2$. —— : RDT results, o : DNS results of Sarkar (1995).

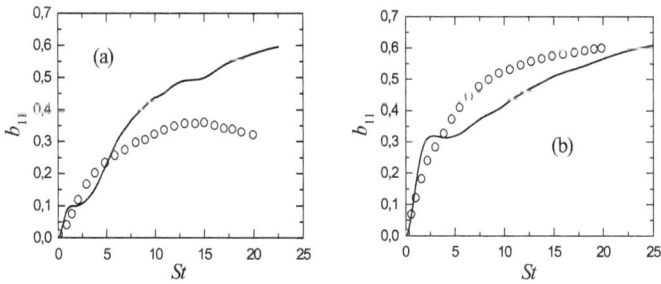

Figure 10. Evolution of b_{11} component anisotropy tensor with (a): $M_{g0} = 2.2$ and (b): $M_{g0} = 13.2$. —— : RDT results, o : DNS results of Sarkar (1995).

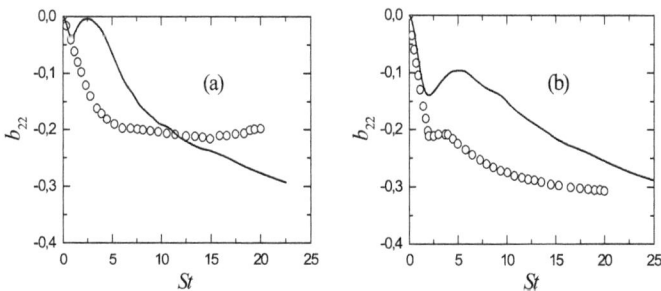

Figure 11. Evolution of b_{22} component anisotropy tensor with (a): $M_{g0} = 2.2$ and (b): $M_{g0} = 13.2$. —— : RDT results, o : DNS results of Sarkar (1995).

In conclusion, RDT is valid for small values of the non-dimensional times St ($St < 3.5$). RDT is also valid for large values of St ($St > 10$) in particular for large values of M_{g0}. This essential feature justifies the resort to RDT in order to critically study equilibrium states in the compressible regime.

4. Various regimes of flow

The various curves of Figure 1(b) permit to determine different regimes of the flow (Riahi & Lili, 2011). This figure shows that there is an increase in the turbulent kinetic energy when M_{g0} increases for various cases considered $A_1,...,A_{10}$. In addition, when initial gradient Mach number increases, we observe a break of slope which is accentuated when the value of M_{g0} becomes more significant. It appears from this figure that the turbulent kinetic energy varies quasi-linearly in cases A_1 and A_2, where M_{g0} takes respectively values 2.7 and 4. These two values of M_{g0} correspond to the incompressible regime. A weak amplification of the turbulent kinetic energy shows that cases A_4 ($M_{g0} = 12$) and A_{10} ($M_{g0} = 66.7$) correspond to the intermediate regime. From $M_{g0} = 16.5$, the regime becomes compressible. Indeed, cases $A_5,...,$ A_{10} show a significant amplification of the total kinetic energy more and more marked when the initial gradient Mach number M_{g0} increases.

These different regimes permit to better understanding compressibility effects on structure of homogeneous sheared turbulence and to analyze the causes of turbulence structure modification generated by compressibility.

Several explanations were proposed these last years to analyze causes of the stabilising effect of compressibility which are still not cleared up.

5. Compressibility effects on structure of homogeneous sheared turbulence

5.1. Introduction

The study of compressibility effects on the turbulent homogeneous shear flow behavior made these last years the objective of several researches as mentioned in the works of Blaisdell et al. (1993 & 1996) and Sarkar et al. (1991 & 1992). DNS developed by Sarkar (1995) show that the temporal growth rate of the turbulent kinetic energy is extensively influenced by compressibility. Simone et al. (1997) identified the new coupling term in the quasi-isentropic RDT equations, which was responsible for long-term stabilization, comparing incompressible and compressible RDT. They concluded that the modification of the compressible turbulence structure is due to linear processes.

The prediction of compressible turbulent flows by rapid distortion theory provides useful results which can be used to clarify the physics of the compressible turbulent flows and to study compressibility effects on structure of homogeneous sheared turbulence. This study is important for the analysis of various physical mechanisms which characteristic the turbulence. Thus, physical comprehension of compressibility effects on turbulence leads it possible to better predict compressible turbulent flows and to improve existing turbulence models. The Helmoltz decomposition of the velocity field in solenoidal and dilatational parts reveals two additional terms in the turbulent kinetic energy budget. Several studies of the behavior of the different terms present in this budget and the Reynolds stress equations show the role of the explicit compressible terms. From Simone et al. (1997) and Sarkar (1995), for the case in which the rate of shear is much larger than the rate of non-linear interactions of the turbulence, amplification of turbulent kinetic energy by the mean shear caused by compressibility is due to the implicit pressure-strain correlations effects and to the anisotropy of the Reynolds stress tensor. These authors also recommend to take into account correctly the explicit dilatational terms such as the pressure-dilatation correlation and the dilatational dissipation. In contrast, the role of explicit terms was over-estimated by Zeman (1990) and Sarkar et al. (1991). These last authors show that both those terms have a dissipative contribution in shear flow, leading to the reduced growth of the turbulent kinetic energy.

The study of the budget behaviors of the turbulent kinetic energy and the Reynolds stress anisotropy by RDT enables to better understand and explain compressibility effects on structure and evolution of a sheared homogeneous turbulence.

5.2. The turbulent kinetic energy equation

In the case of homogeneous turbulence, the turbulent kinetic energy is written (Simone, 1995) as:

$$\frac{d}{dt}(\frac{q^2}{2}) = P - \varepsilon_s - \varepsilon_d + \Pi_d, \tag{22}$$

in which $P = -\overline{Su_1u_2}$ is the rate of production by the mean flow and $\Pi_d = \overline{pu_{i,i}}$ the pressure-dilatation correlation. $\varepsilon_s = v\,\overline{\omega_i\omega_i}$ and $\varepsilon_d = \frac{4}{3}v\,\overline{u_{i,i}u_{i,i}}$ are respectively the solenoidal and the dilatational parts of the turbulent dissipation rate given that ω_i is the fluctuating vorticity and $u_{i,i}$ denotes the fluctuating divergence of velocity. ε_s represents the turbulent dissipation arising from the traditional energy cascade which is solenoidal, ε_d represents the turbulent dissipation arising from dilatational regimes. Note that the last two terms on the right-hand side in equation (22) do not appear when the flow is incompressible. The explicit/energetic approach is embodied in the modeling of ε_d and Π_d done by Zeman (1990), Sarkar et al. (1991) and others.

5.3. Results

Figures 12(a), (b), (c) show the budget of the turbulent kinetic energy for three values of initial gradient Mach number (respectively 1, 12 and 66.7) which describe the various regimes of the flow (Riahi & Lili, 2011). It will be shown from Figures 12(b), (c) that the production P and the pressure-dilatation Π_d intervene significantly in this budget. In the compressible regime [Fig. 12(c)], the production is dominating and the pressure-dilatation Π_d remains with relatively low values (practically null until $St = 1$). Figure 12(a) shows the results for $M_{g0} = 1$ and $M_{t0} = 0.1$ (case A₀). In this case, it is still the production which is dominating. The explicitly compressible terms which are the dilatational dissipation rate ε_d and the pressure-dilatation Π_d are not negligible in spite of the small value of the initial gradient Mach number M_{g0}. Consequently, we can affirm that "the incompressible behavior" cannot be obtained in the borderline case of low values of M_{g0}; that's why we preferred to give the results for $M_{g0} = 1$ and not for $M_{g0} = 2.7$ and 4 with an aim of better approaching the "incompressible limit".

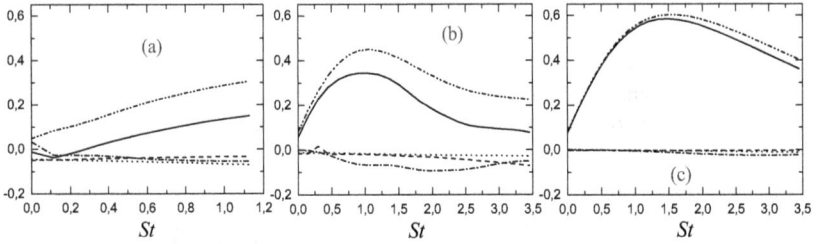

Figure 12. The turbulent kinetic energy budget (a) case A₀ ($M_{g0} = 1$), (b) case A₄ ($M_{g0} = 12$) and (c) case A₁₀ ($M_{g0} = 66.7$). —— : $\dfrac{d}{dt}(\dfrac{q^2}{2}) = P - \varepsilon_s - \varepsilon_d + \Pi_d$, ---- : rate of the turbulent production (P) , ⋯ : rate of the solenoidal dissipation ($-\varepsilon_s$), --- : rate of the dilatational dissipation ($-\varepsilon_d$), -⋅- : pressure-dilatation correlation term (Π_d) .

In conclusion, compressibility affects more the turbulent production term which represents the dominating parameter in the turbulent kinetic energy budget. In the compressible regime, the production becomes preponderant. This property is already observed in the analysis of the budget of the Reynolds stress equations related to $\overline{u_1^2}$ and $\overline{u_1 u_2}$ (Riahi et al., 2007).

In Figures 13(a), (b), (c), (d) are represented the different components of the anisotropy tensor b_{ij}. These figures show obviously a remarkable property concerning the anisotropy. In fact, it appears clearly that the anisotropy increases with M_{g0} i.e. with compressibility and that it is responsible of the behavior of $\overline{u_1^2}$ which becomes dominating compared to $\overline{u_2^2}$ and $\overline{u_3^2}$ in the compressible case ($M_{g0} = 66.7$). As an indication, $\dfrac{\overline{u_1^2}}{\overline{u_2^2}} = 12.62$ and $\dfrac{\overline{u_1^2}}{\overline{u_3^2}} = 13.11$ for $St = 3.5$.

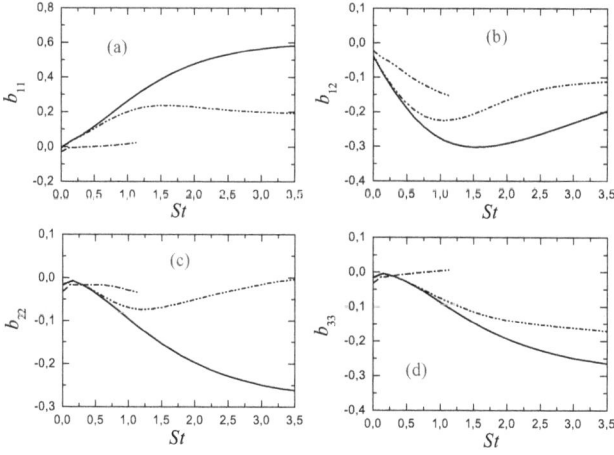

Figure 13. Components evolution of the anisotropy tensor (a) b_{11}, (b) b_{12}, (c) b_{22} and (d) b_{33}.
$-\cdot-$: case A_0 $(M_{g0} = 1)$, $-\cdot\cdot$: case A_4 $(M_{g0} = 12)$, $\underline{\quad\quad}$: case A_{10} $(M_{g0} = 66.7)$.

6. Equilibrium states of homogeneous compressible turbulence

RDT is also used to determine equilibrium states in the compressible regime (Riahi & Lili, 2011) in particular for large values of St $(St > 10)$ and M_{g0}. Evolution of the relevant component of Reynolds stress anisotropy tensor b_{12} is presented in Figure 14 for different values of the initial gradient Mach number M_{g0} (66.7, 200, 500 and 1000). Numerical results show that, for large values of St, b_{12} is independent of the initial turbulent Mach number M_{t0} as shown in Figure 14(a) $(M_{t0} = 0.25)$ and in Figure 14(b) $(M_{t0} = 0.6)$. As one can remark also from these results that b_{12} reaches its stationary value all the more quickly as M_{g0} is large $(M_{g0} = 1000)$. At this stage, we study equilibrium states corresponding to $M_{g0} = 1000$, value which we assimilate to M_{g0} arbitrarily large $(M_{g0}$ infinity). We present now some properties relative to these equilibrium states corresponding to pressure-released regime. The case corresponding to pressure-released limit has been discussed by Cambon et al. (1993), Simone et al. (1997) and Livescu et al. (2004). We begin by presenting equilibrium values of Reynolds stress anisotropy components: $(b_{11})_\infty = 0.66$, $(b_{22})_\infty = (b_{33})_\infty = -0.33$ and $(b_{12})_\infty = 0$. These values confirm the independence of the anisotropy tensor with M_{t0} in the pressure-released regime; calculation gives effectively these values for $M_{t0} = 0.25$, 0.4, 0.5 and 0.6. Pressure-released state corresponds thus to a particular state of one-component turbulence in the direction of the shear (x_1 direction): $\left(\overline{u_1^2}\right)_\infty = q^2$, $\left(\overline{u_2^2}\right)_\infty = 0$ and $\left(\overline{u_3^2}\right)_\infty = 0$.

Another interesting property of the pressure-released regime is the quadratic increase of $\overline{u_1^2}$ as shown in Figure 15 related to $M_{g0} = 10^6$ and $M_{t0} = 0.25$ (Riahi & Lili, 2011). This property has been established by Livescu et al. (2004) as analytical RDT solution in the pressure-released limit.

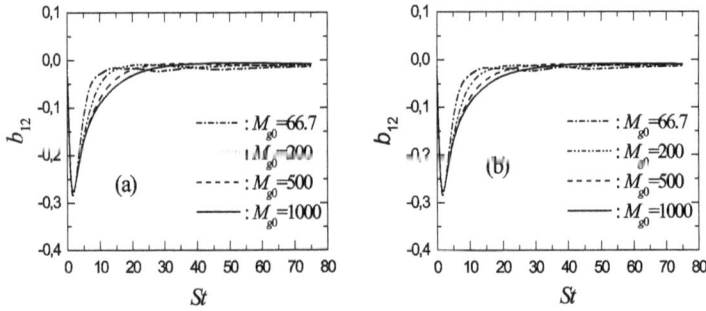

Figure 14. Evolution of the relevant component of the Reynolds stress anisotropy tensor b_{12} for various values of initial gradient Mach number M_{g0} with (a): $M_{t0} = 0.25$ and (b): $M_{t0} = 0.6$.

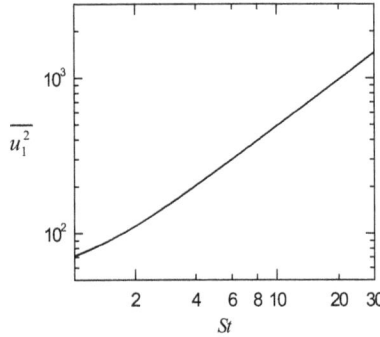

Figure 15. Quadratic evolution of $\overline{u_1^2}$.

Table 2 lists equilibrium values of the normalized dissipation rate due to dilatation $\left(\dfrac{\varepsilon_d}{\varepsilon_s + \varepsilon_d} \right)_\infty$, the normalized pressure-dilatation correlation $\left(\dfrac{\overline{pd}}{Sq^2} \right)_\infty$ and the normalized pressure variance $\left(\dfrac{\overline{p^2}}{\overline{\rho}^2 q^4} \right)_\infty$ for various values of initial turbulent Mach number M_{t0} (Riahi & Lili, 2011). From this table, we deduce the dependence of all these parameters with M_{t0}.

Figure 16 shows variation of the equilibrium normalized pressure variance $\left(\dfrac{\overline{p^2}}{\overline{\rho}^2 q^4} \right)_\infty$ with various initial turbulent Mach number ($M_{t0} = 0.25, 0.4, 0.5$ and 0.6 ($M_{t0} = 0.6$ is relatively high initial turbulent Mach number)) (Riahi & Lili, 2011). From this figure, it can be seen that this parameter can be written as $\left(\dfrac{\overline{p^2}}{\overline{\rho}^2 q^4} \right)_\infty = (M_{t0})^\alpha$ where $\alpha = 5.8$.

M_{t0} Equilibrium parameters	0.25	0.4	0.5	0.6
$\left(\dfrac{\varepsilon_d}{\varepsilon_s+\varepsilon_d}\right)_\infty$	0.0410	0.0277	0.0230	0.0197
$\left(\dfrac{\overline{pd}}{Sq^2}\right)_\infty$	0.000387	0.000311	0.000280	0.000257
$\left(\dfrac{\overline{p^2}}{\overline{\rho}^2 q^4}\right)_\infty$	0.00937	0.00278	0.00156	0.000973

Table 2. Equilibrium values of the normalized dissipation rate due to dilatation, the normalized pressure-dilatation and the normalized pressure variance for various initial turbulent Mach number M_{t0}.

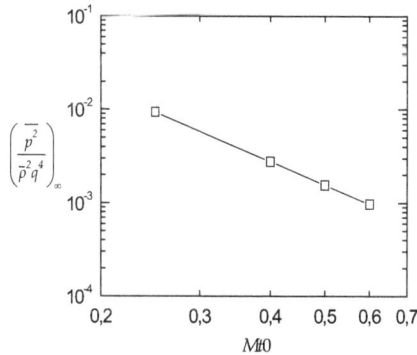

Figure 16. Variation of the equilibrium normalized pressure variance with various initial turbulent Mach number (M_{t0} = 0.25, 0.4, 0.5 and 0.6).

7. Turbulence model to be tested

7.1. Introduction

RDT is an efficient and accurate method for testing linear turbulence models. The Fujiwara and Arakawa model (1995) has been proposed in literature for pressure-dilatation correlation studying compressible turbulence. In order to verify its capacity to describe the homogeneous compressible flow, since it was established to this objective, three cases of the initial gradient Mach number (M_{g0} = 1, 12 and 66.7) are considered. The evaluation of this model, in the different regimes of flow, stays in the field of RDT validity (Riahi & Lili, 2011). Its detailed linear form is provided below.

7.2. Fujiwara and Arakawa model

The model suggested by Fujiwara and Arakawa (1995) for the pressure-dilatation correlation takes into account the compressible part (dilatational) of the kinetic energy (q_d^2). Linear part of this model has the following form:

$$\Pi_d = C_1 \bar{\rho} q^2 b_{ij} \sqrt{\frac{q_d^2}{q^2}} \frac{\partial U_i}{\partial x_j},$$

where $C_1 = 0.3$ and ε is the turbulent dissipation rate. b_{ij} is the Reynolds anisotropy tensor.

7.3. Results

Comparison between RDT and Fujiwara and Arakawa (1995) model results concerning the pressure-dilatation correlation term Π_d are represented in Figures 17(a), (b), (c). As one can remark from these results, that the contribution of Π_d obtained by this model is negligible. Except in the case of the intermediate regime (M_{g0} = 12), the Fujiwara and Arakawa model does not represent an appreciable discrepancies with the RDT results.

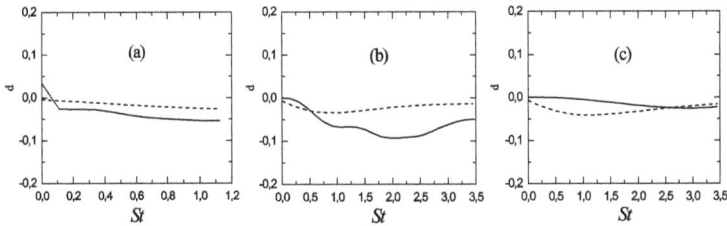

Figure 17. Evolution of the pressure dilatation correlation term Π_d (a) case A0 (M_{g0} = 1), (b) case A4 (M_{g0} = 12) and (c) case A10 (M_{g0} = 66.7). ———— : RDT results, – – – : Fujiwara and Arakawa model.

8. Conclusion

Rapid distortion theory (RDT) is a computationally viable option for examining linear compressible flow physics in the absence of inertial effects. Evolution of compressible homogeneous turbulence has been described completely by finding numerical solutions obtained by solving linear double correlations spectra evolution. Numerical integration of these equations has been carried out using a second-order simple and accurate scheme. This numerical method has proved more stable and faster and allows in particular to obtain accurately the asymptotic behavior of the turbulence parameters (for large values of St) characteristic of equilibrium states. In this chapter, RDT code developed by authors solves linearized equations for compressible homogeneous shear flows (Riahi & Lili, 2011). It has been validated by comparing RDT results with direct numerical simulation (DNS) of Simone et al. (1997) and Sarkar (1995) for various values of initial gradient Mach number M_{g0} which is the key parameter controlling the level of compressibility. The study of the behavior of the non-dimensional turbulent kinetic energy $q^2(t)/q^2(0)$ permit to determine various regimes of flow. This study allows to check relevance of an incompressible regime for low values of initial gradient Mach number M_{g0}, of an intermediate regime for moderate values of M_{g0} and of a compressible regime for high values of M_{g0}. Agreement between RDT and DNS of Simone et al. (1997) and Sarkar (1995) is obtained for small values of the non-dimensional times St ($St <$ 3.5). This agreement gives new insight into compressibility effects and reveals the extent to which linear processes are responsible for modifying the structure of compressible turbulence. The behavior analysis of the various terms presented in the turbulent kinetic energy budget shows that compressibility affects more the turbulent production which becomes preponderant in the compressible regime. This property is already observed in the analysis of the budget of the Reynolds stress equations related to $\overline{u_1^2}$ and $\overline{u_1 u_2}$. Another important property reveals that anisotropy increases with compressibility.

Agreement of RDT with DNS (Simone et al., 1997) found for large values of St ($St >$ 10) in particular for large values of M_{g0} which allows to determine equilibrium states in the compressible regime. Evolution of the relevant component of the Reynolds stress anisotropy tensor b_{12}, for different values of initial gradient Mach number M_{g0} (66.7, 200, 500 and 1000) and for large values of St, shows that b_{12} is independent of the initial turbulent Mach number M_{t0} (which is a parameter characterizing the effects of compressibility). In addition, b_{12} becomes stationary all the more quickly as M_{g0} is large ($M_{g0} = 1000$). Considering equilibrium states associated to this value of M_{g0} as representative of equilibrium states related to infinite M_{g0} (pressure-released regime), equilibrium values of the Reynolds stress anisotropy components $(b_{11})_\infty$ and $(b_{22})_\infty$ are independent of the initial turbulent Mach number M_{t0}. In contrast, equilibrium values of the normalized dissipation rate due to dilatation, the normalized pressure-dilatation correlation and the normalized pressure variance are dependent of M_{t0}. Variation of the equilibrium normalized pressure variance with various initial turbulent Mach number ($M_{t0} = 0.25, 0.4, 0.5$ and 0.6) can be written as $\left(\dfrac{\overline{p^2}}{\overline{\rho}^2 q^4} \right)_\infty = (M_{t0})^\alpha$ where $\alpha = 5.8$.

In conclusion, after a critical analysis, we were able to justify that RDT permits to well identify compressibility effects in order to develop models taking them into account satisfactorily. The analysis of rapid distortion theory showed that it is possible to better understand the compressible turbulent flows. In addition, RDT is valid to predict asymptotic equilibrium states for compressible homogeneous sheared turbulence for large values of initial gradient Mach number (pressure-released regime). RDT is an efficient method for testing linear contribution of turbulence models.

Author details

Mohamed Riahi and Taieb Lili

Faculté des Sciences de Tunis, Département de Physique, Laboratoire de Mécanique des Fluides, Campus Universitaire, Manar II, Tunis, Tunisia

Abbreviations

RDT   Rapid Distortion Theory
DNS   Direct Numerical Simulation

Nomenclature

M_g, M_{g0}	gradient Mach number, initial gradient Mach number
M_t, M_{t0}	turbulent Mach number, initial turbulent Mach number
S	magnitude of the mean velocity gradient
u_i	velocity fluctuation
p	pressure fluctuation
$\bar{\rho}$	mean density
ν	kinematic viscosity
λ_{ij}	mean velocity gradient
a	mean sound speed
Δt	time-step size
$q^2/2$	turbulent kinetic energy
$b_{ij} = \dfrac{\overline{u_i u_j}}{q^2} - \dfrac{\delta_{ij}}{3}$ anisotropy tensor	
ε_0	initial total (solenoidal and dilatational) dissipation rate of turbulent kinetic energy
R_{et0}	initial turbulent Reynolds number
r_0	initial rapidity of the shear
γ	ratio of specific heats

9. References

Blaisdell, G.A.; Mansour, N.N. & Reynolds, W.C. (1993). Compressibility effects on the growth and structure of homogeneous turbulent shear flow. *Journal of Fluid Mechanics*, vol. 256, pp. 443-485.

Blaisdell, G. A.; Coleman, G. N. & Mansour, N. N. (1996). Rapid distortion theory for compressible homogenous turbulence under isotropic mean strain. *Physics of Fluids*, vol. 8, pp. 2692-2705.

Coleman, G. N. & Mansour, N. N. (1991). Modeling the rapid spherical compression of isotropic turbulence, *Physics of Fluids*, vol. A 3, pp. 2255-2259.

Cambon, C.; Coleman, G.N. & Mansour, N.N. (1993). Rapid distortion analysis and direct simulation of compressible homogeneous turbulence at finite Mach number. *Journal of Fluid Mechanics*, vol. 257, pp. 641-665.

Durbin, P.A. & Zeman, O. (1992). Rapid distortion theory for homogeneous compressed turbulence with application to modelling. *Journal of Fluid Mechanics*, vol. 242, pp. 349-370.

Fujiwara, H. & Arakawa, C. (1995). Modeling the pressure-dilatation correlation in isotropic and homogeneous sheared turbulence. *Proc. 10th Symposium on Turbulent Shear Flows*, vol. 26, pp. 1-6.

Hunt, J. C. R. & Carruthers, D. J. (1990). Rapid distortion theory and the problems of turbulence. *Journal of Fluid Mechanics*, vol. 212, pp. 497-532.

Jacquin, L.; Cambon, C. & Blin, E. (1993). Turbulence amplification by a shock wave and rapid distortion theory. *Physics of Fluids*, vol. A 5, pp. 2539-2550.

Livescu D. & Madnia, C. K. (2004). Small Scale Structure of Homogeneous Turbulent Shear Flow. *Physics of Fluids*, vol. 16, pp. 2864-2876.

Riahi, M.; Chouchane, L. & Lili, T. (2007). A study of compressibility effects on structure of homogeneous sheared turbulence. *The European Physical Journal Applied Physics*, vol. 39, Issue: 1, pp. 67-75.

Riahi, M. (2008). Modélisation et simulation d'une turbulence homogène compressible. *Thèse de Doctorat, Faculté des Sciences Mathématiques Physiques et Naturelles de Tunis*, Tunisie.

Riahi, M.; & Lili, T. (2011). Equilibrium states of homogeneous sheared compressible turbulence, *AIP Advances*, vol. 1, Issue: 2, pp. 022117-022117-11, ISSN 2158-3226.

Savill, A. M. (1987). Recent Developments in Rapid-Distortion Theory. *Annual Review of Fluid Mechanics*, vol. 19 pp. 531-573.

Sarkar, S.; Erlebacher, G.; Hussaini, M. & Kreiss, H.O. (1991). The analysis and modelling of dilatational terms in compressible turbulence. *Journal of Fluid Mechanics*, vol. 227, pp. 473-493.

Sarkar, S. (1992). The pressure-dilatation correlation in compressible flows, *Physics of Fluids*, vol. A 4, pp. 2674-2682.

Sarkar, S. (1995). The stabilizing effect of compressibility in turbulent shear flow. *Journal of Fluid Mechanics*, vol. 282, pp. 163-186.

Simone, A.; Coleman, G.N. & Cambon, C. (1997). The effect of compressibility on turbulent shear flow: a rapid-distortion-theory and direct-numerical-simulation study. *Journal of Fluid Mechanics*, vol. 330, (July 1996), pp. 307-338.

Simone, A. (1995). Etude théorique et simulation numérique de la turbulence compressible en présence de cisaillement ou de variation de volume à grande échelle. *Thèse de Doctorat, Ecole Centrale de Lyon*, France.

Tavoularis, S. & Corrsin, S. (1981). Experiments in nearly homogeneous turbulent shear flow with a uniform mean gradient temperature. *Part I. Journal of Fluid Mechanics*, vol. 104, pp. 311-347.

Zeman, O. (1990). Dilatation dissipation: The concept and application in modeling compressible mixing layers. *Physics of Fluids*, vol. 2, pp. 178-188, ISSN 0899-8213.

Industrial Applications

Chapter 5

Numerical Simulation
of Fully Grouted Rock Bolts

Hossein Jalalifar and Naj Aziz

Additional information is available at the end of the chapter

1. Introduction

This chapter describes the application of numerical modelling to civil and mining projects, particularly rock bolting, developing a Final Element (FE) model for the bolt, grout, rock, and two interfaces under axial and lateral loading, verifying the model, analysing the stress and strains developed in the bolt and surrounding materials.

Numerical methods are the most versatile computational methods for various engineering disciplines because a structure is discritised into small elements and the constitutive equations that describe the individual elements and their interactions are constructed. Finally, these numerous equations are solved together simultaneously using computers. The results from this procedure include the stress distribution and displacement pattern within a structure. Numerical modelling includes analytical techniques such as finite elements, boundary elements, distinct elements, and other numerical approaches that depend upon the material. The finite element method "FEM" is considered to evaluate the behaviour of materials and their interactions in a fully grouted bolt which is installed in a jointed rock mass. The simulations were carried out by ANSYS code.

2. FE in ANSYS

ANSYS is a powerful non-linear simulation tool, *Bhashyam.G.R* (2002).The ANSYS software is a commercial FE analysis programme, which has been in use for more than thirty years, *Pool et al.* (2003). The software can analyse the stress and strain built up in a variety of problems, especially designing roof bolts and long wall support systems.

The original code developed around a direct frontal solver has been expanded over the years to include full featured pre and post processing capabilities which support a comprehensive list of analytical capabilities including linear static analysis, multiple non-linear analyses, modal analysis, contact interface analyses and many other types.

In this chapter only structural analysis is considered. Structural analyses are available in the ANSYS Multiphysics, ANSYS Mechanical, ANSYS Structural, and ANSYS Professional programmes only. Statistical analysis is used to determine displacement and stress and strain under static loading conditions (both linear and non-linear statistical analyses). Non-linearity can include plasticity, stress stiffening, large deflection, large strain, hyper-elasticity, contact surfaces, and creep behaviour.

3. A review of numerical modelling in rock bolts

A number of computer programmes have been developed for modelling civil and geo-technical problems. Some of them can be partially used to design and analyse roof bolting systems. It is noted that 3D software is necessary to simulate the whole characters of a model, such as modelling the joints, bedding planes, contact interface and failure criterion. Several numerical methods are used in rock mechanics to model the response of rock masses to loading and unloading. These methods include the method (FEM), the boundary element method (BEM), finite difference method (FDM) and the discrete element method (DEM).

A number of studies were carried out on bolt behaviour in the FE field, including those by *Coats and Yu* (1970), *Hollingshead* (1971), *Aydan* (1989), *Saeb and Amadei* (1990), *Aydan and Kawamoto* (1992), *Swoboda and Marence* (1992), *Moussa and Swoboda* (1995), *Marence and Swoboda* (1995), *Chen et al.* (1994, 1999, 2004), and *Surajit* (1999).

One of the earliest attempts to use standard FEs to model the bolt and grout was done by *Coats and Yu* (1970). The study was carried out on the stress distribution around a cylindrical hole with the FEmodel either in tension or compression. It was found that the stress distribution was a function of the bolt and rock moduli of elasticity. The presence of grout between the bolt and the rock was not considered and there was no allowance for yielding. The analysis was only carried out in linear elastic behaviour with two-phase materials, which limited the model. *Hollingshead* (1971) solved the same problem using a three phase material (bolt-grout and rock) and allowed a yield zone to penetrate into the grout using an elastic, perfectly plastic criterion, according to the Tresca yield criterion, for the three materials (Figure 1). How the interface behaved was not considered in the model.

John and Dillen (1983) developed a new one-dimensional element passing through a cylindrical surface to which elements representing the surrounding material are attached (Figure 2). They considered three important modes of failure for fully grouted bolts, a bi-linear elasto-plastic model for axial behaviour, elastic- perfectly plastic, and residual plastic model for bonding material, was assumed. Although this model eliminated many previous limitations and agreed with the experimental results, it neglected rock stiffness and in-situ stress around the borehole. They claimed that critical shear stress occurred at the grout - rock interface, which is not always the case in the field or laboratory. Aydan (1989) presented a FE model of the bolt. He assumed that a cylindrical bolt and grout annulus is connected to the rock with a three-dimensional 8-nodal points.

Two nodes are connected to the bolt and six to the rock mass. The use of boundary element and FE techniques to analyse the stress and deformation along the bolt was conducted by Peng and Guo (1992) (Figure 3). The effect of the face plate was replaced by a boundary element. The effect of reinforcement because of the assumption of perfect bonding was overestimated.

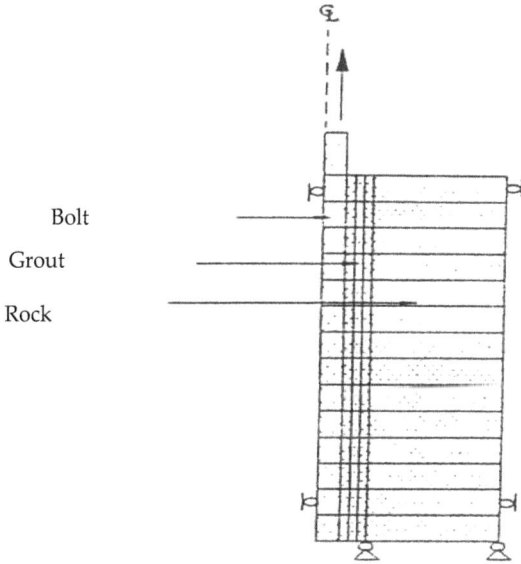

Bolt

Grout

Rock

Figure 1. FE Simulation of bolted rock mass (after Hollingshead, 1971)

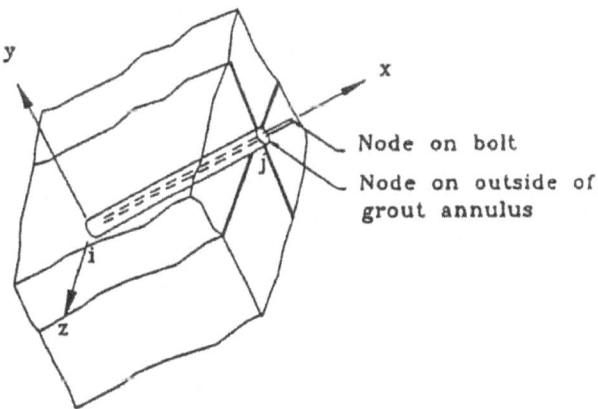

Node on bolt

Node on outside of grout annulus

Figure 2. Three-Dimensional rock bolt element (after John and Dillen, 1983)

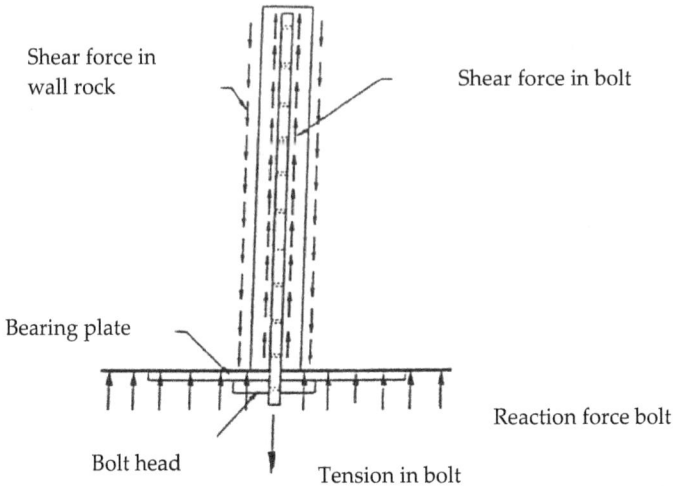

Figure 3. Bolt-Rock interaction model (after Peng and Guo, 1992)

Stankus and Guo (1996) investigated that in bedded and laminated strata, point anchor and fully grouted bolts are very effective, especially if quickly installed at high tension after excavation. They used three lengths 3300, 2400, and 1500 mm and three tensions, 66, 89, and 110 kN and found that:

- Bolts with higher pre-tension induce a smaller deflection
- The longer the bolt, the larger the load,
- In bolts with the same length and high tension, there is small deflection,
- Large beam deflection was observed in long bolts and small deflection in short bolts.

They developed a method for achieving the optimum beaming effect (OBE). However there were some assumptions in their methodology such as, the problem with the gap element, which is not flexible for any kind of mesh, especially with thin grout. Many relevant parameters about the contact interface cannot be defined in gap element. All materials were modelled in the elastic region.

Marence and Swoboda (1995) developed the Bolt Crossing Joint (BCJ) element that connects the elements on both sides of the shear joint. It has two nodes, one each side of the discontinuity. The model cannot predict the de-bonding length along the bolt, grout interface and hinge point position.

It was realised that to further facilitate data analysis and the stress and strain build up along a bolt surrounded by composite material and their interaction, a powerful computer simulation was needed. FE modelling is considered to be the only tool able to accomplish this goal. There is still a lack of an adequate global models of grouted bolts to analyse bolt behaviour properly, particularly at the contact interfaces.

In this chapter, three-dimensional formulations and non-linear deformation of rock, grout, bolt, and two interfaces are taken into account in the reinforced system. A description of the numerical model developed is presented below.

4. Materials design model

The FE method is the most suitable computational method to evaluate the real behaviour of the bolt, grout, and surrounding rock when there are composite materials with different interfaces. A three dimensional FE model of a reinforced structure subjected to shear loading was used to examine the behaviour of bolted rock joints. Three governing materials (steel, grout, and concrete) with two interfaces (bolt-grout and grout-concrete) were considered. To create the best possible mesh, symmetry rules should be applied. To reduce computing demand and time (when a fine mesh is used) the density of the mesh has been optimised during meshing. The division of zones into elements was such that the smallest elements were used where details of stress and displacement were required. The process of FE analysis is shown in Figure 4.

4.1. Modelling concrete and grout

Care was taken to develop the best model for concrete and grout that could offer appropriate behaviour. 3D solid elements, Solid 65 that has 8 nodes was used with each node having three translation degrees of freedom that tolerates irregular shapes without a significant loss in accuracy. Solid 65 is used for the 3-D modeling of solids with or without reinforcing bars (rebar). The geometry and node locations for this type of element are shown in Figure 5 a. The solid element is capable of plastic deformation, cracking in tension, crushing in compression, creep non-linearity, and large deflection geometrical non-linearity, and also includes the failure criteria of concrete Fanning (2001), Feng et al. (2002) and Ansys (2012). Concrete can fail by cracking when the tensile stress exceeds the tensile strength, or by crushing when the compressive stress exceeds the compressive strength. A FE mesh for concrete is shown in Figure 5 b. Figure 6 shows the FE mesh for grout. Due to symmetry only a quarter of the model needed to be treated.

4.2. Modelling the bolt

The steel bar, which resists axial and shear loads during loading, due to rock movement, is the main element within the rock bolt system,. The steel bar was modelled appropriately, particularly with regard to the type of element designed and bolt behaviour, in the linear and non-linear region. 3D solid elements, solid 95 with 20 nodes, was used to model the steel bar, with each node having three translation degree of freedom. The approach adopted is to reveal that the experimentally verified shear resistance of fully grouted bolt can be investigated by numerical design. Elastic behaviour of the elements was defined by Young's Modulus and Poisson's ratio of various materials. The stress, strain relationship of steel is assumed as the bi-linear kinematic hardening model and the modulus of elasticity of strain hardening after yielding, is accounted as a hundredth of the original one, *Cha et al.* (2003),

Hong et al. (2003) and *Abedi et al.* (2003). Figure 7 displays the solid 95 elements and FE mesh for bolt.

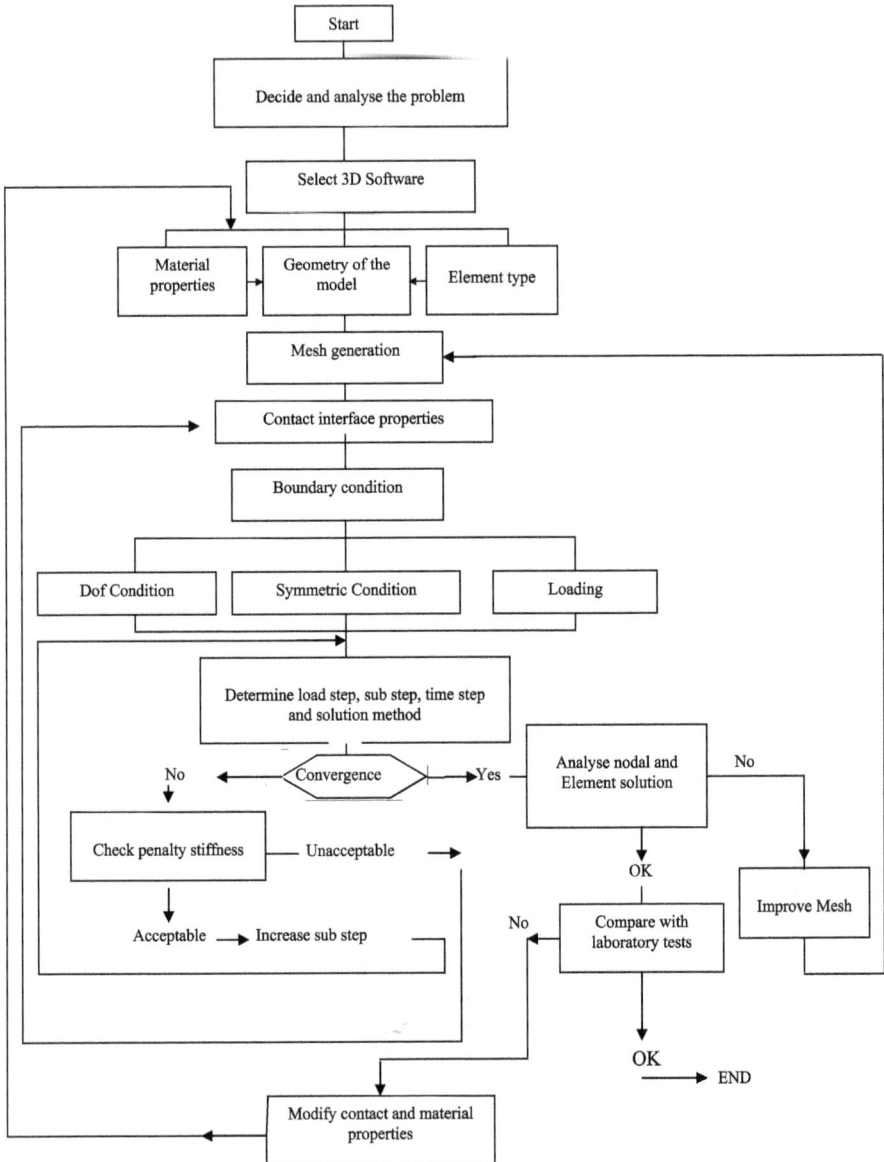

Figure 4. The process of FE simulation (Dof = degrees of freedom)

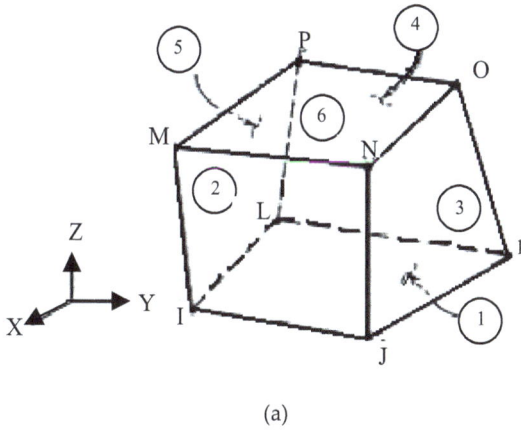

(a)

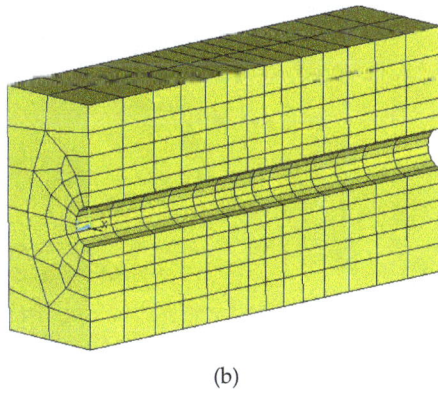

(b)

Figure 5. (a) 3D Solid 65 elements; (b) Concrete mesh

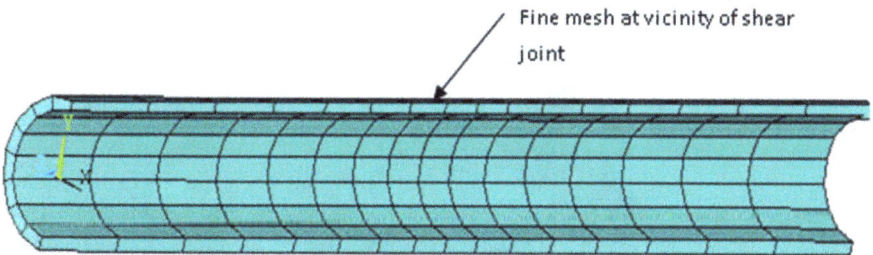

Figure 6. FE mesh for grout

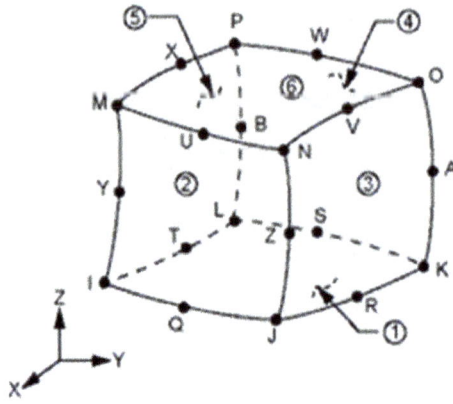

(a)

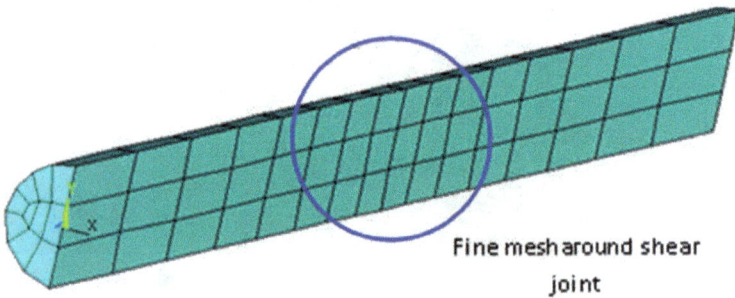

Fine mesh around shear
joint

(b)

Figure 7. (a) 3D Solid 95 elements (b) FE mesh for bolt

4.3. Contact interface model

The main difficulties with numerically simulating a reinforced shear joint are the bolt- grout
and grout-rock interfaces. An important parameter controlling the load transfer from the
bolt to the rock through resin is bond behaviour between the interfaces. If they are not
designed properly it is difficult to understand their behaviour, when and where de-bonding
occurs, how a gap is created between the interfaces, and how the load is transferred. Thus
the contact interfaces were designed to act realistically. To study the stress, strain generation
through numerical modelling, it is very important to model the interfaces accurately, *Pal et
al.* (1999). *Ostreberge* (1973) also emphasised the bond strength between two adjacent

mediums for an accurate load transfer. *Nietzsche and Hass* (1976) proposed a model for bolt, grout-rock that assumed a linear elastic behaviour for all materials, and perfect bonding for all contact interfaces (bolt- grout and grout- rock). It has to be noted that perfect bonding, particularly between the bolt-grout interface could not be considered to be the right behaviour, because there is no cohesion strong enough between them. In addition, there are large stresses and strains concentrated near the shear joints, which restrict perfect bonding. The interface between the grout and concrete was considered as standard behaviour where normal pressure changes to zero when separation occurs. As found from laboratory results, a low cohesion (150 kPa) was adopted for the contact interface, which was determined from the test results under constant normal conditions.

3D surface-to-surface contact element (contact 174) was used to represent contact between the target surfaces (steel-grout and rock - grout). This element is applicable to 3D structural contact analysis and is located on the surfaces of 3D solid elements with mid-side nodes. This contact element is used to represent contact and sliding between 3-D "target" surfaces (Target 170) and a deformable surface, is defined by this element. The element is applicable to three-dimensional structural and coupled thermal structural contact analysis. This element is also located on the surfaces of 3-D solid or shell elements with mid-side nodes. It has the same geometric characteristics as the solid or shell element face to which it is connected. Contact occurs when the element surface penetrates one of the target segment elements on a specified target surface. The contact elements themselves overlay the solid elements describing the boundary of a deformable body and are potentially in contact with the target surface. This target surface is discritised by a set of target segment elements (Target 170) and is paired with its associated contact surface via a shared real constant set. Figure 8 displays the target 170 geometry.

4.4. 3D geometrical model

An actual 3D geometrical model was created to simulate the rock-bolt- grout behaviour and their interactions. The model bolt core diameter (D_b) of 22 mm and the grouted cylinder (D_h) of 27 mm diameter had the same dimensions as those used in the laboratory test. Due to the symmetry of the problem, only one fourth of the system was considered. Figure 9 shows the geometry of the FE model with mesh generation.

5. Verification of the model

A numerical representation model for a fully grouted reinforcement bolt was developed and its validity assessed with laboratory data conducted in a variety of rock strengths and pre-tension loadings. A comparison of experimental results with numerical simulations showed that the model can predict the interaction between bolt, grout, and concrete, and how the interfaces behave. The consistency of the experimental observations with a numerically design model is presented by typical shear load, shear displacement curves shown in Figure 10. It is clear that when the strength of the concrete was doubled there was a twofold reduction in shear displacement.

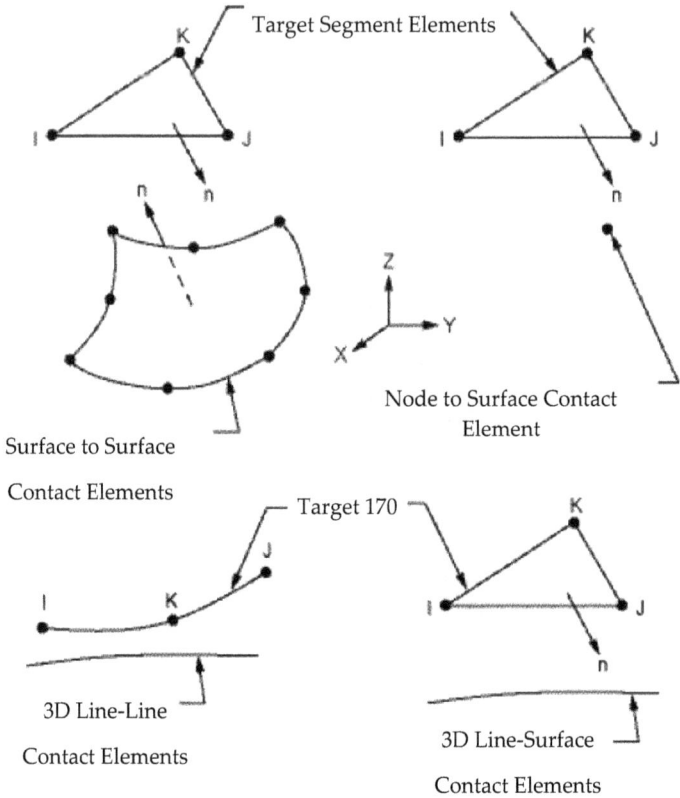

Figure 8. Target 170 geometry (Ansys 2012)

6. Modelling bolts under lateral loading

An extensive series of laboratory tests to analyse the bending behaviour of fully grouted bolts in different strength rock, bolt pre-tension and thickness of resin were carried out. Three governing materials (steel, grout, and rock) with two interfaces (bolt-grout and grout-rock) were considered for 3D numerical simulation.

By this three dimensional FEM, the characteristics of elasto - plastic materials and contact interfaces are simulated. Numerical modelling in different strength rock (20, 40, 50 and 80 MPa) and different pre-tension loads (0, 20, 50, and 80 kN) were carried out and the results were analysed. As the output results were large, only the main results of 0 and 80 kN pre-tension are presented here.

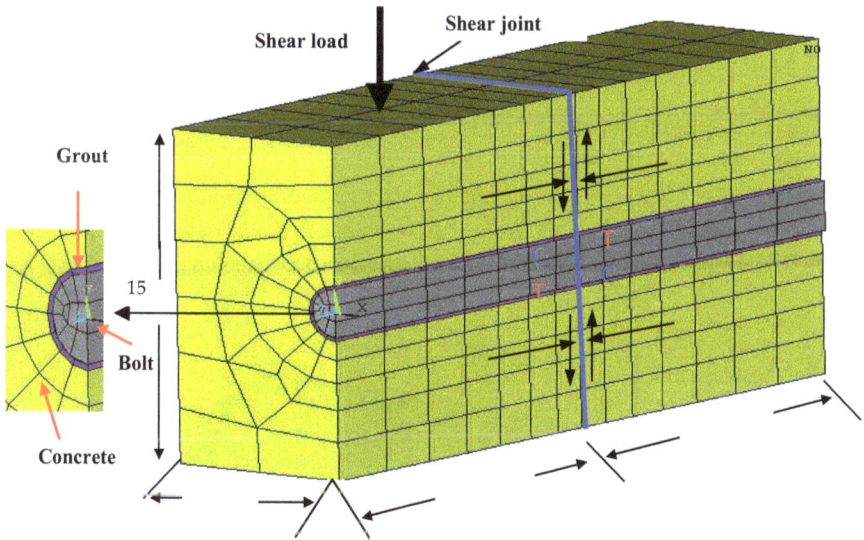

Figure 9. Geometry of the model and mesh generation

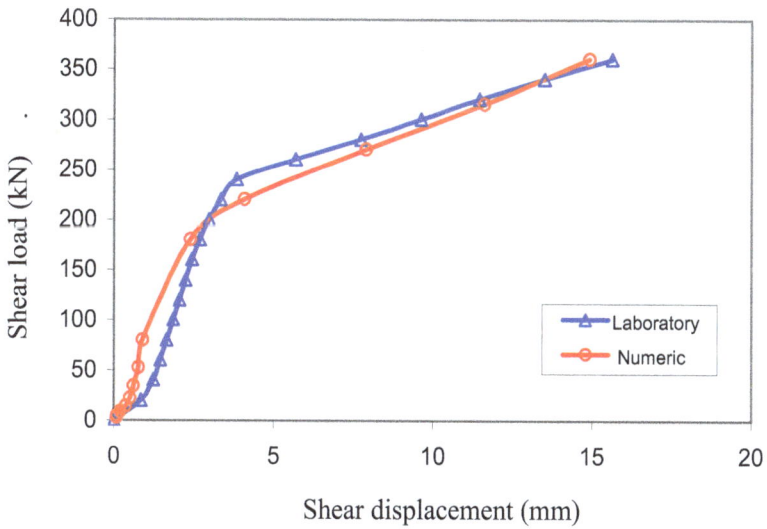

Figure 10. Load-deflection in 80 kN pretension bolt load and 40 MPa concrete

6.1. Bolt behaviour

6.1.1. Stresses developed along the bolt

When a beam with a straight longitudinal axis is loaded laterally, its longitudinal axis is deformed into a curve, and the resulting stresses and strains are directly related to the deflection curve, which is affected by the surrounding materials. Figure 11 shows a quarter of the model with induced loads along the shear joint.

When the beam was bent there was deflection and rotation at each point. The angle of rotation α is the angle between the bolt axis and the tangent to the deflection curve, shown as point o. α was measured for the bolts tested. The deflection trend in 20 MPa concrete is shown in Figure 12.

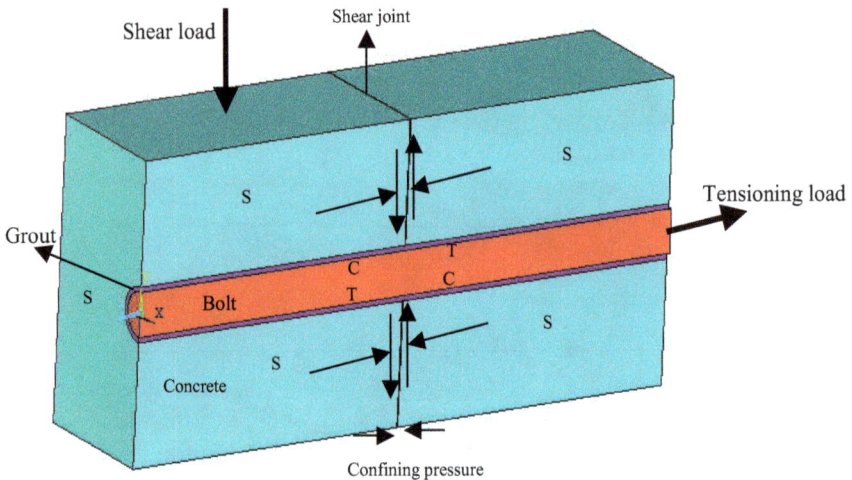

Figure 11. Numerical model (s = symmetric planes, c = compression zone, T = tension zone

Also to find the relationship between deflection and each point along the axis of the bolt, raw output data from the numerical simulation were classified and entered as input data to Maple software. Equation 3 and Figure 13 were established.

$$\tan\alpha = \frac{dv}{dx}, \tag{1}$$

$$\alpha = \arctan\frac{dv}{dx} \tag{2}$$

$$U_y = -40.76 + 26\,Arc\tan(e^{(0.05x-7.2)}) \tag{3}$$

where;

- U_y = Shear displacement (mm)
- x = Distance from the bolt centre to the end (mm), from A to B.

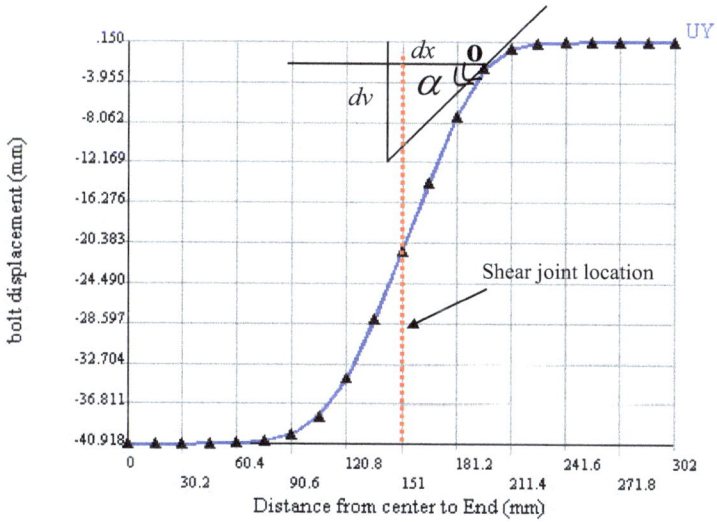

Figure 12. Bolt displacement in 20 MPa, without Pre-tension

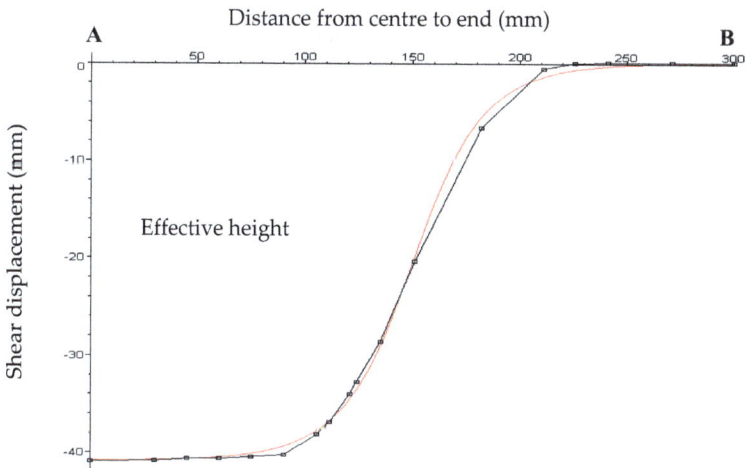

Figure 13. Shear displacement as a function of bolt length sections in 20 MPa concrete

The relationship between vertical displacement at the bolt-joint intersection and hinge point is:

$$Uy \text{ (hinge)} = (0.15\text{-}0.2) \, Uy \text{ (joint)}$$

Which is consistent with the laboratory results. Figure 14 shows the bolt deflection in 40 MPa concrete.

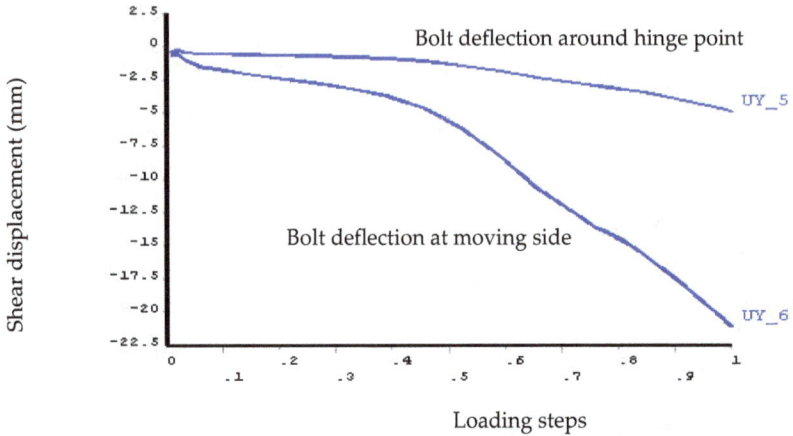

Figure 14. Bolt deflection at the moving side and hinge point versus loading process, in 40 MPa concrete without pre-tension load

Figure 15 shows the contours of stress developed along the bolt in 20 MPa concrete, where the stress in the top part of the bolt and towards the perimeter are tensile and compressive at the centre. However, the stress conditions at the lower half section of the bolt are reversed. In addition, the shape of the bolt between the hinges can be considered as linear. The rate of stress changes in the post failure region is plotted in Figure 16.

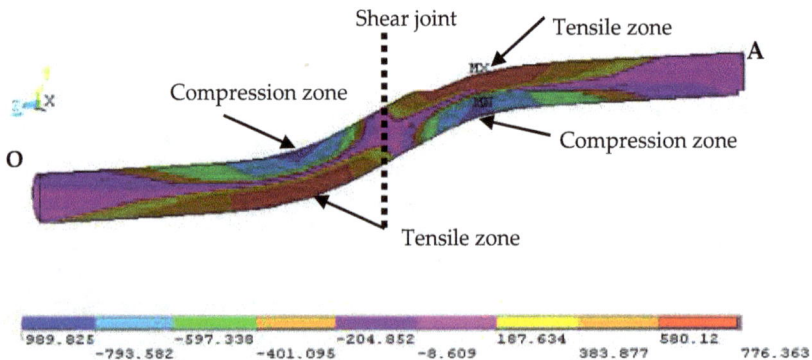

Figure 15. Stress built up along the bolt axis in 20 MPa concrete without pre-tension

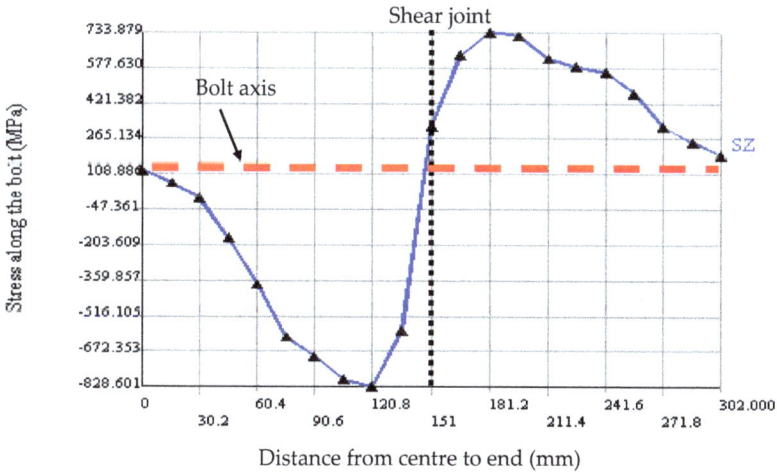

Figure 16. Trend of stress built up along the bolt axis 20 MPa concrete with 80 kN pre-tension

It can be seen that induced stresses at these tensile and compression zones are high and the bolt appears to be in a state of yield. At the two hinges the yield limit of the bolt is reached quickly. However, a further increase in the shear load has no apparent influence on the stress built up at the hinge point. From this stage afterwards, only tensile stresses are developed and expanded between the hinge points, and may lead the bolt to fail at some distance between the hinge points located near the shear joint, as the maximum stress and strain occurs between them.

From analysing the results in different pre-tension loads it was found there are no significant changes in induced stresses along the bolt with an increase in pre-tension load in the tension zone. However there is a slight reduction in compressive stress with an increasing pre-tension load. Induced stresses are higher than the yield point and less than the maximum tensile strength of the steel bolt in both situations (with and without pre-tension in all strength concrete). Moreover, in different strength concrete it was observed that the strength of the concrete affects shear displacement and bolt contribution. However there were no meaningful changes in induced stress beyond the yield point along the bolt axis with increasing rock strength but stress was reduced slightly with high pre-tension loading and strength of concrete. The Von Mises stress trend along the bolt axis perpendicular to the shear joint in 20 MPa concrete is plotted in Figure 17. Comparing the results in 20 MPa concrete with and without pre-tension, Von Mises stress decreased slightly, with an increase in bolt pre-tension. However, this difference is insignificant.

Figure 18 shows the concentration of shear stress along the bolt and the rate of change along the axis is shown in Figure 19. Figure 20 shows the trend of shear stress along the length of the bolt in one side of the joint surrounded with soft concrete.

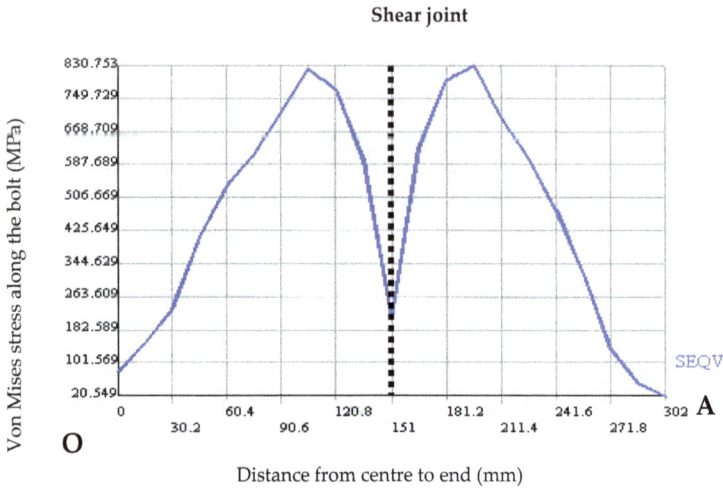

Figure 17. Von Mises stress trend in 20 MPa concrete without pre-tension

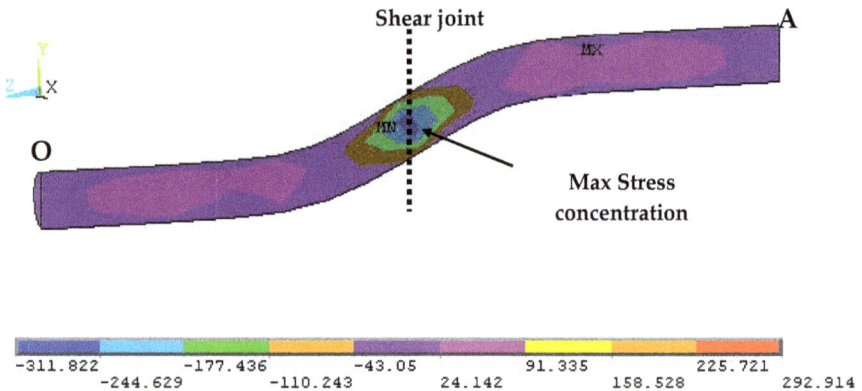

Figure 18. Shear stress contour in the concrete 20 MPa without pre-tension

As it shows the maximum shear stress is concentrated in the vicinity of the joint plane, and according to structural analysis, the bending moment at this point is zero. These stress slowly increase, beginning with plastic deformation, and end with a stable situation. The shear stress reduces dramatically from the shear joint towards the bolt end. This trend reaches zero at the hinge point. In the two hinges, the yield limit of the steel is reached quickly, at about 0.3 P and 0.4 P in concrete 20 and 40 MPa respectively, (P is the maximum given applied load). Further increase in the shear force has no apparent influence on stress in the hinges. The distance between the hinge points is reduced as the strength of the concrete is increased.

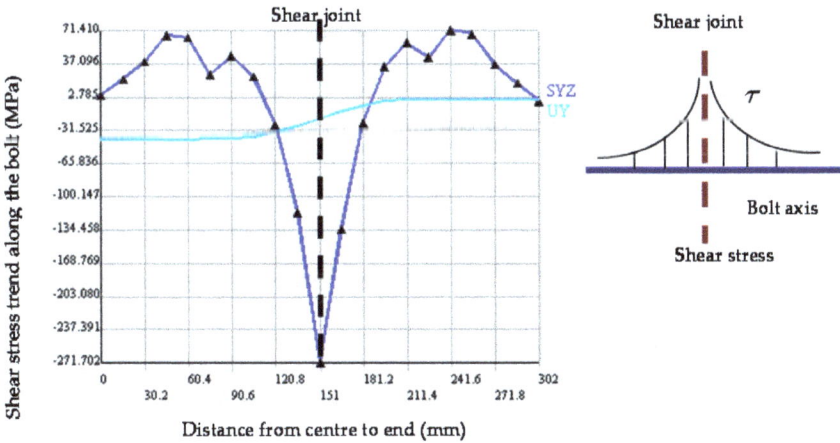

Figure 19. The rate of shear stress change along the bolt axis in concrete 20 MPa without pre-tension

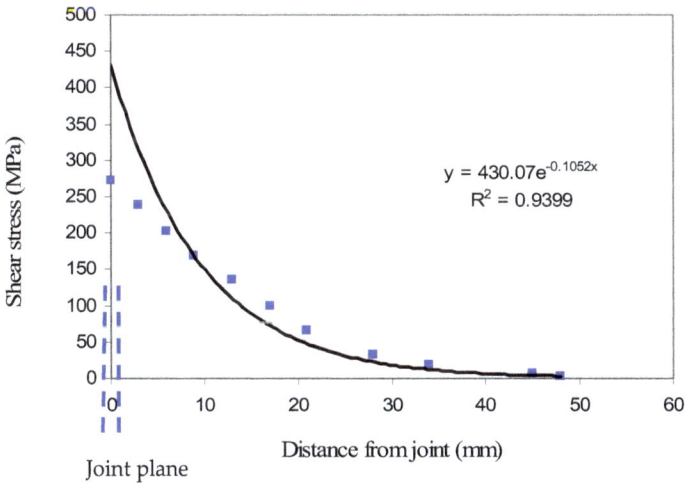

Figure 20. The rate of shear stress along the bolt axis in concrete 20 MPa without pre-tension in one side of the joint plane

Figure 21 shows the trend of changes in shear stress profile with the shear stress tapering off to a stable state past the yield point. It shows the shear stress trend is not exceeded during further loading after the yield point.

Eventually, a combination of this stress with induced tensile stress at the bolt - joint intersection leads the bolt to fail. By increasing the initial tensile load on the bolt, the shear stress was decreased, which was seen in different strength concrete but there was no significant changes with increasing shear load after the yield point. Any reduction in shear stress causes an increase in the resistance of the bolt to shear. It can be noted that the shear stress increased slightly with an increasing strength of concrete.

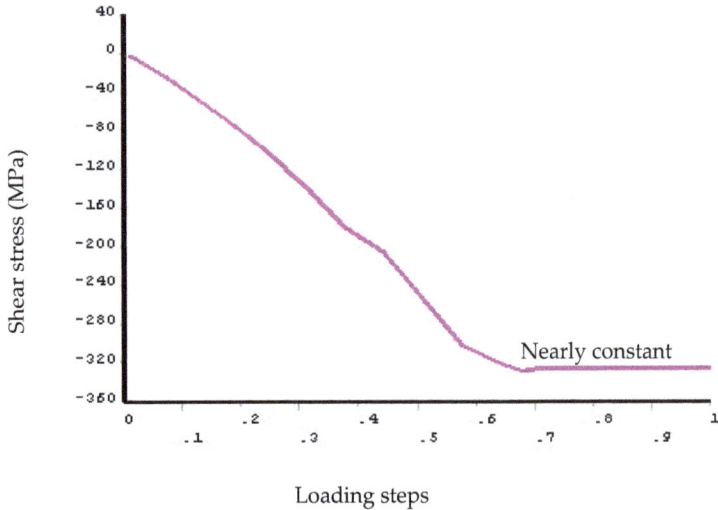

Figure 21. Shear stress trend in bolt –joint intersection in concrete 20 MPa at post failure region without pre-tension load

6.1.2. Strain developed along the bolt

Strain was generated along all the surrounding materials as the shear load increased, particularly along the axis of the bolt. As deflection increased, plastic strain is induced in the critical locations in all three materials (bolt - resin and concrete). Figure 22 shows the location of maximum plastic deformation along the bolt while bending. It shows there are two hinge points around the shear plane approximately 50 mm from the shear joint in 20 MPa concrete.

However an increasing pre-tension load has not affected hinge point distances, which are around 2.3 D_b (D_b is bolt diameter). This value in the laboratory test is around 44 mm that is 2 D_b. The strain and the rate of strain changes along the bolt in 20 MPa concrete are shown in Figures 23 and 24.

As Figure 23 shows that the outer layer of the bolt yielded, whereas the middle section remained in an elastic state.

Shear load

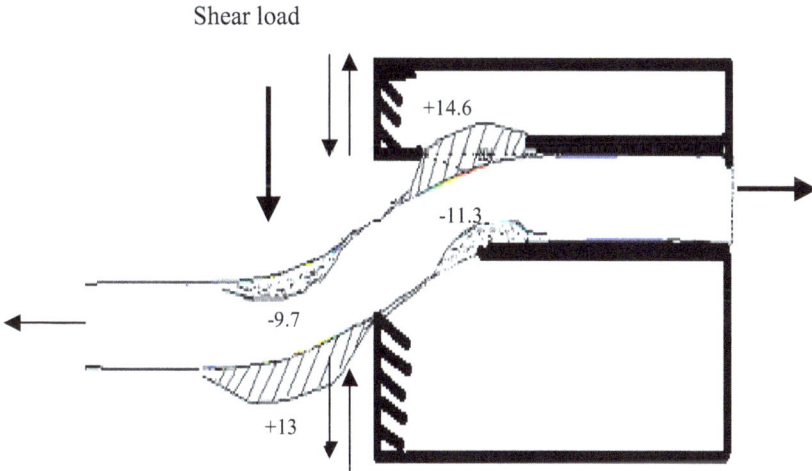

Figure 22. Deformed bolt shape in post failure region in 20 MPa concrete

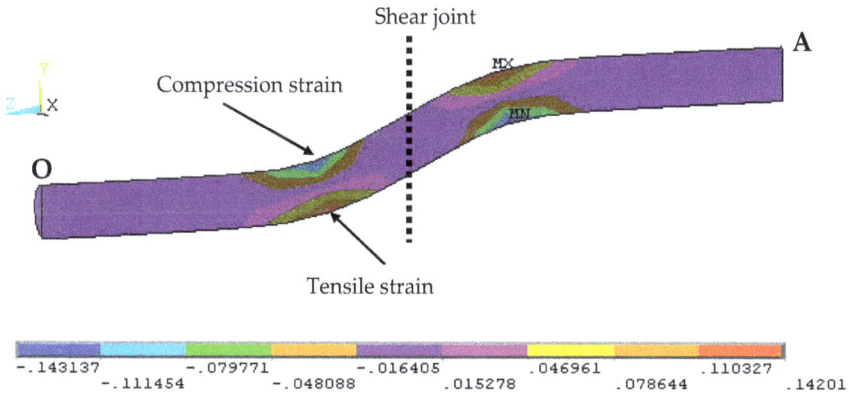

-.143137		-.079771		-.016405		.046961		.110327	
	-.111454		-.048088		.015278		.078644		.14201

Figure 23. Plastic strain contour along the bolt axis in concrete 20 MPa without pre-tension

Figure 25 shows the beginning of plastic strain during shearing and a trend of strain developing as a function of load stepping. It notes that both the tensile and compression strain around the bolt started approximately 27-30 % after loading began and increased with an increasing shearing load. However, the rate of increase in the tensile zone is higher than the compression zone. It also showed these strains appeared in the early stage of loading with a small displacement (around 3 mm), which increased with increase in shear deflection.

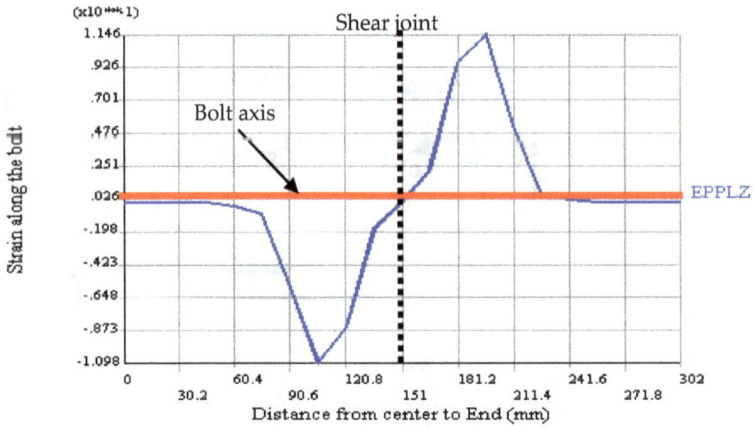

Figure 24. Strain trend along the bolt axis in concrete 20 MPa without pre-tension in upper fibre of the bolt

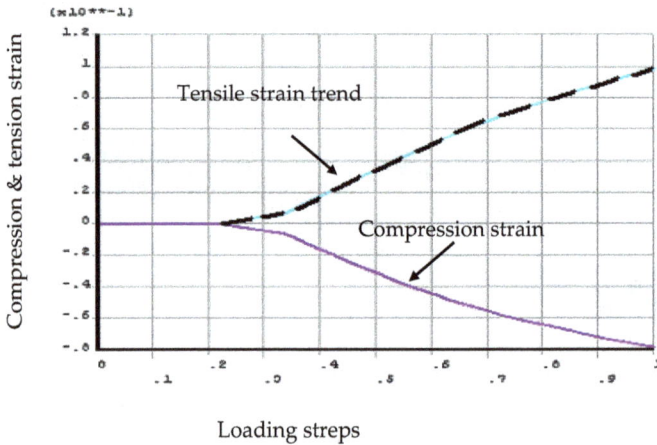

Figure 25. Yield strain trend as a function of time stepping concrete 20 MPa in 20 kN pre-tension load

With an increase in loading, shear displacement was increased. There was a significant increase in shear displacement after 35% of loading time. Bending of the bolt is predominant at a low loading time. plastic strain begins at the hinge point around 35 % of loading. A comparison of the data (with and without pre-tension) shows that the intensity of the strain along the axis of the bolt is slightly reduced with an increase in pre-tension load. However the affected area in the tensile zone expands towards the shear joint. The strains in the compression and tension zones were reduced in higher strength concrete.

6.2. Concrete behaviour

6.2.1. Stress developed in concrete

The behaviour of the centre concrete under shear load in double shearing assembly was analysed in different strength concrete and different pre-tension loads. During shearing the middle part of the assembled system was displaced downwards with increasing shear load. Figure 26 shows the deflection rate after failure. Reaction forces are developed during the middle concrete block displacement, which increased in critical locations (at the vicinity of the shear joint), affected by the bolt. The reaction forces induce and propagate stress and strain in sheared zones. Figure 27 shows the high-induced stress near the shear joint as the maximum reaction forces are expected there. When induced stress is larger than the ultimate stress the concrete will be crushed. Figure 28 displays the rate of induced stress at the interface near the shear joint. It shows that induced stresses are much higher than the compressive strength, and the concrete at this location would be severely crushed. From the figure it can be seen that the high stress is approximately 60 mm from the shear plane. At an early stage of loading, the concrete was crushed and stresses propagated throughout, with bolt yield to start at around 2 mm from the edge of the intersection. Beyond this point stresses increased quickly near the joint intersection and reaction zones. Induced stresses near the shear joints were reduced slightly with increase in the pre-tension load on the bolt. In addition the trend of induced stresses and strains built up along the concrete interface in 40 MPa concrete was the same as with 20 MPa concrete. However, the value of stresses and strains were slightly reduced in higher strength concrete.

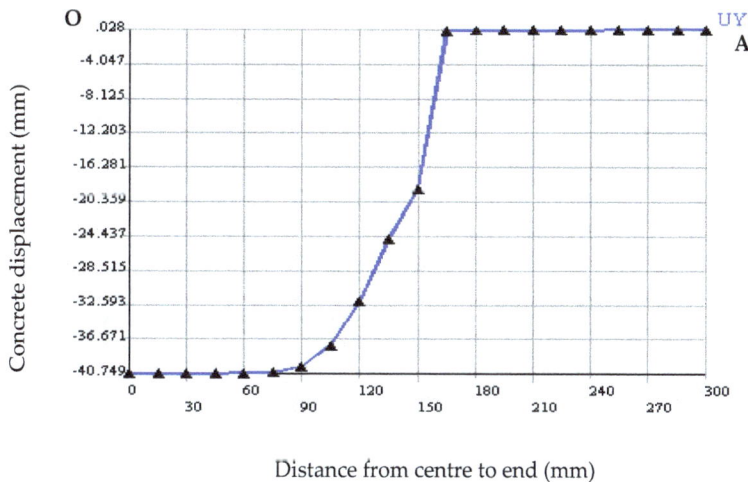

Distance from centre to end (mm)

Figure 26. Concrete displacement in non-pretension condition in 20 MPa concrete

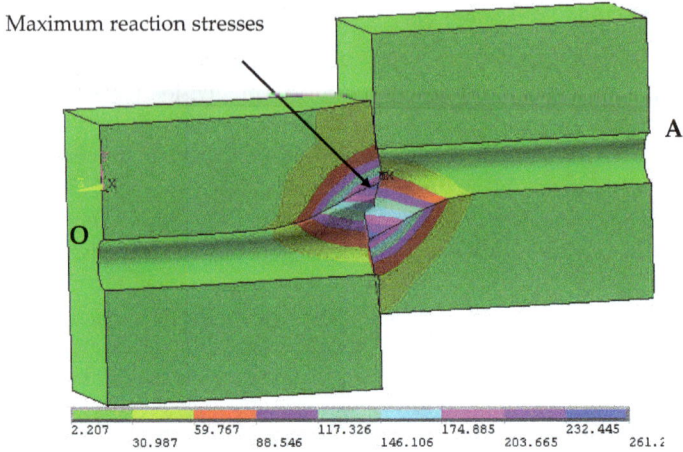

Figure 27. Yield stress induced in 20 MPa concrete without pre-tension condition

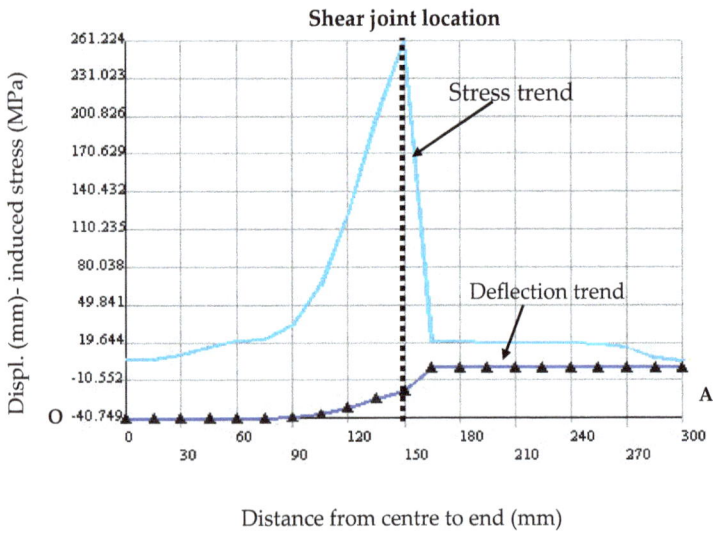

Distance from centre to end (mm)

Figure 28. Induced stress and displacement trend in 20 MPa concrete without pre-tension

6.2.2. Strain developed in concrete

The highest level of induced stress was near the shear joint, so it is expected that strain would be highest around this zone. Figure 29 shows the induced strain contours at the high pressure zone. Figure 30 shows induced strain in terms of loading time in grout and concrete. It shows that the strain generation begins in the concrete before it is seen in the resin grout because lower strength concrete is one third the strength of grout. There is an approximate exponential relationship in the strain trend as loading increases. After 20% of loading steps, plastic strain is induced along the contact interface near the shear joint. This value in soft concrete (20 MPa) is at an earlier stage, which is around 15% of loading step. This shows the strain built up along the axis of the bolt is lower than in the shear direction.

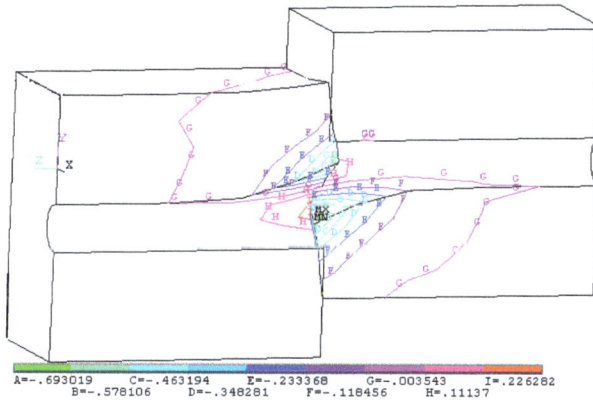

Figure 29. Strain contours in 20 MPa concrete without pre-tension

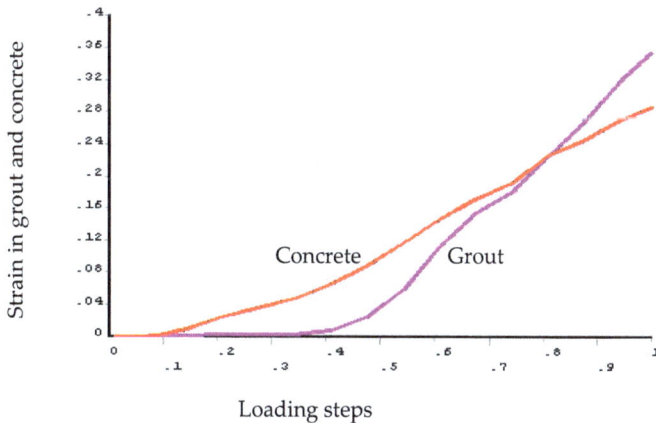

Figure 30. Induced strain in concrete 20 MPa in grout and concrete versus loading without a pre-tension and 27 mm diameter hole

A comparison of induced strain along the joint interface with and without pre-tension found that the strain in the shear direction is reduced (around 15%) with increasing pre-tension. In the axial and shear direction strain was concentrated near the shear joint.

Figure 31 shows the deformation behaviour of both concrete medium and bolt. Plastic deformation of concrete occurs nearly 15 % of the maximum shear load while the deformation of the bolt occurs at 33% of the loading steps. From the graphs it can be inferred that in very low values of bolt deflection and time steps, fractures happen in the concrete, which is in the elastic range of the bolt. Any further increase in shearing does not influence the stress at the hinge points, however induced stress in the concrete blocks causes extensively fractures and eventually leads to failure.

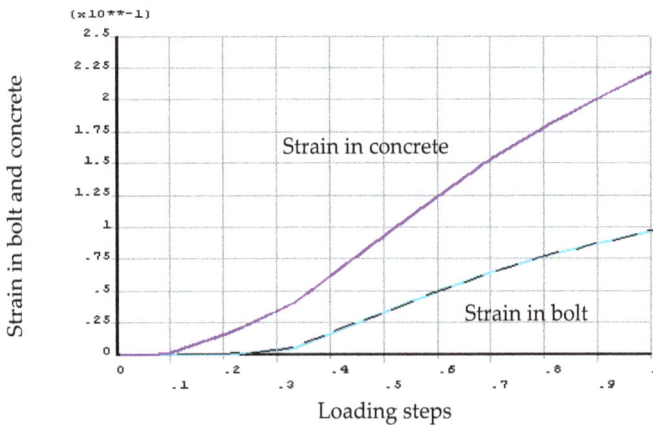

Figure 31. Induced strain in concrete and bolt as a function of loading steps in 20 MPa concrete with 80 kN pre-tension

6.3. Grout behaviour

6.3.1. Stress in grout

It is known that grout bonds the shanks to the ground making the bolt an integral part of the rock mass itself. Its efficiency depends on the shear strength of the bolt - grout, and grout - rock interface. Figure 32 shows the contours of induced stress through the resin layer surrounded by 20 MPa concrete, without pre-tension. It was revealed that the induced stress exceeded the uniaxial compressive strength of the grout near the bolt - joint intersection which crushed the grout in this zone. It shows that the value of induced stress in the grout near the shear joint is much higher than the uniaxial strength, and grout in this location can be crushed. The broken sample showed that the grout was crushed around this zone. The damaged area on the upper side of the grout was approximately 60 mm from the shear joint. Figures 33 and 34 show the gap formation after bending in the numerical and laboratory methods respectively. It is noted that the induced stresses were slightly reduced as the pre-tension increased (nearly 10 %). However, it shows they are slightly expanded.

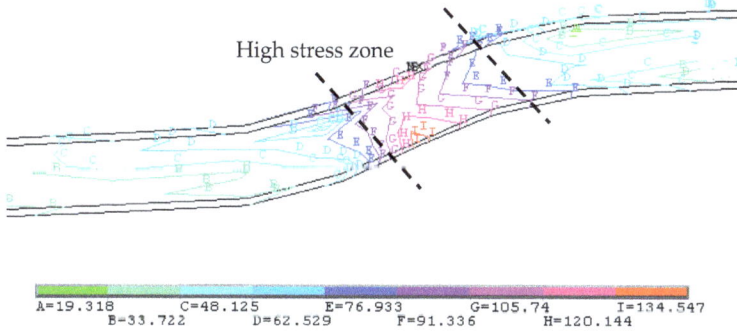

Figure 32. Maximum induced stress contours in grout layer without pre-tension and 20 MPa

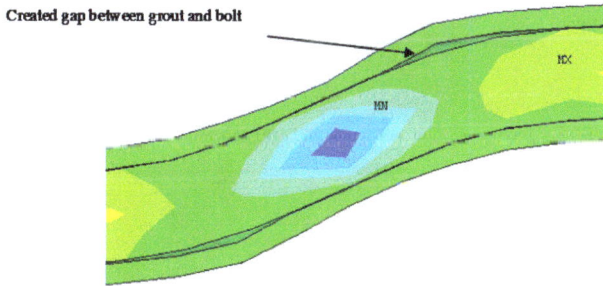

Figure 33. Gap formation in post failure region in 20 MPa concrete in the Numerical simulation

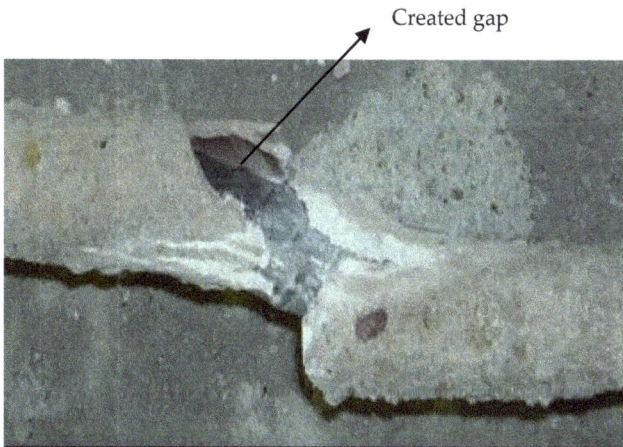

Figure 34. Gap formation in post failure region in 20 MPa concrete in the laboratory test

6.3.2. Strain in grout

While shearing takes place, strains are induced through the grout near the shear joint and reaction zones. The strain in the grout was around ten times greater than the linear region at critical zones. This means that the grout in those areas had broken off the sides that were in tension. The rate of induced strain along the grout in an axial direction is shown in Figure 35.

A comparison of the strain along the joint interface in the grout showed that it decreased between 3% and 5% in the compression and tension zones with increasing pre-tension to 80 kN, which is due to higher shear resistance and lower lateral displacement. It was also found that the grout layer at the bolt - joint intersection will start to crush after slight movement along the joint, which causes plastic strain in the grout layer.

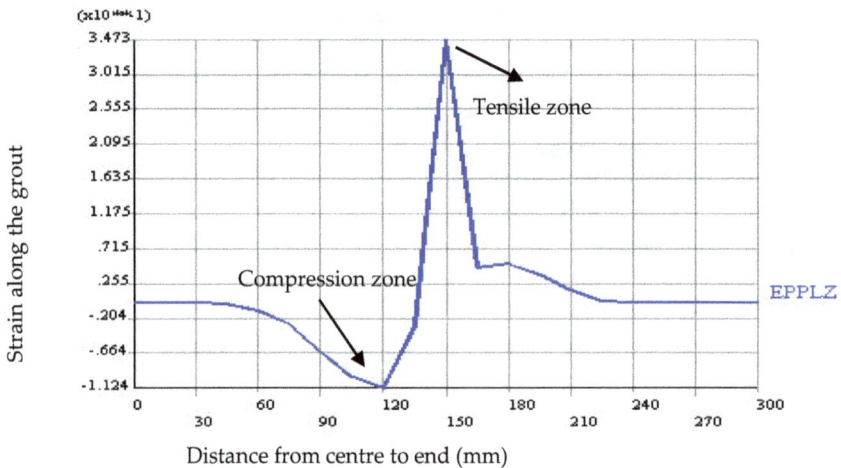

Figure 35. The rate of induced strain along the grout layer without pre-tension in an axial direction

In high strength concrete induced stress was reduced slightly and pre-tension reduces induced stresses along the bolt - grout interface.

From the results at contact pressure in the bolt-grout-concrete it was found that there is an exponential relationship between contact pressure and loading process at the bolt - grout interface, which started after around 15% of the loading process. However, the contact pressure trend in the concrete - grout interface was formed by 2 parts. From the beginning to around 15% of the loading, there is an approximate linear relation followed by an exponential relationship till the end of the load stepping process.

7. Bolt modelling under axial loading

A numerical model was developed to investigate the contact interface behaviour during shearing under pull and push tests. The same 3D solid elements and surface-to-surface contact elements were used to simulate grout and steel. The numerical simulation of the cross section of the bolt and its ribs was complicated, and is almost impossible with the range of software available in the market today. However an attempt was made to model the bolt profile configurations by taking into account the realistic behaviour of the rock - grout and grout - bolt interfaces based on laboratory observations. To achieve this end, the coordinates of all nodes for all materials were defined then all these co-ordinates were inter-connected to form elements, which were extruded in several directions to obtain the real shape of the bolt.

Figure 36 shows the FE mesh. Figure 37 shows the bolt under pull test. Two main fractures were produced as a result of shearing the bolt from the resin. The first one begins at the top of the rib at an angle of about 53^0 running almost parallel to the rib, and the second one has an angle of less than 40^0 from the axis of the bolt. When these fractures intersect they cause the resin to chip away from the main body because it is overwhelmed by the surface roughness of the rib while shearing. Internal pressure produced by the profile irregularities of the bolt induces tangential stress in the grout. The grout fractures and shears when the induced stress exceeds the shearing strength, allowing the bolt to slide easily along the sheared and slikenside fractures in the grout interface.

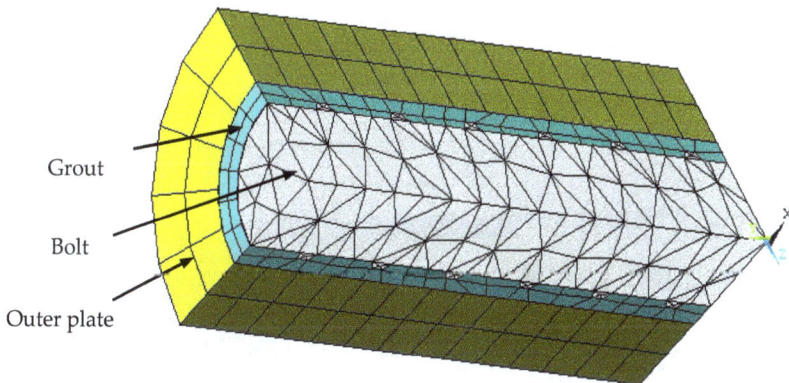

Figure 36. FE mesh: a quarter of the model

7.1. Bolt behaviour

From the simulations it was found that there will be an increase in grout - bolt surface de-bonding, and this decrease in diameter due to Poisson's effect in the steel, contributes to an axial elongation of about 0.084 mm at the top collar where the load is applied. This value in push test is around 0.05 mm as shown in Figure 37.

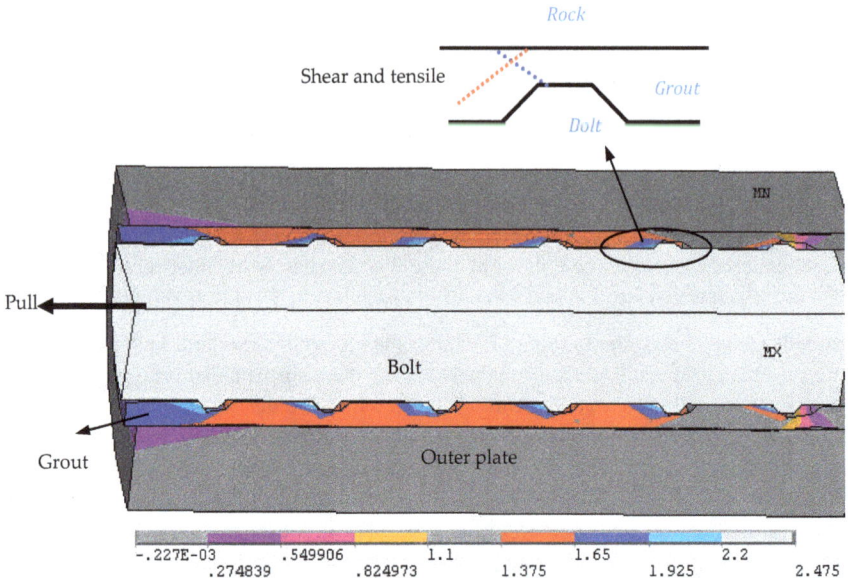

Figure 37. The bolt movement in pulling test

Figure 38 show the maximum induced strain near the applied load position in both the pull and push results. The strain is around the elastic strain and therefore the bolt is unlikely to yield.

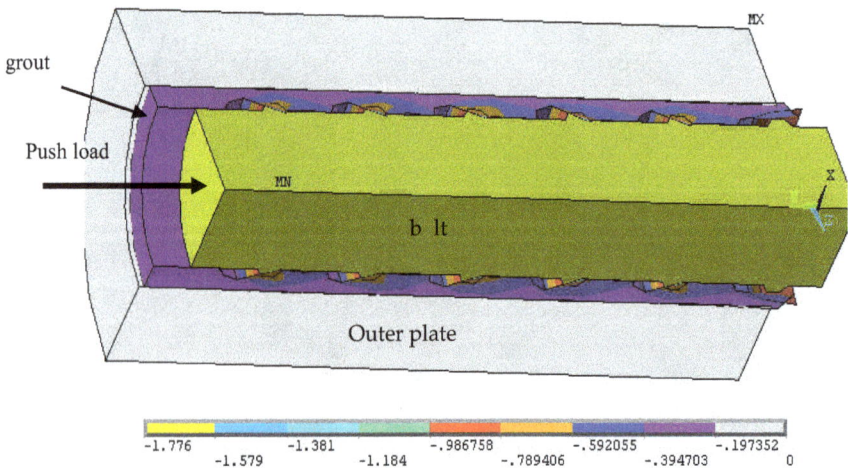

Figure 38. Bolt displacement contour in Bolt Type T1 in case of push test

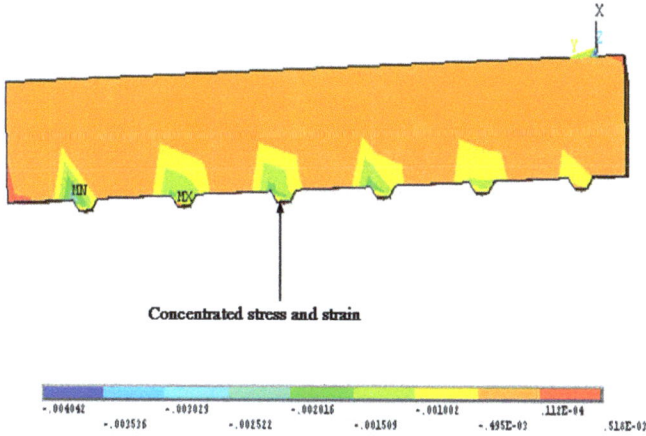

Figure 39. Shear strain in bolt ribs in push test

Maximum tensile stress along the bolt is 330 MPa. This is one half of the strength of the elastic yield point of 600 MPa. This means the bolt behaves elastically and is unlikely to reach a yield situation. Axial stress developed along the bolt is given by:

$$\sigma_t = \frac{4T}{\pi D_b^{\,2}} \tag{4}$$

and

$$T = \frac{\pi D_b^{\,2} * \sigma_t}{4} \tag{5}$$

Where, σ_t is the tensile stress, T is the axial load, D_b is the bolt diameter and σ_y is the yield strength of the bolt. The bolt behaves elastically as long as the following expression is satisfied:

$$\sigma_t < \sigma_y \tag{6}$$

So in this situation with failure along the bolt-grout interface will not yield.

7.2. Grout behaviour

The behaviour of interface grout annulus is assumed to be elastic, softening, residual, plastic flow type. This behaviour was developed by Aydan (1989), and is given as:

$$\tau = G\gamma \quad \tau < \tau_{max} \tag{7}$$

$$\tau = \tau_{max} - \frac{\gamma - \gamma_{max}}{\gamma_r - \gamma_{max}}(\tau_{max} - \tau_r) \qquad (8)$$

$$\tau = \tau_r \qquad (9)$$

where;

- G = Shear modulus of grout interface
- γ = Shear strain at any point in the interface
- γ_r = Shear strain at residual shear strength
- γ_{max} = Shear strain at peak shear strength
- τ_r = Residual shear strength of the interface
- τ_{max} = Peak shear strength of interface
- τ = Shear stress at any point in interface

The grout material is in elastic conditions if the following expression is satisfied;

$$T_t < T_y \qquad (10)$$

where;

- T_t = Actual bond stress in the grout
- T_y = Yield stress of the grout in shear

From the strain generated along the grout interface it was found that the surface of the grout was disturbed by shear stress induced at the interface and this strain is higher than the elastic strain that damaged the grout at the contact surface. Figure 39 shows the shear stress contour at the grout interface. The whole contact area of the grout was affected by the shear stress and consequently the induced shear strain dominated. The maximum bonding stress was approximately 38% of the uniaxial compressive strength of the resin grout. The stress produced along the grout contact interface was greater than the yield strength of the grout of 16 MPa, and beyond the yield point only a slight increase in load is enough to damage the whole contact surface. Shear displacement increased as a result bonding failure. The shear stress at the bolt - grout interface can be calculated by Equation (11), which agrees with the results from the numerical simulation.

Thus,

$$\tau = \frac{f}{A} = \frac{\sigma \pi D^2}{8\pi r l} = 23.2 MPa \qquad (11)$$

where;

- τ = Shear stress in the grout - bolt interface (MPa)
- f = Axial force in the bolt (kN)

- A = Contact interface area (mm²)
- D = Bolt diameter (mm)

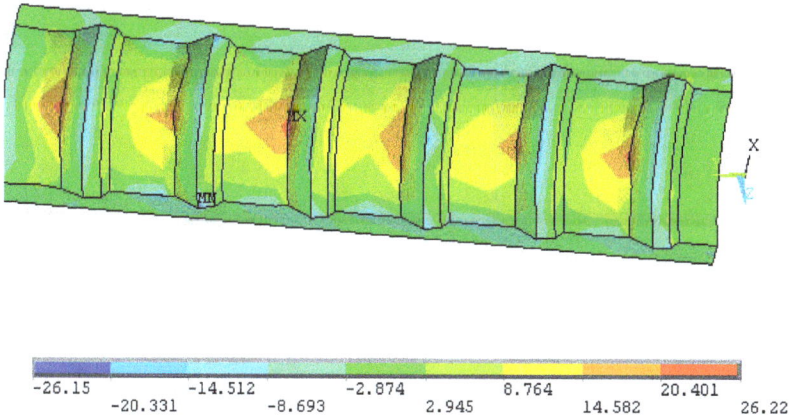

Figure 40. Shear stress contours along the grout interface

Using the *Farmer* (1975) equation the shear strength was equal to 27 MPa.

$$\frac{\tau}{\sigma} = 0.1e^{-(\frac{0.2x}{a})} \tag{12}$$

where;

- τ = Shear stress along the bolt grout interface
- σ = Axial stress
- a = Bolt radius

During shearing the outer plate of the bolt was influenced by the stresses and strains of the resin. From the analyses it was found that induced stress along the surface of the outer plate was insignificant at about 30 % of the yield stress, which is not sufficient to cause the outer plate to yield. In addition, grout de-bonding occurred around 50 to 60 kN at different levels of applied load.

8. Summary

Numerical analysis of the grout – concrete - bolt interaction has demonstrated that:

- There were no significant changes in induced stresses along the bolt with increasing pre-tension load, particularly in the tension zone. However, there was a small reduction in compression stress.
- The yield limit of the bolt at the hinge point depends on the strength of the concrete. In 20 MPa concrete the yield limit was 0.3P and in 40 MPa concrete it increased to 0.4P. A

further increase in the shear force has no apparent influence on stress at the hinges. The distance between the hinge points reduced with increasing strength of concrete.

- The strength of the concrete greatly affects shear displacement and bolt contribution. However, no significant change was observed in the induced stresses beyond the yield point along the axis of the bolt with increasing concrete strength.
- The maximum shear stress was concentrated near the bolt - joint intersection.
- There was an exponential relationship between the shear stress and distance from the shear joint.
- The shear stress was not exceeded during further loading after the yield point. Eventually, a combination of this stress with induced tensile stress at the bolt - joint intersection caused the bolt to fail.
- Shear stress at the bolt - joint intersection increased slightly with an increasing strength of concrete.
- There was no significant change in the hinge point distances with an increase in bolt pre-tension.
- There was a significant increase in shear displacement beyond 35% of the loading step, which is the likely yield point.
- The strain in the shear direction along the concrete was reduced (around 15%) with increasing the pre-tension loading. In both axial and shear directions the strain concentrated near the shear joint.
- The induced stresses exceeded the uniaxial compressive strength of the grout near the bolt - joint intersection, crushing the grout.
- The damaged area in the upper side of the grout was approximately 60 mm from the shear joint.
- Induced stress along the grout was reduced by increasing the pre-tension load nearly 10%. However they have expanded slightly.
- The strain was decreased by around 3% and 5% in the compression and tension zones where the bolt pre-tension load increased to 80 kN.
- Failure of the bolt - resin interface occurred by the grout shearing at the profile tip in contact with the resin.
- Numerical simulation provided an opportunity to better understand the stress and strains generated as a result of the bolt - resin interface shearing. Such an understanding is supported both analytically and by simulation.
- Findings from the experimental test agreed with the numerical simulations and analytical results.

Author details

Hossein Jalalifar
Shaihid Bahonar University of Kerman-Iran

Naj Aziz
Wollongong University- Australia

9. References

[1] Pool.G, Cheng.Y and Mandel.J (2003). Advancing analysis capabilities in Ansys through solver technology. Electronic transactions on numerical analysis 15: 106-121.

[2] Bhashyam.G.R (2002). Ansys Mechanical-A powerful nonlinear simulation tool, 2004. Ansys,Inc. 275 Technology Drive Canonsburge,PA 15317.

[3] Coates.D.F and Yu.Y.S (1970). Three dimensional stress distribution around a cylindrical hole and anchor. Proceeding of 2nd Int. Cong. Rock Mechanics.175-182.

[4] Hollingshead.G.W (1971). Stress distribution in rock anchors. Canadian Geotechnical Journal 8: 588-592.

[5] Aydan.O (1989). The stabilisation of rock engineering structures by rock bolts. Geotechnical Engineering.Thesis. Nagoya, Nagoya: 202.

[6] Saeb.S and Amadei.B (1990). Finite element implementation of a new model for rock joints. Int. Symp. of Rock Joints.707-712.

[7] Aydan, O. and T. Kawamoto (1992). Shear reinforcement effect of rockbolts in discontinious rock masses. Int.Sym. of Rock support in mining and underground construction, Canada.483-489.

[8] Swoboda.G and Marence.M (1992). Numerical modelling of rock bolts in interaction with fault system. Int.Sym. of Numerical modelling in Geomechanics.729-737.

[9] Moussa.A and Swoboda.G (1995). Interaction of rock bolts and shotcrete tunnel lining. Int.Sym. of Numerical models in Geomechanics.443-449.

[10] Marence.M and Swobodea. M (1995). Numerical model for rock bolt with consideration of rock joint movements. Rock Mechanics & Rock Enginnering 28.(3): 145-165.

[11] Chen, S. H. and G. N. Pande (1994). Rheological model and finite element analysis of jointed rock masses reinforced by passive, fully-grouted bolts. International Journal of Rock Mechanics and Mining Science & Geomechanics Abstracts 31.(3): 273-277.

[12] Chen, S. H. and P. Egger (1999). Three dimensional elasto-viscoplastic finite element and its application. Int. J. for Numerical and analytical methods in geomechanics 23: 61-78.

[13] Chen, S. H., S. Qiang and S. F. Chen (2004). Composite element model of the fully grouted rock bolt. Rock mechanics and rock engineering 37.(3): 193-212.

[14] Surajit, P. and G. Wije Wathugala (1999). disturbed state model for sand-geosynthetic interfaces and application to pull-out tests. Int. J. for Numerical and analytical methods in geomechanics 23: 1873-1892.

[15] John.C.M and Van Dillen.D.E (1983). Rock bolts: A new numerical representation and its application in tunnel design. Proc.24th U.S. Symp. Rock Mech., Texas A&M University.13-25(Cited in Moosavi 1994).

[16] Peng, S. and S. Guo (1992). An improved numerical model of grouted bolt-roof rock interaction in underground openings. Rock support in mining and underground construction, Canada.67-74.

[17] Stankus.J.C and Guo.S (1996). Computer automated finite element analysis-A powerful tool for fast mine design and ground control problem diagnosis and solving. 5th Conference on the use of computer in the coal industry, West Virginia-USA.108-115.

[18] Fanning, P. (2001). Nonlinear models of reinforced and post tensioned concrete beams. Structural engineering: 111-119.

[19] Feng.P, Lu.X.Z and Ye.L.P (2002). Experimental research and finite element analysis of square concrete column confined by FRP under uniaxial compression. 17th Australian Conference on the Mechanics of Structures and Materials, Gold Coast, Australia.60-65.

[20] Ansys. 2012. Manual.

[21] Cha.E.J, Choi.S.M and Kim.Y (2003). A moment-rotation curve for CFT square column and steel beams according to reliability analysis. Int. Conference on Advances in Structures, Sydney-Australia.943-950.

[22] Hong.S.D, Choi.S.M and Kim.Y (2003). A moment-rotation curve for CFT square columns and steel beams. Int. Conference on Advances in Structures, Sydney-Australia.951-956.

[23] Abedi, K., H. Afshin and A. Ferdousi (2003). Investigation into the behaviour of a novel steel section for concrete filled tubular columns under axial and cyclic loadings. Advances in structures, Australia. 891-897.

[24] Pal.S and Wathugala.G.W (1999). Distribution state model for sand-geosynthetic interfaces and application to pull out tests. Int.J. for Numerical and Analytical Methods in Geomechanics 23: 1873-1892.

[25] Ostreberge. J.O and Gill. S.A (1973). load transfer mechanism for piers. Proceedings of the 9th Canadian rock mechanics symposium, Montreal , Canada, Mines Brance, Departement of Energey, Mines and Resources.235-262.

[26] Nitzsche. R. N and Haas. C. J (1976). Installation induced stresses for grouted roof bolts. International Journal of Rock Mechanics and Mining Science & Geomechanics Abstracts 13.(1): 17-24.

[27] Farmer, I. W. (1975). Stress distribution along a resin grouted rock anchor. Rock mechanics and mining science 12: 347-351.

Chapter 6

Numerical Methods for Analyzing the Transients in Medium Voltage Networks

Dumitru Toader, Stefan Haragus and Constantin Blaj

Additional information is available at the end of the chapter

1. Introduction

The analysis of transient regimes in electric networks is a complex problem due to great number of elements, some nonlinear, as well as, due to the nonsinusoidal variation of currents and voltages. Numerical simulation can solve properly such a problem

There are available a lot of simulation programs for transients. One of the first of them, dedicated to electro energetic systems, was initiated by H.W.Dommel. Electromagnetic Transient Program (EMTP) was bought by Bonneville Power Administration (USA) (Dommel, 1995). Later on, from this initial program were developed several versions, such as MicroTrans, EMTP-RV, ATP (Alternative Transients Program), PSCAD (Chuco, 2005; DeCarlo & Lin, 2001).

Comparing some of the main simulation programs is not very conclusive, because each one of them are used in some versions trying to solve a large group of problems (Iordache & Mandache, 2004; Istrate et al., 2009).

Interface becomes more and more friendly, the library becomes larger and the facilities for creating own models are simpler to use. The speed of calculations and the storage capacity do no more represent a problem. It is possible now to implement some complicated algorithms able to solve problems with high speed in the variation of variables, able to solve all kinds of nonlinearities or to handle eigenvalues highly distanced one from another (large stiffness ratio dynamical systems).

An important criterion for the selection is represented by the integration of complex dynamical systems like rotating electric machines, relays, power electronic devices, FACTS, controllers etc. in the simulation. Some of the programs, depending on the way that were conceived, allow a better integration of these subsystems.

Choosing one or other of the simulation programs depends on the previous experience of the user, as well as, on the manner the program responds to the specific demands (Chuco, 2005; Danyek et al., 2002; Foltin et al., 2006; Karlsson, 2005; Rashid & Rashid, 2006).

No mater what program is used, the numerical analysis of transients eventually must be able to solve a set of differential – algebraic equations that models the physical system. By accounting the stage the numerical integration takes place at, there are two main classes of simulation programms.

One class consists of those programs that make the integration at element level, meaning that for each step of time discretization the differential equations associated to dynamic elements are transformed in finite difference relations. Several ways of approximation can be used, more frequently the trapeze rule being used. All these algebraic relations are then assembled and as a result an algebraic system of equations is obtained and consequently solved using specific algorithms. This procedure is repeated at each iteration step, the parameters involved in the equations being modified by the results obtained at the previous step. Usually the integration step is fixed, but if this is required by a lack of convergence, the integration step can be split in half and the computation process is resumed. This method is known under the name of the implicit integration method and is used in programs such as SPICE and EMTP (Blume, 1986).

A second class of programs uses a two step procedure. In the first step the mathematic model is expressed as a system of first order differential equations, known as the system's state equations. During the next step this system of equations is integrated using algorithms with fixed step or with variable steps, depending on the systems particularities. The advantage of this method, called the state variables method, is to treat in the same way, in a unitary manner, electric networks, electric machines, drives, control devices or any other device that allows state equations. A representative of this method is SimPowerSystems™ which is an extension of Simulink® with tools for modeling and simulating of electrical power systems. It provides models of many components used in these systems, including three-phase machines, electric drives, and libraries of application-specific models such as Flexible AC Transmission Systems (FACTS) and wind-power generation. Harmonic analysis, calculation of Total Harmonic Distortion (THD), load flow, and other key power system analyses are automated (Mathworks®).

The precision of the results obtained by integration of the equations describing the simulated system is quite remarkably high. It is obvious that none of the simulation programs, no mater which of the methods is using, cannot be more precise than the mathematic models used for the simulated components. The accuracy of the values of the parameters of these models is of high importance.

2. The state-variable method for electrical circuits

Medium voltage electric networks are in fact complex circuits mainly made of power sources, resistors, inductors, capacitors and, sometimes, other electric components.

Analyzing transient regimes in such networks is the same as the analysis of any other complex electric circuit (Dessaint, et al. 1999; Mandache & Topan, 2009; SIMULINK, 1997).

Generally, an electric circuit, no mater how complex it is, is described by an algebraic linear system of equations, obtained by applying the Kirchhoff's laws. This system of equations reflects the circuit's topology. To complete the model, the voltage-current equation at the terminals of each element of circuit (called also constitutive relations) must be added.

The inductors and the capacitors are described by constitutive relations in which are involved the derivatives of currents, respectively voltages. As a result, the mathematic model of an electric circuit consists of a system of algebraic and differential equations (DAE).

In the electric circuit theory it is demonstrated that each current or voltage can be expressed as function of inductors currents and capacitors voltages. By other words if the inductors currents and the capacitors voltages are known, all remaining currents and voltages are uniquely determined and because of this they are called state-variables.

The state-variable equations are obtained by elimination of the algebraic equations in the initial DAE.

The modelization of an electric circuit made using state-equations has several important advantages.

First of all, electric circuits can be integrated with other dynamic systems, of completely different physical nature, as long as the last ones are described also by state-equations.

By the other hand, the procedures for integrating state-equations have a well-established theoretic support, being the domain of interest for quite a long period of time for famous mathematicians.

Finally, but not last, the method can be applied to linear circuits, as well as, to nonlinear circuits.

Of course there is a price to be paid for all these advantages: a certain difficulty in the elimination of the algebraic equations from the system.

In order to give an example for the basics of the state-variable method, we will analyze the simple RLC series connection circuit supplied by an ideal voltage source $u(t)$. The Kirchhoff equations and the constitutive relations for such a circuit are, respectivelly:

$$
\begin{aligned}
i_R(t) &= i_L(t) = i_C(t), \\
u(t) &= u_R(t) + u_L(t) + u_C(t), \\
u_R(t) &= Ri(t), \\
u_L(t) &= L\frac{di_L(t)}{dt}, \\
i_C(t) &= C\frac{du_C(t)}{dt}.
\end{aligned}
\tag{1}
$$

It is customary to write the last two equation in the form

$$\frac{di_L(t)}{dt} = \frac{1}{L}u_L(t),$$
$$\frac{du_C(t)}{dt} = \frac{1}{C}i_C(t). \tag{2}$$

With the remaining equations $u_L(t)$ and $i_C(t)$ are obtained as functions of $i_L(t)$, $u_C(t)$:

$$u_L(t) = u(t) - Ri_L(t) - u_C(t),$$
$$i_C(t) = i_L(t). \tag{3}$$

By the substitution in (2) it follows

$$\frac{di_L(t)}{dt} = \frac{1}{L}u(t) - \frac{R}{L}i_L(t) - \frac{1}{L}u_C(t),$$
$$\frac{du_C(t)}{dt} = \frac{1}{C}i_L(t), \tag{4}$$

or, in matrix form,

$$\frac{d}{dt}\begin{bmatrix} i_L(t) \\ u_C(t) \end{bmatrix} = \begin{bmatrix} -\frac{R}{L} & -\frac{1}{L} \\ \frac{1}{C} & 0 \end{bmatrix}\begin{bmatrix} i_L(t) \\ u_C(t) \end{bmatrix} + \begin{bmatrix} -\frac{1}{L} & 0 \end{bmatrix}\begin{bmatrix} u(t) \\ 0 \end{bmatrix}. \tag{5}$$

This is a first order differential equations system, with the variables $i_L(t)$ and $u_C(t)$.

If the initial conditions $i_L(0)$ and $u_C(0)$ are known, then, according to the uniqueness theorem for the solution of a first order differential equation, the solution $i_L(t)$ and $u_C(t)$, $t \geq 0$, exists and is unique.

Once the solutions for the variables $i_L(t)$ and $u_C(t)$ are available, immediately all the other unknowns are solved. Therefore, $i_L(t)$ and $u_C(t)$ are the state-variables of the circuit, and (5) is the matrix state equation of the circuit.

Generally, for a linear electric circuit, the matrix state equation is of the type (6):

$$\frac{dx(t)}{dt} = \mathbf{A}x(t) + \mathbf{B}u(t), \tag{6}$$

where $x(t)$ is the column matrix of the inductor currents and of the capacitor voltages (state-vector),

$u(t)$ is the column matrix of independent supply sources (sources vector), and, A and B are matrix depending on the circuits topology and of its parameters.

The total number of the state variables is equal with the sum of the number of the inductors and capacitors from the circuit, representing the order of the circuit.

All the other unknown variables are represented by the column matrix $y(t)$ (also called output vector) represented by a linear combination of the state-vector and the sources vector):

$$y(t) = Cx(t) + Du(t),\qquad(7)$$

C and D being matrix of the same origin as A and B.

The equation given by (6) describes the model, in state-space, of a linear electric circuit.

It is possible, as the result of using ideal elements in the modelization, to occur capacitive loops or inductive sections.

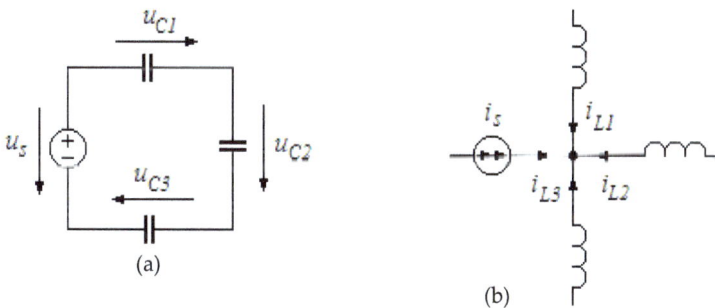

Figure 1. Capacitive loop and, respectively, inductive section

A capacitive loop (see Fig. 1a) is a loop consisting only from capacitors and, possibly, independent voltage sources.

In this case the voltages at the terminals of capacitors from the loop are no more linearly independent variables, because the second Kirchhoff's law gives:

$$-u_S + u_{C1} + u_{C2} + u_{C3} = 0.\qquad(8)$$

One, arbitrarily selected, of the capacitor voltages can be expressed as a linear combination of the other voltages.

Such a situation is present also for inductive sections.

An inductive section is represented in Fig.1b, being composed by the convergence in a node of ideal inductors and, possibly, independent current sources. In this case the first Kirchhoff's law states that

$$-i_S + i_{L1} + i_{L_2} + i_{L_3} = 0.\qquad(9)$$

Again anyone of the state variables can be expressed as a function of the others.

The order of an RLC circuit , meaning the number of linearly independent variables (and by consequence the number of state-equations) is equal to $n_{LC} - n_{bC} - n_{sL}$:

n_{LC} is the total number (sum) of inductors and capacitors

n_{bC} is the number of the capacitive loops

n_{sL} is the number of inductive sections.

The programs performing the analysis make the detection of capacitive loops and of the inductive sections presents in the electric circuit. Usually, it is recommended to introduce a very small value resistor in series connection with one of the elements of the capacitive loop, or to connect a high value resistor in parallel to one of the elements of the inductive section. By this the circuits differs from the original, but the difference in currents values and voltages values are insignificant

The method of state variables can be naturally extended for nonlinear circuits. In this case the circuit equations are of the form

$$f_R\left(u_R, i_R\right) = 0,$$
$$u_C = \frac{dq}{dt}; f_C(u_C, q) = 0, \tag{10}$$
$$i_L = \frac{d\Phi}{dt}; f_L(i_L, \Phi) = 0,$$

with f_R, f_C, f_L being the characteristics of nonlinear elements and, q and Φ, the electric charge of the capacitor, respectively the magnetic flux trough the inductor.

The procedure for obtaining the state equations is similar as for linear circuits, with the difference that the state variables are q and Φ. These variables are present in the dynamic constitutive equations.

The variables u_C and i_L are obtained as functions of q and Φ and afterwards :

$$\frac{dx}{dt} = F\left(x, t\right), \tag{11}$$

In (11) x is the column matrix of state variables (state vector), and F is a matrix depending on the circuits topology and on the nonlinear characteristics.

If for linear circuits, the circuits equations can be always reduced to state equations, for certain nonlinear circuits it is not possible to allow state equations (Hasler & Neirynck, 1985).

Anyhow, for a modelization not excessively idealized and a reasonable choice of state variable it is possible to obtain always state equations. For nonlinear state equations even the problem of existing or not of a solution might occur.

State equations for linear circuits have exact analytically determined solution, available in any book of college mathematics. If the order of the circuit is higher than three, analytic solution implies calculations so complicated that this method becomes inefficient. For this kind of situations a large offer of numeric integrations methods is available embedded in the commercial programs of numeric simulation.

3. Simulation of electric circuits using PSPICE

The levels of the analysis of electric circuits performed by PSPICE are, mainly, the following (Radoi, 1994; Vladimirescu, 1999):

a. Solving a linear electric circuit, invariable with time,
b. Solving a nonlinear electric circuit,
c. Solving a time variable electric circuit.

For solving the electric circuit problems PSPICE uses the branch current method. Each element of the circuit is considered being placed between two nodes. The nodes are numbered from "0", the reference node, to N and the position of an element is given by the numbers of its terminals. A very simple circuit example is given in Fig. 2.

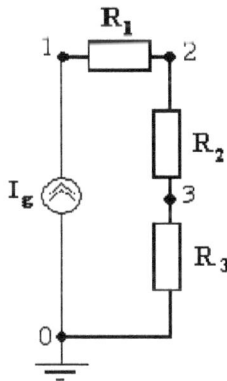

Figure 2. Simple circuit for SPICE application

$$
\begin{cases}
\dfrac{V_0 - V_3}{R_3} = -I_g & \text{for node 0,} \\[2ex]
\dfrac{V_1 - V_2}{R_1} = I_g & \text{for node 1,} \\[2ex]
\dfrac{V_1 - V_2}{R_1} = \dfrac{V_2 - V_3}{R_2} & \text{for node 2,} \\[2ex]
\dfrac{V_2 - V_3}{R_2} = \dfrac{V_3 - V_0}{R} & \text{for node 3.}
\end{cases}
\tag{12}
$$

These equations can be written in matrix form also as (13), where $[I]$ is the column matrix of the sources for all the nodes, in the same order as the unknown nodes potentials in matrix $[V]$. $[G]$ is the square conductance matrix.

The potential of the reference node, "0", is taken with "zero" value so all the other N-1 nodes potentials remain as unknown values in the equation system

With $V_0 = 0$, system (13) is transformed into a system (14) with N-1 equations.

In this very simple example remain only 3 unknown values of nodes potentials (14).

$$
\begin{bmatrix} -I_g \\ +I_g \\ 0 \\ 0 \end{bmatrix} =
\begin{bmatrix}
\dfrac{1}{R_3} & 0 & 0 & -\dfrac{1}{R_3} \\
0 & \dfrac{1}{R_1} & -\dfrac{1}{R_1} & 0 \\
0 & -\dfrac{1}{R_1} & \dfrac{1}{R_1}+\dfrac{1}{R_2} & -\dfrac{1}{R_2} \\
-\dfrac{1}{R_3} & 0 & -\dfrac{1}{R_2} & \dfrac{1}{R_2}+\dfrac{1}{R_3}
\end{bmatrix}
\cdot
\begin{bmatrix} V_0 \\ V_1 \\ V_2 \\ V_3 \end{bmatrix}
\quad or, \; [I] = [G] \cdot [V] \;(13)
$$

$$
\begin{bmatrix} +I_g \\ 0 \\ 0 \end{bmatrix} =
\begin{bmatrix}
\dfrac{1}{R_1} & -\dfrac{1}{R_1} & 0 \\
-\dfrac{1}{R_1} & \dfrac{1}{R_1}+\dfrac{1}{R_2} & -\dfrac{1}{R_2} \\
0 & -\dfrac{1}{R_2} & \dfrac{1}{R_2}+\dfrac{1}{R_3}
\end{bmatrix}
\cdot
\begin{bmatrix} V_1 \\ V_2 \\ V_3 \end{bmatrix}.
\qquad (14)
$$

The system (14) can be solved using the Gaussian elimination method (Ross, 2004).

An equivalent method that can be used for solving the system is the LU factorization, by decomposing the [G] matrix.

Matrix [G]=[L]·[U], with [L] being a matrix with "1" on the principal diagonal, and [U] is a matrix having only "0" below the principal diagonal. This method is more advantageous than the "classic" Gauss elimination method, mainly when the equation [I] = [G]·[V] has to be solved several times: several sources [I] for the same conductance matrix. This is the situation for solving our problem using PSPICE, so that the LU factorization method shall be used.

In the analysis of dynamical circuits either the sources, voltage or current sources, have time variation, either the topology of the circuit is changing. If the passive elements are resistors, linear or nonlinear, using a time step Δt for the discretization of the time variation (sinusoidal, exponential, linear) of the sources. For each time moment a set of values [V(t)] is calculated and memorized. These values can be used in further graph representations or printed.

The presence of inductors and/or capacitors implies replacing them with equivalent elements.

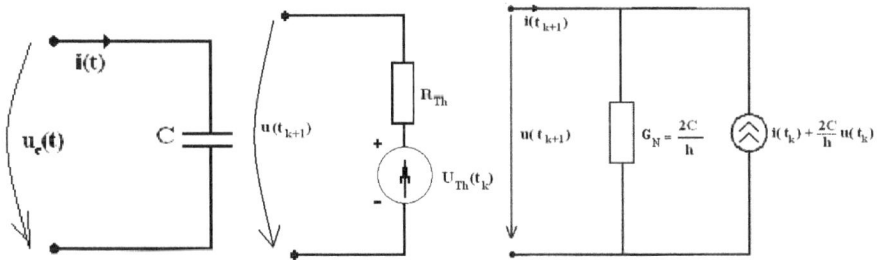

Figure 3. Thevenin and Norton generators for a capacitor , when trapeze method is used

For an iterative calculation process the capacitor can be replaced by a Thevenin or a Norton generator. The Norton generator presented in the Fig. 3 has the parameters calculated according to the trapeze method of integration. The method using the Euler's regressive algorithm and the Gear's second order algorithm can be used also, but the equivalences (parameters of the Norton or Thevenin generators) are different from those presented in Fig. 3.

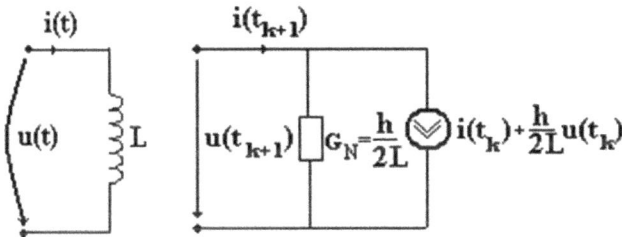

Figure 4. Norton equivalent generator representation of a inductor, when trapeze method is used

In a similar manner for an inductor can be used a Norton equivalent generator or a Thevenin equivalent generator, with parameters depending on the method used for solving the problem. Taking the nodes potentials as unknown values it is more suitable to use the Norton generator equivalent to the inductor in a dynamic circuit.

The transient regime analysis is the most important analysis that can be performed using PSPICE.

When the transient regime begins with non zero initial conditions this fact can be introduced in the program by the option UIC (use initial conditions).

The situation of the modelization of a complex network is shown in the Fig. 5.

The transient regime is triggered by closing the switch, meaning that phase one is grounded, as if a fault phase to ground is produced.

Resistor R_{17} has a value that is corresponding to the conditions of the fault.

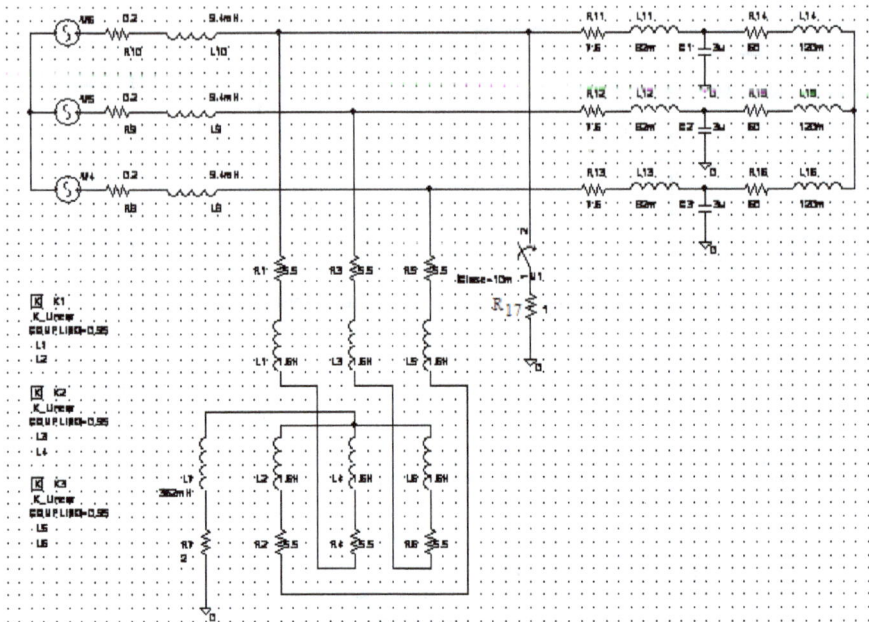

Figure 5. MVN represented as a complex circuit in SPICE drawing

Currents and voltages can be represented, as on a "soft" oscilloscope, after solving the transient regime, using the elements of the V[t] matrix, calculated for each time step, and the resulting, calculated, currents values.

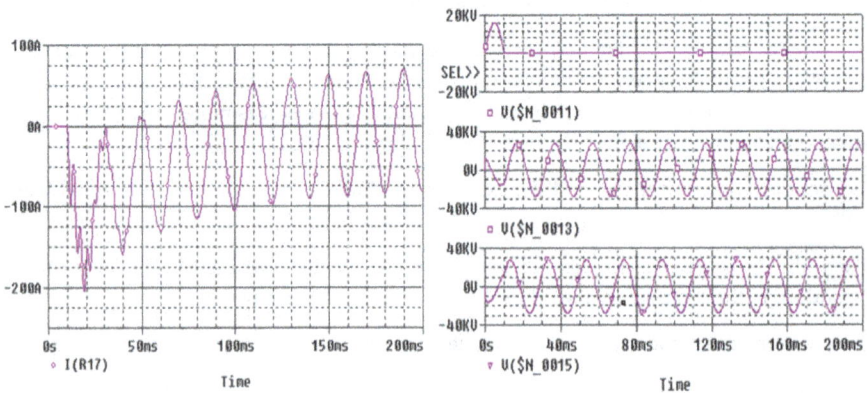

Figure 6. Current and voltages representation as SPICE output oscillogrames

4. Numeric simulation of transients triggered by faults in medium voltage networks

Faults in medium voltage networks (MVN) generate transient regimes with duration depending on the networks parameters, on the contact type at the fault place and on the default type itself. The numeric simulation performed in this chapter is based on PSPICE and the transient regimes are caused by simple and double grounding faults, as well as, by the fault produced by a broken conductor, grounded towards the consumer. A fault of the type broken conductor grounded towards the source can be treated similar as the simple grounding fault, with an insignificant error, so shall be no more analyzed in the followings (Sybille et al., 2000).

The elements that are important in the analysis are:

- the initial phase α of the faulty phase,
- the electric resistance at the fault, grounding point,
- the method used for grounding the neutral of the network,
- the functioning regime of the MVN.

4.1. The structure of MVN

The single line diagram of the MVN whose behavior is studied in case of transient caused by faults is presented in Fig. 7.

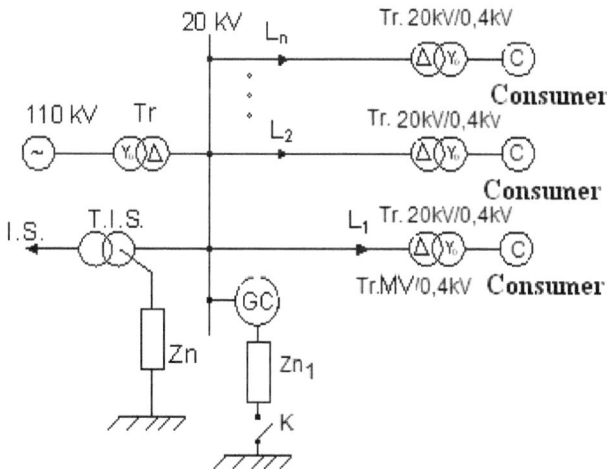

Figure 7. Single Line Diagram of the MVN

The MVN in Fig. 7 contains the following elements:

- The source, the 110kV line supplying the transformer 110kV/20kV, considered as being of infinite power supply and internal impedance zero,

-   The transformer (Tr) 110kV/20kV star connection with neutral point on 110kV side and delta connection on the 20kV side,
-   The lines of 20kV ($L_1 ... L_n$),
-   The transformers TR.20kV/0.4kV supplying low voltage consumers, transformers having star or delta connection on the 20kV side and star with null or zigzag with null on the low voltage side,
-   Low voltage consumers (C),
-   The transformer for internal services (TIS) ; this transformer is used also for creating the artificial null of the 20kV network (has zigzag connection with null on the 20kV side and star with null on the low voltage side),
-   The grounding inductor (GC) , having zigzag with null and creates the artificial null of the 20kV network,
-   The impedance for treatment of the null (Z_n), impedance that might be a compensation inductor, a resistor or might be missing in the case of isolated null,
-   The resistor for the null's treatment (Z_{n1}),
-   The switch for the connection to the ground of the resistor Z_{n1}.

The model is designed with the possibility of changing the values of parameters.

4.2. Numerical simulation of some fault types

The numeric simulator of the three phased circuit corresponding to a MVN is shown in Fig.8. The PSPICE program is used to perform the modelization. Several types of faults are simulated using the model:

-   simple grounding fault,
-   double grounding fault,
-   broken conductor and grounded on the consumer side.

The MVN where the faults are modelized has the following characteristics:

-   nominal line voltage 20kV,
-   S = 25MVA, the apparent power of the transformer 110kv/MV,
-   the capacitive current of the MVN is 100A,
-   the current of the two faulty lines , 8A each,
-   the apparent power of the consumers is 80% of the nominal one and is uniformly distributed on the MV lines.

The simulator contains three lines, denoted with "a", "b" and "c" , so that to each index of parameter such as resistance or inductance is added a, b or c. The line "c" is considered as the healthy line and is the equivalent of all the 20kV lines without fault (Bucatariu et al., 2009; Drapela, 2009)

The line "a" is provided with the possibility of the simulation of a simple grounding fault, by connecting Vs_a. The same line "a" might simulate a fault of the type broken conductor (by opening the switch VSGOL). Even more, by opening the switch VSGOL and closing, in the same time, the switch VSₐ it is possible to simulate a fault of the type broken conductor and grounded on the consumer's side.

The line "b" can simulate a simple grounding fault by closing the switch VS_b.

The fault of the type double grounding is simulated by closing the switch VSₐ connected to the line "a" and closing the switch VS_b connected to the line "b", as well.

If a fault of the type double grounding on the same line is wanted to be simulated, the switch VS_b should be moved from line "b" to line "a".

The parameters of the Earth (resistance, inductance) are encompassed in the parameters of the electric line.

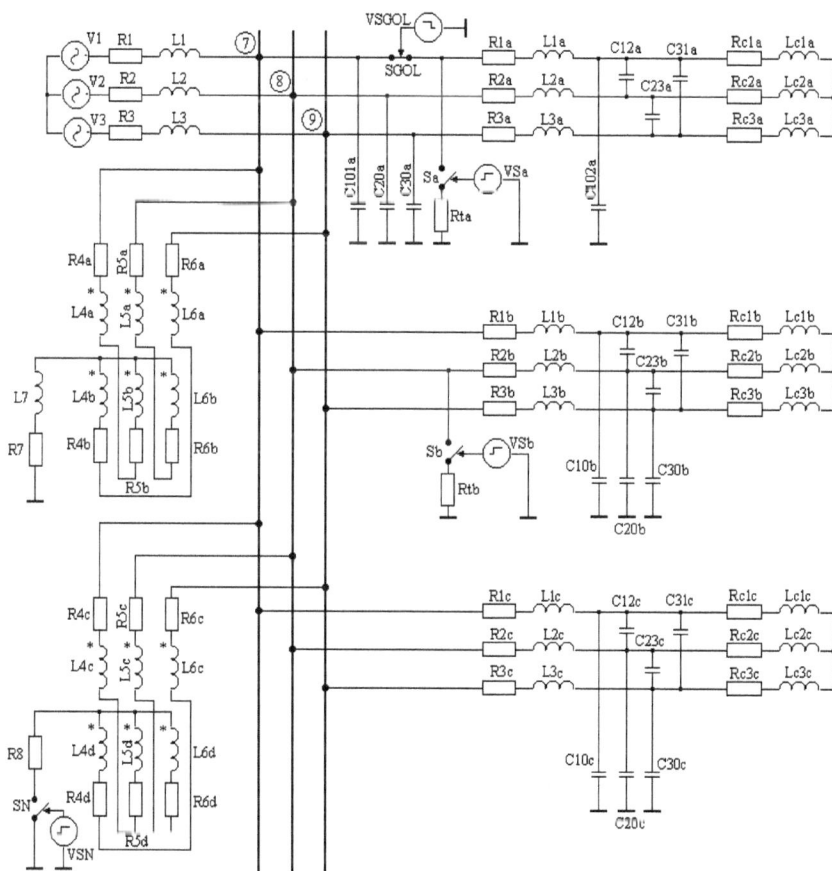

Figure 8. Three phased circuit simulating a MVN

4.2.1. The single phase-to-ground fault

Using the numerical simulator a multitude of situation could be analyzed.

One of those situations is the one when the MVN has the null grounded trough compensation inductor and the MVN functions under resonance conditions, respectively overcompensated 10%.

In these circumstances the evolution of voltages and currents during the transient is analyzed according to:

-   the initial phase of the faulty phase (α) in the very moment when the fault is produced
-   the electric resistance at the fault place (R_t)
-   the way the null is treated
-   the functioning regime (resonance, under or over compensated)

The first oscillogram, left side, is corresponding to the currents and the second corresponding to the voltages.

In these oscillograms I(S) represents the time variation of the current at the fault place, the zero sequence current of the faulty line is denoted as (I(R1a)+I(R2a)+I(R3a))/3, the zero sequence current is (I(R1b)+I(R2b)+I(R3b))/3 on the healty line, and the current trough the element for grounding the neutral point is I(L7).

In the oscillograms of the voltages , V(7), V(8) and V(9) represent, respectively, the voltages of the three phases and the zero sequence voltage of the MVN bars is represented by (V(7) + V(8) + V(9))/3.

In figures 9....20 the results correspond to the 10% overcompensated MVN with the neutral point grounded with compensation inductor. Only two values, 0 and 90°, of the initial phase of the voltage of the faulty line are taken into account. An intermediate value of 45° for this angle does not modify significantly the transient of the simple grounding fault.

From the oscillograms it can be observed that the initial phase of the faulty phase has great importance in the evolution of voltages and currents during the transient regime caused by the simple gounding fault.

Knowing, as accurate as possible, the evolution in time of the voltages gives the possibility of a good design for the MVN, strong enough to resist to the maximal applied voltage.

The maximal values of the currents are important for the correct evaluation of the mechanical forces acting on lines conductors, on the insulators and on the bars of MVN from the transformer station 110kV/MV.

The way that the electric contact is established, R_t , at the fault place, is very important in the transient evolution of the voltages and currents. If the resistance is over 100Ω the transient vanishes rapidly, in less than 5ms, and the maximal values of voltages and currents are not significant. This conclusion results by comparing oscillograms from Fig. 9 to Fig. 19, and, respectively, from Fig. 10 to Fig. 20.

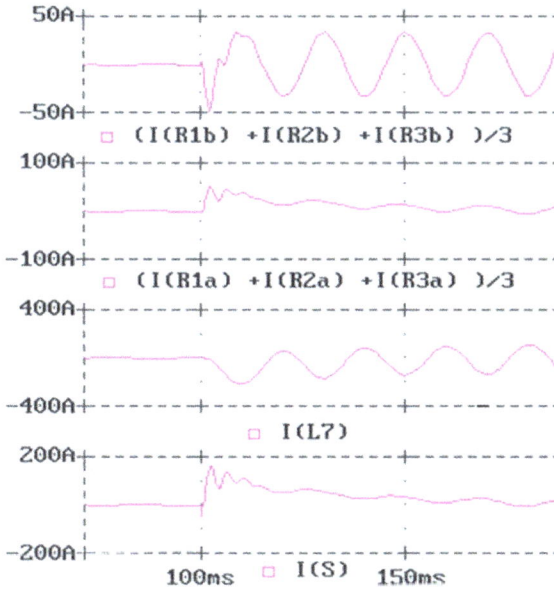

Figure 9. Time variation of the currents when MVN is 10% overcompensated and $R_t = 1\Omega$, $\alpha = 0°$

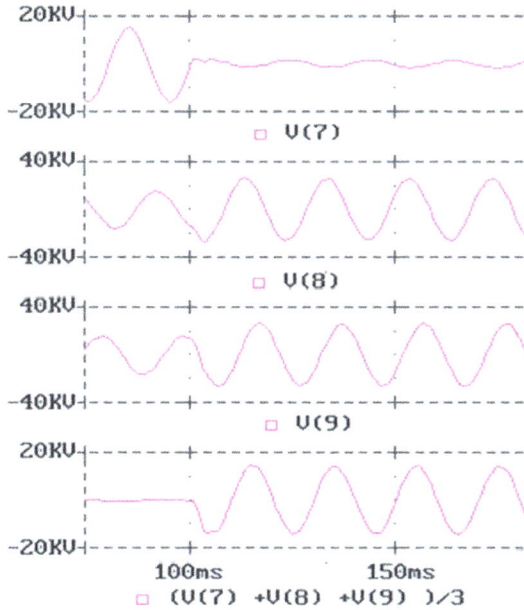

Figure 10. Time variation of the voltages when MVN is 10% overcompensated and $R_t = 1\Omega$, $\alpha = 0°$

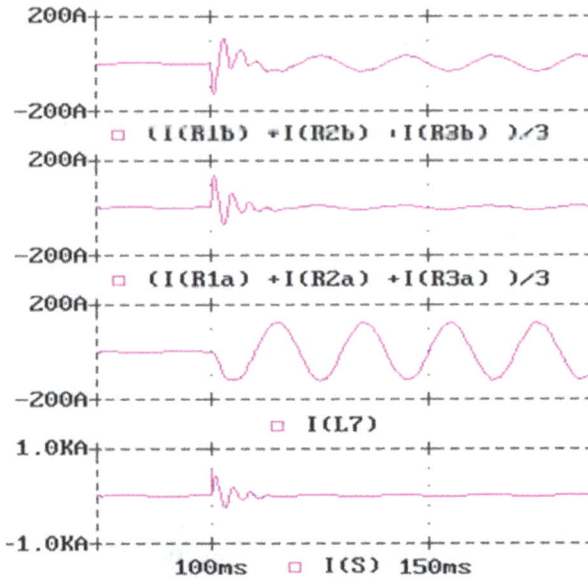

Figure 11. Time variation of the currents when MVN is 10% overcompensated and $R_t = 1\Omega$, $\alpha = 90°$

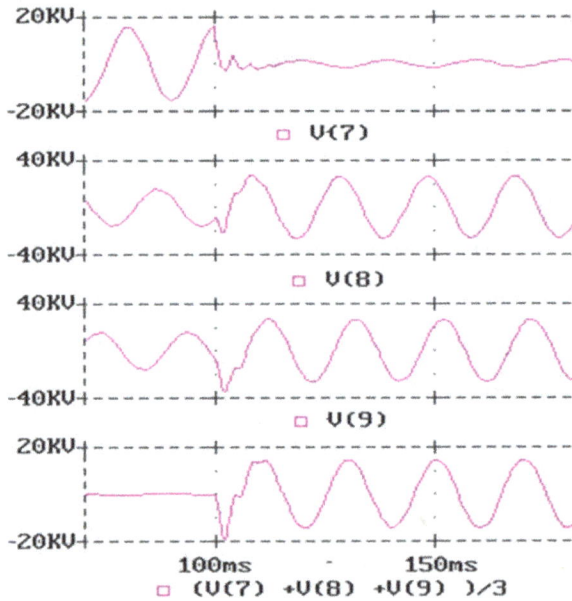

Figure 12. Time variation of the voltages when MVN is 10% overcompensated and $R_t = 1\Omega$, $\alpha = 90°$

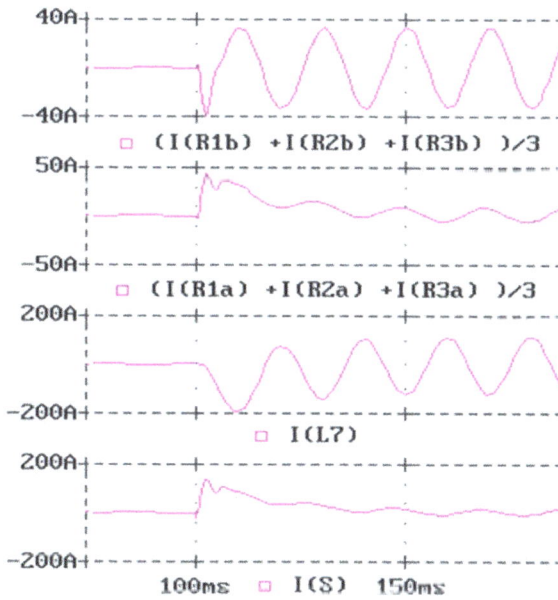

Figure 13. Time variation of the currents when MVN is 10% overcompensated $R_t = 10\Omega$, $\alpha = 0°$

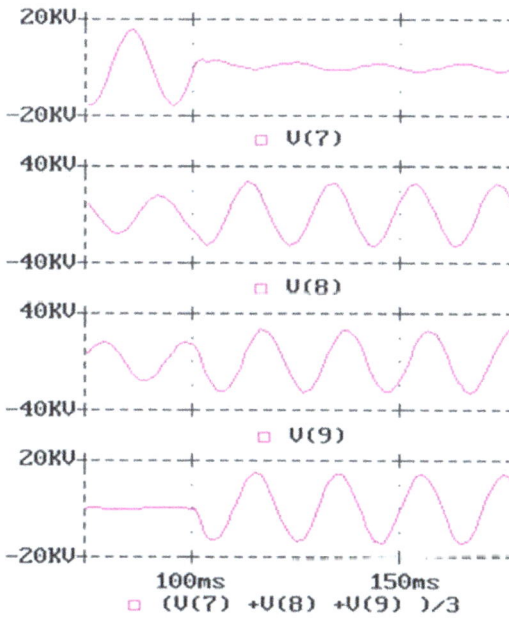

Figure 14. Time variation of the voltages when MVN is 10% overcompensated and $R_t = 10\Omega$, $\alpha = 0°$

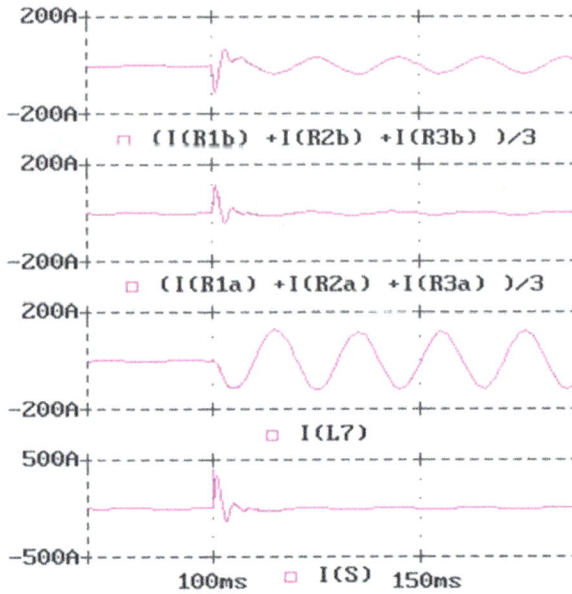

Figure 15. Time variation of the currents when MVN is 10% overcompensated $R_t = 10\Omega$, $\alpha = 90°$

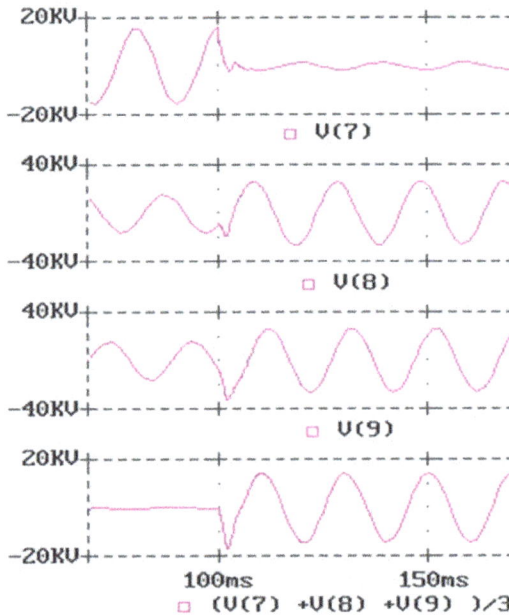

Figure 16. Time variation of the voltages when MVN is 10% overcompensated and $R_t = 10\Omega$, $\alpha = 90°$

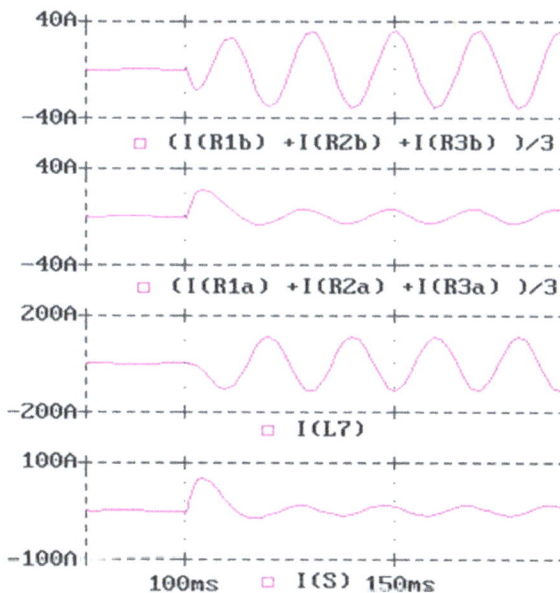

Figure 17. Time variation of the currents when MVN is 10% overcompensated R_t = 100Ω, $\alpha = 0°$

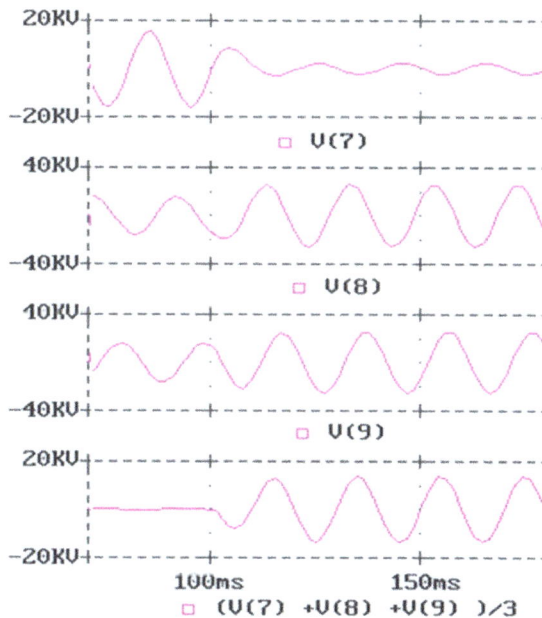

Figure 18. Time variation of the voltages when MVN is 10% overcompensated and R_t = 100Ω, $\alpha = 0°$

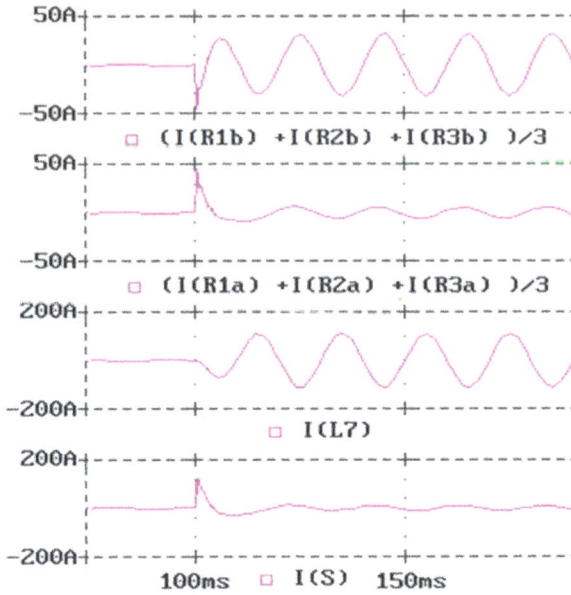

Figure 19. Time variation of the currents when MVN is 10% overcompensated $R_t = 100\Omega$, $\alpha = 90°$

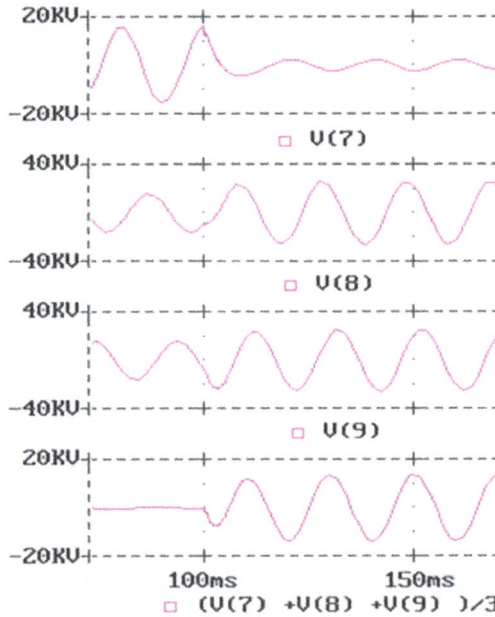

Figure 20. Time variation of the voltages when MVN is 10% overcompensated and $R_t = 100\Omega$, $\alpha = 90°$

If the MVN is functioning at resonance, the same simulation results are presented and, in addition, also the 45° value for the initial phase of the faulty phase voltage is presented.

The oscillograms show that the differences between the 10% overcompensated regime and the resonant regime are not significant.

If the functioning regime is far from being at resonance the transient regim produced by simple grounding fault is very short and the variation of voltages and currents is less important.

The maximal value of the current at fault place is obtained for 90° (comparing α values 0°, 45°, respectively 90°).

For $R_t = 1\Omega$ at $\alpha = 45°$ the current is twice its value at 0° and at 90° the current at the fault place is four times greater than the same current if $\alpha = 0°$.

If R_t becomes 10Ω the conclusions regarding the dependence of the current at the fault place on the phase α remain the same (higher α, higher value of the current) but with a decrease of about 25% of the maximal value of the current for a 10 times increase of the resistance at the fault place.

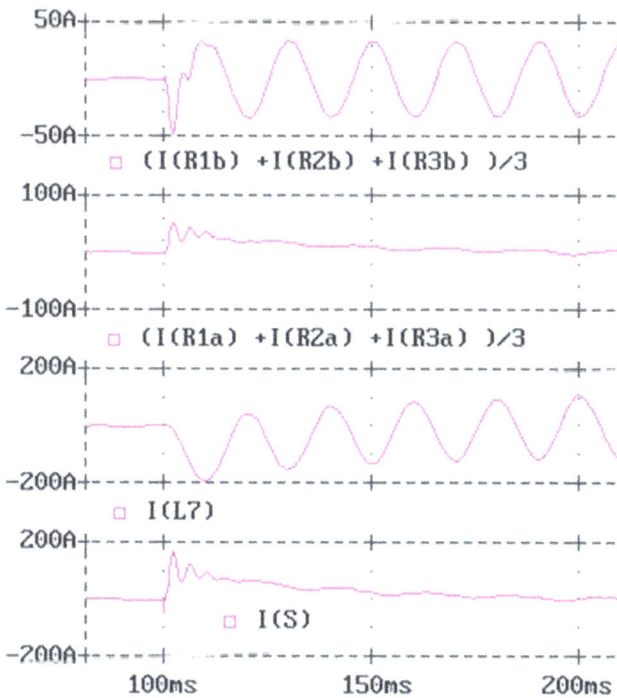

Figure 21. Time variation of the currents when MVN is at resonance , $R_t = 1\Omega$, $\alpha = 0°$

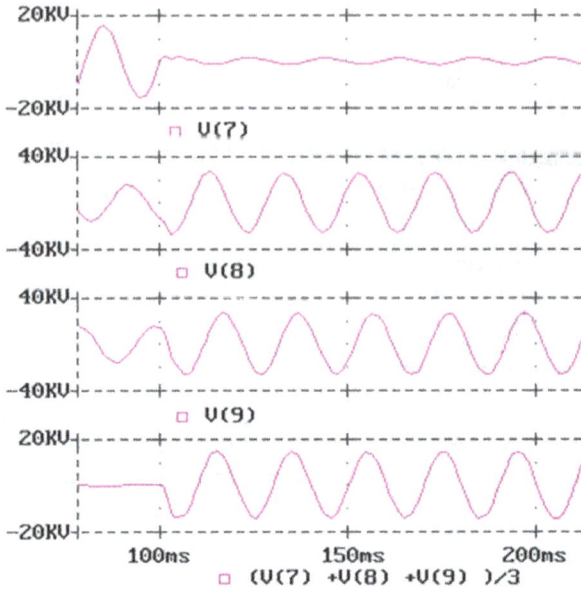

Figure 22. Time variation of the voltages when MVN is at resonance, $R_t = 1\Omega$, $\alpha = 0°$

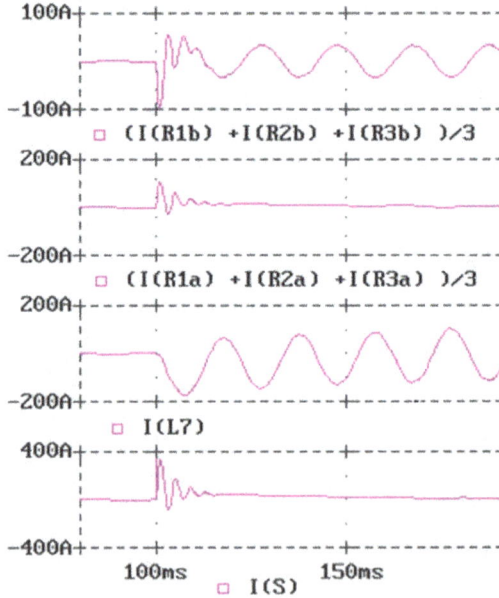

Figure 23. Time variation of the currents when MVN is at resonance , $R_t = 1\Omega$, $\alpha = 45°$

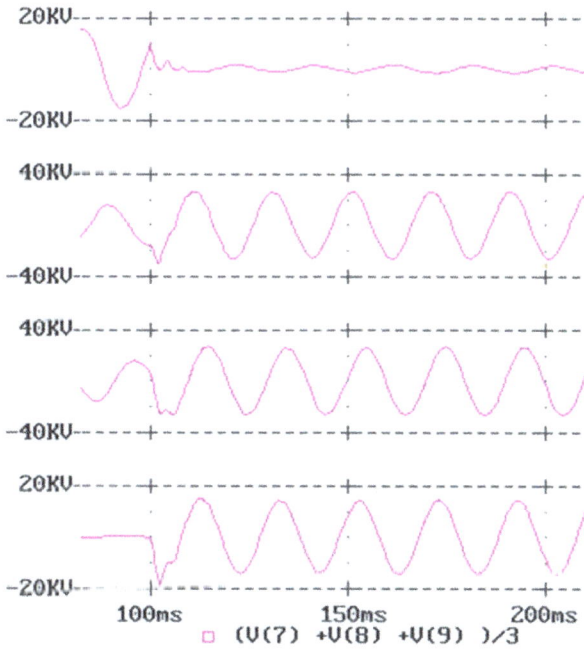

Figure 24. Time variation of the voltages when MVN is at resonance, $R_t = 1\Omega$, $\alpha = 45°$

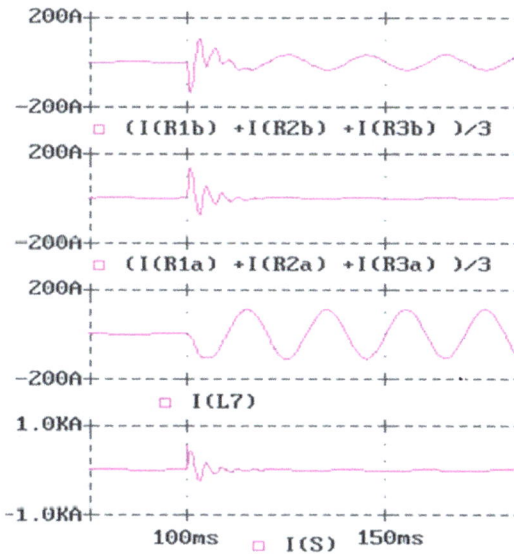

Figure 25. Time variation of the currents when MVN is at resonance, $R_t = 1\Omega$, $\alpha = 90°$

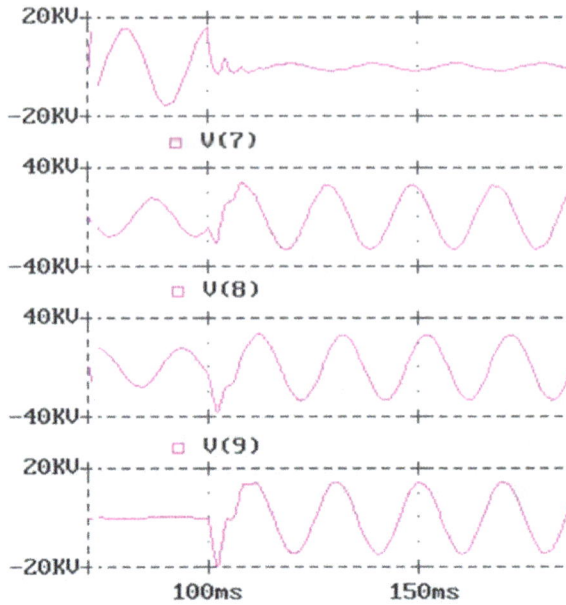

Figure 26. Time variation of the voltages when MVN is at resonance, $R_t = 1\Omega$, $\alpha = 90°$

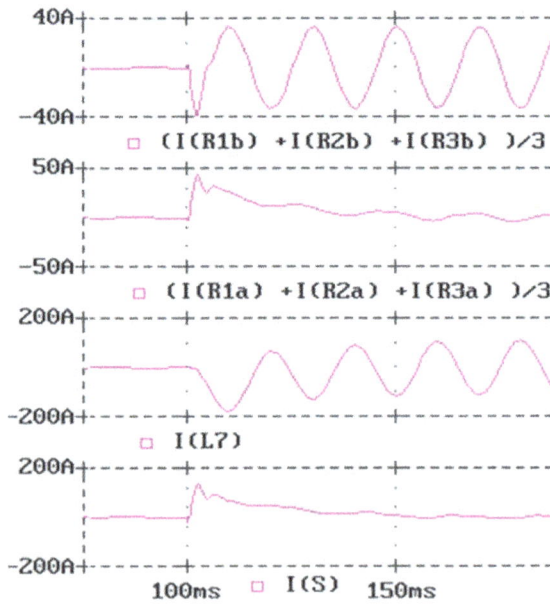

Figure 27. Time variation of the currents when MVN is at resonance , $R_t = 10\Omega$, $\alpha = 0°$

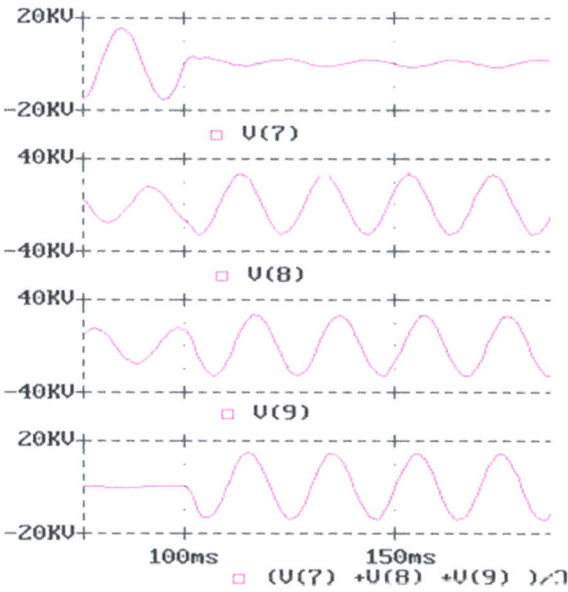

Figure 28. Time variation of the voltages when MVN is at resonance, $R_t = 10\Omega$, $\alpha = 0°$

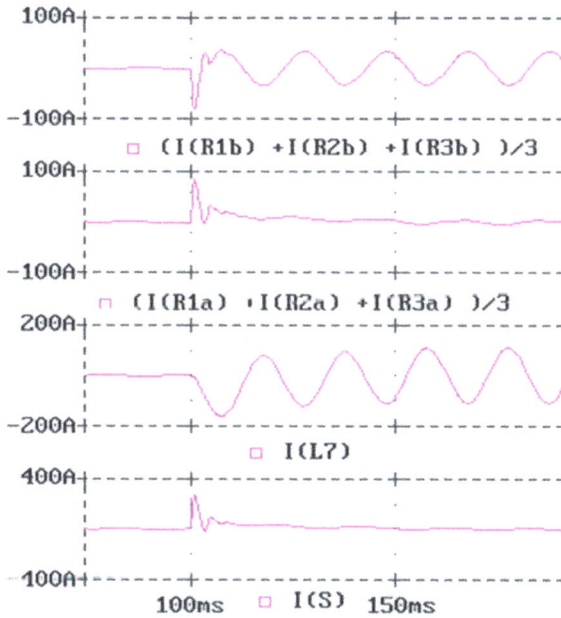

Figure 29. Time variation of the currents when MVN is at resonance , $R_t = 10\Omega$, $\alpha = 45°$

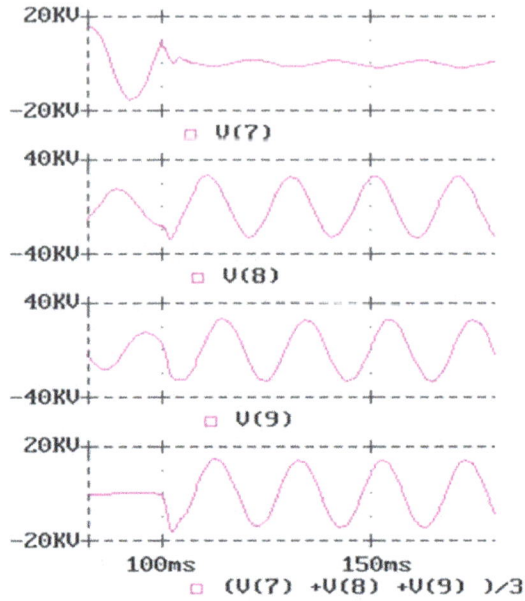

Figure 30. Time variation of the voltages when MVN is at resonance, $R_t = 10\Omega$, $\alpha = 45°$

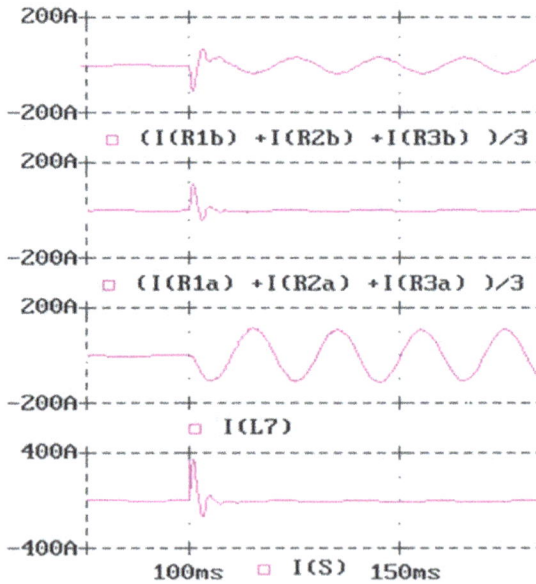

Figure 31. Time variation of the currents when MVN is at resonance , $R_t = 10\Omega$, $\alpha = 90°$

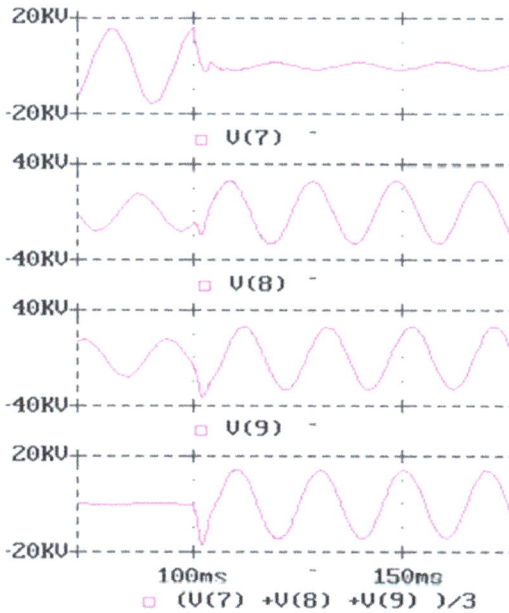

Figure 32. Time variation of the voltages when MVN is at resonance, $R_t = 10\Omega$, $\alpha = 90°$

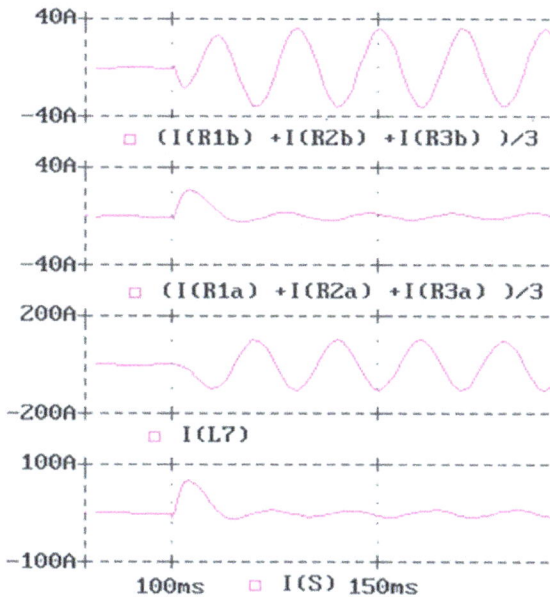

Figure 33. Time variation of the currents when MVN is at resonance, $R_t = 100\Omega$, $\alpha = 0°$

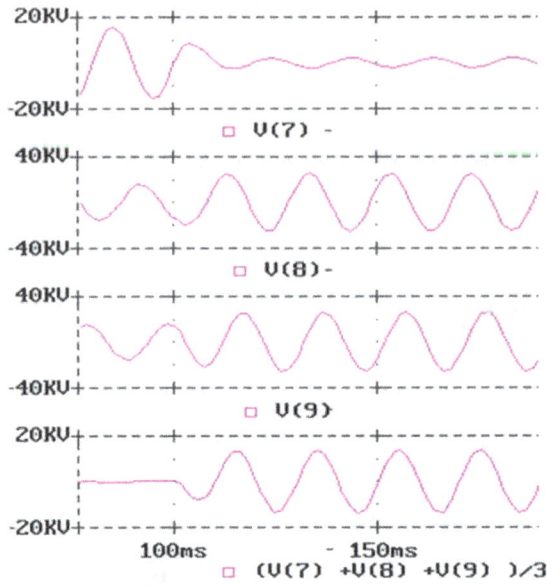

Figure 34. Time variation of the voltages when MVN is at resonance, $R_t = 100\Omega$, $\alpha = 0°$

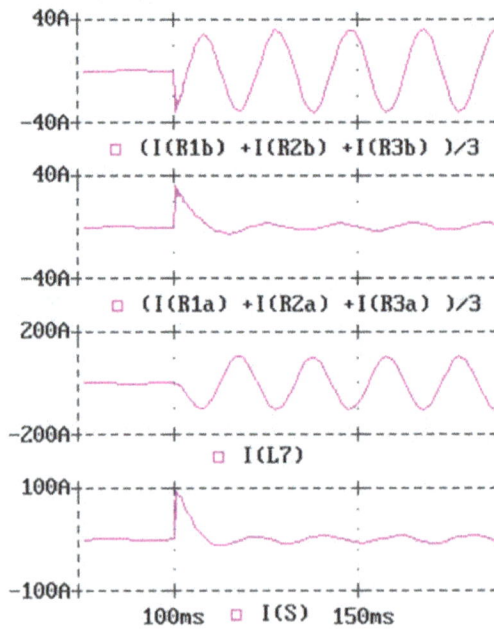

Figure 35. Time variation of the currents when MVN is at resonance , $R_t = 100\Omega$, $\alpha = 45°$

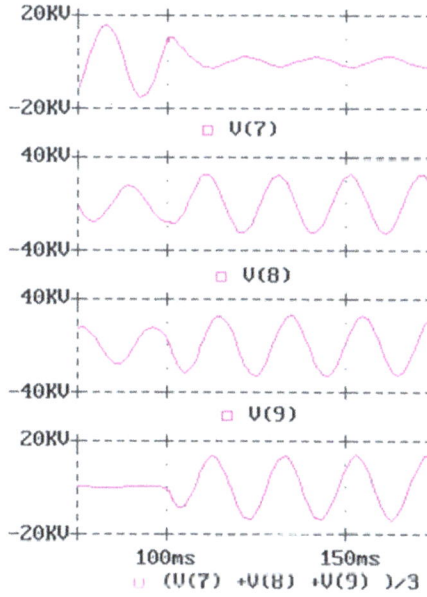

Figure 36. Time variation of the voltages when MVN is at resonance, $R_t = 100\Omega$, $\alpha = 45°$

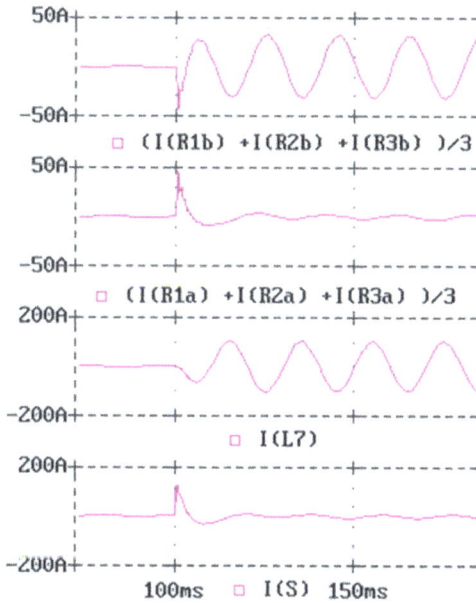

Figure 37. Time variation of the currents when MVN is at resonance , $R_t = 100\Omega$, $\alpha = 90°$

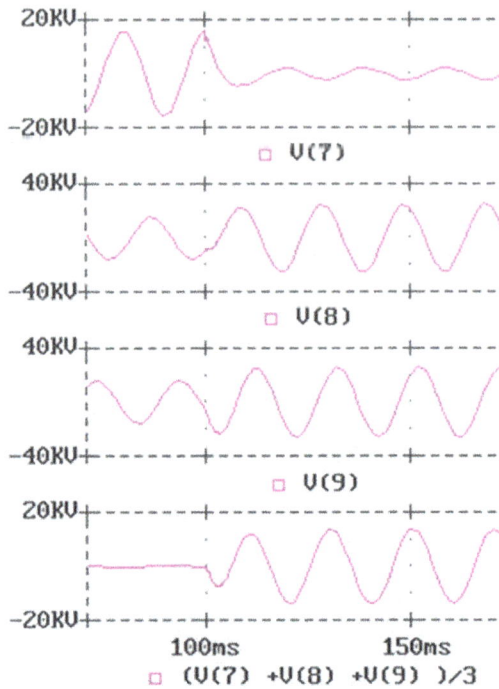

Figure 38. Time variation of the voltages when MVN is at resonance, $R_t = 100\Omega$, $\alpha = 90°$

4.2.2. The double phase-to-ground fault

The same simulation model from Fig. 8 is used with the switch VSa closed, and after 20ms the switch VSb is connected also. VSGOL remains in „closed" position.

The double phase to ground fault was simulated for neutral point grounded by compensation inductor and, respectively, by resistor.

In both cases R_t was considered consecutively with 1Ω, 100Ω, 1000Ω trough VSa switch and with 1Ω trough VSb .

In both cases $\alpha = 90°$ was considered, but also $\alpha = 0°$ was imposed for the situation with compensation inductor.

Also the situation with isolated neutral point was simulated, with same R_t values , but only with $\alpha = 90°$. (Curcanu et al., 2006).

The same symbols are used in oscillograms for the currents, just in addition the index "a" is used for the first faulty line (simulated by switch VSa) and the index "b" for the second faulty line (VSb). For the voltages the symbols are quite the same as those used previously.

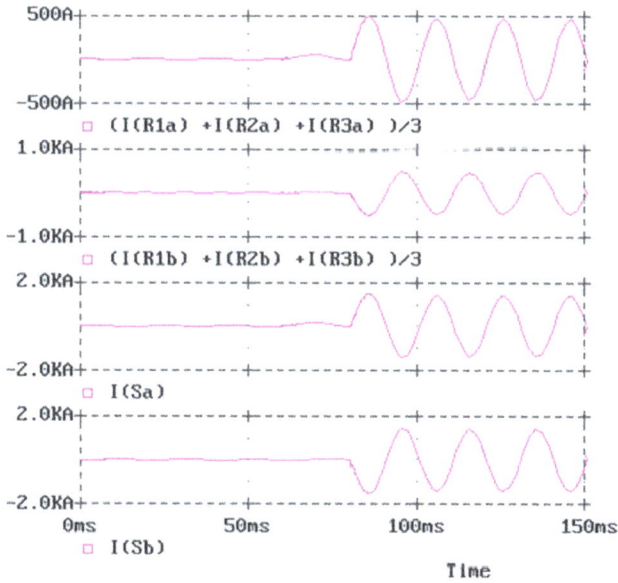

Figure 39. Time variation of the currents when MVN is at resonance , $R_t = 1\Omega$, $\alpha = 0°$

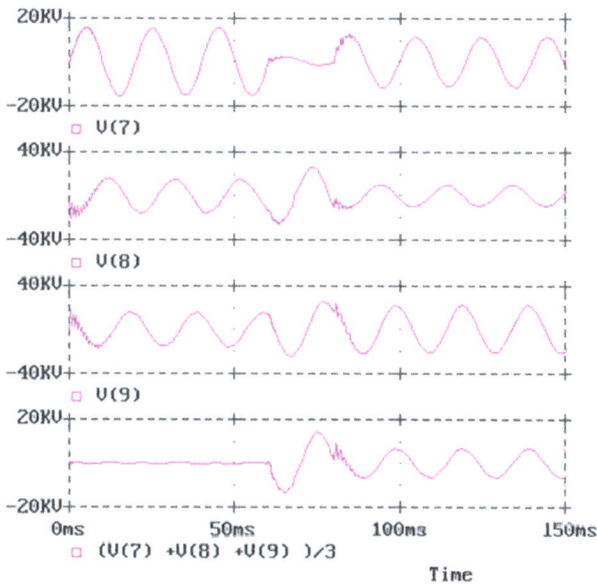

Figure 40. Time variation of the voltages when MVN is at resonance, $R_t = 1\Omega$, $\alpha = 0°$

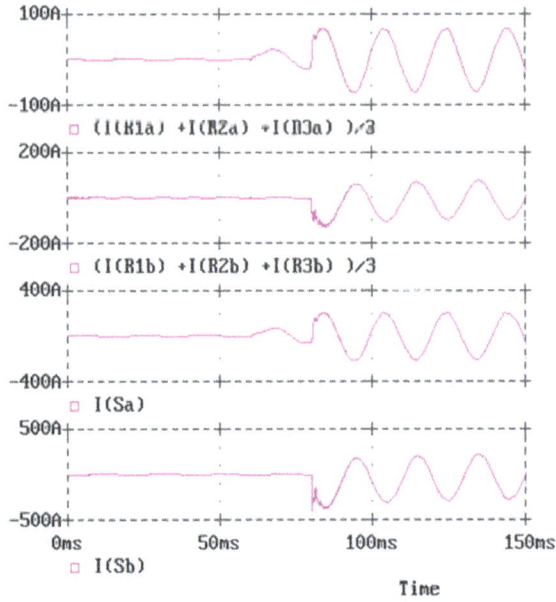

Figure 41. Time variation of the currents when MVN is at resonance , $R_t = 100\Omega$, $\alpha = 0°$

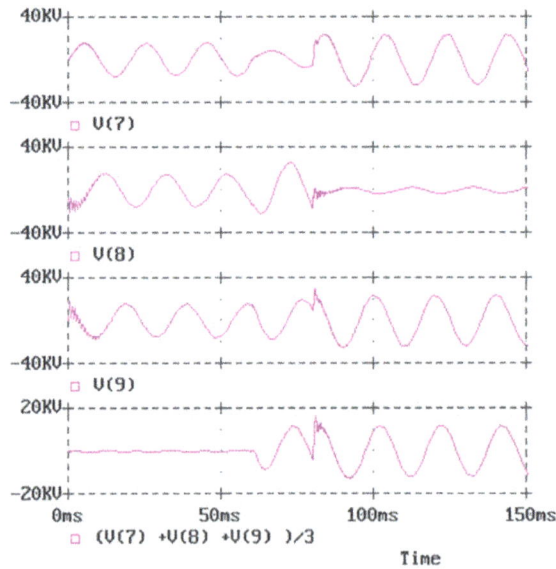

Figure 42. Time variation of the voltages when MVN is at resonance, $R_t = 100\Omega$, $\alpha = 0°$

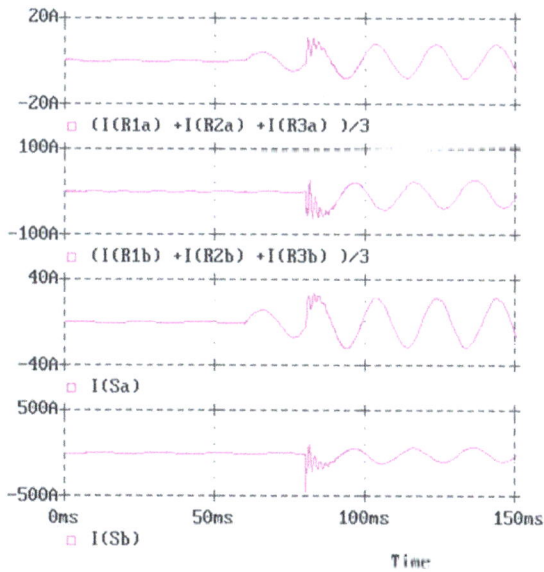

Figure 43. Time variation of the currents when MVN is at resonance , $R_t = 1000\Omega$, $\alpha = 0°$

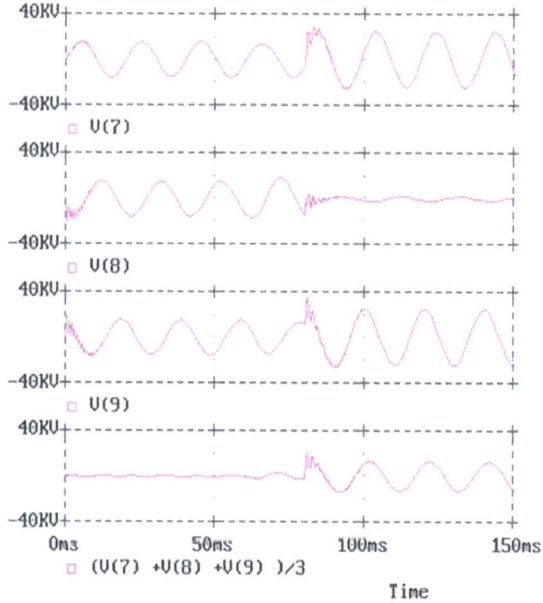

Figure 44. Time variation of the voltages when MVN is at resonance, $R_t = 1000\Omega$, $\alpha = 0°$

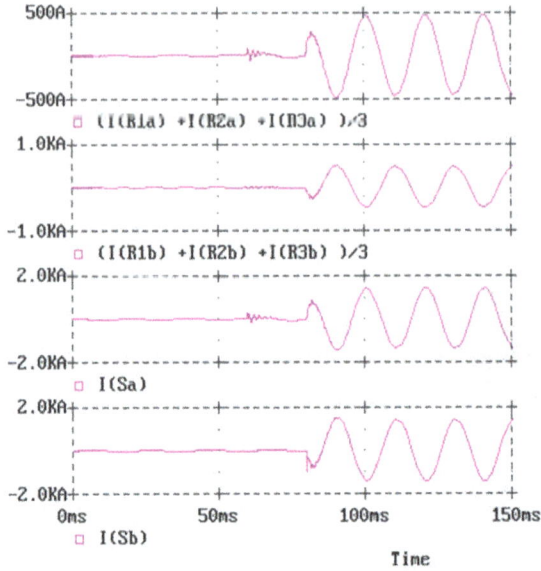

Figure 45. . Time variation of the currents when MVN is at resonance , $R_t = 1\Omega$, $\alpha = 90°$

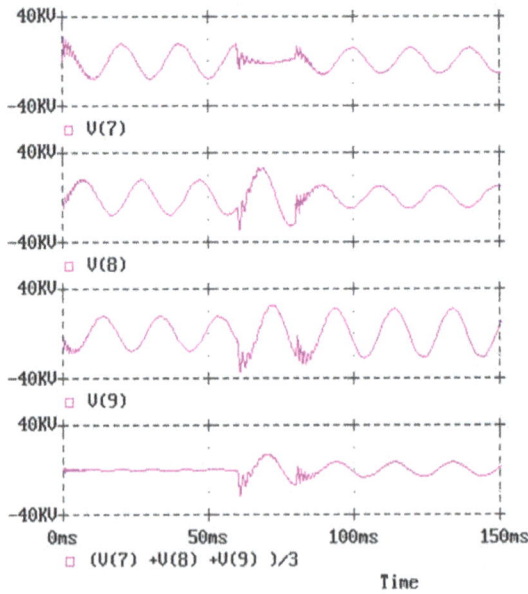

Figure 46. Time variation of the voltages when MVN is at resonance, $R_t = 1\Omega$, $\alpha = 90°$

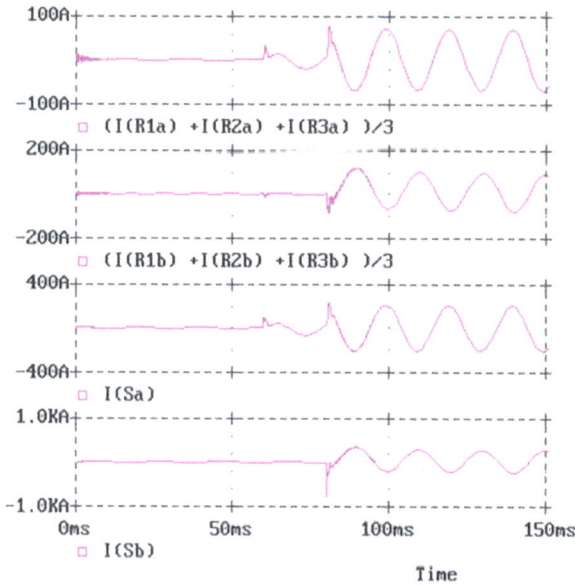

Figure 47. Time variation of the currents when MVN is at resonance , $R_t = 100\Omega$, $\alpha = 90°$

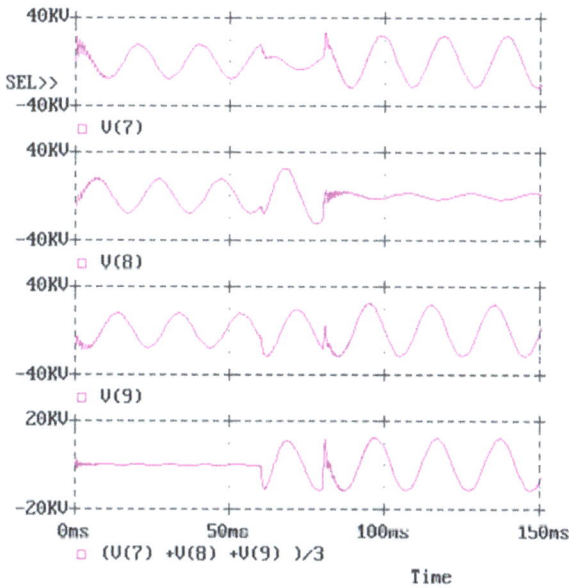

Figure 48. Time variation of the voltages when MVN is at resonance, $R_t = 100\Omega$, $\alpha = 90°$

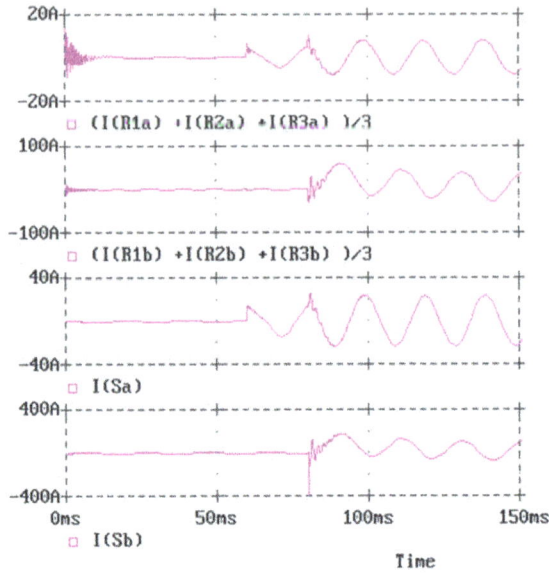

Figure 49. Time variation of the currents when MVN is at resonance , $R_t = 1000\Omega$, $\alpha = 90°$

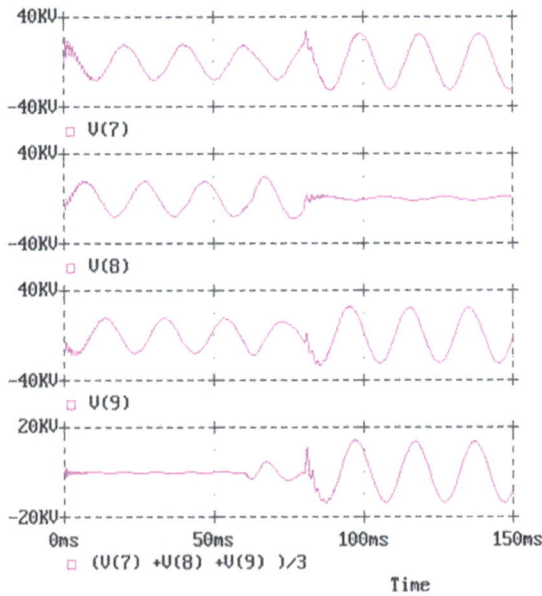

Figure 50. Time variation of the voltages when MVN is at resonance, $R_t = 1000\Omega$, $\alpha = 90°$

The double phase to ground fault has less important transient effects than the single phase to ground fault.

When the simple phase to ground fault occurs than the value of initial phase α is importans, but its variation is totaly insignifiant in the double phase to ground fault. The value of the resistance at the fault place is important in both fault types, the maximal value of the fault currents and the zero sequence currents depending strongly on the electric resistance at the fault place.

The oscillograms for the MVN grounded trough resistor are presented in Fig. 51 ... Fig. 54. The fault is simulated with the switch VSa and for $\alpha = 90°$ the transient is most important.

The current flowing trough the grounding resistor is calculated also and this is important because the value of this current is taken into account for the protection.

The resistance at the fault place is simulated with the switch VSa.

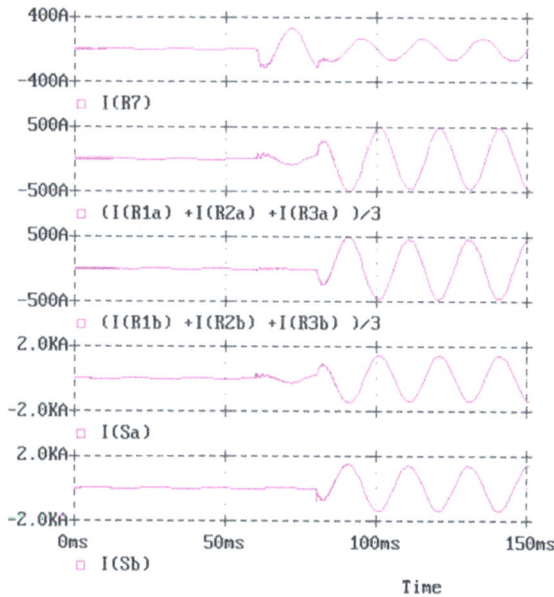

Figure 51. Time variation of the currents when MVN is grounded trough resistor, $R_t = 1\Omega$, $\alpha = 90°$

If the resistance at the fault place is increasing from 1Ω to 100Ω the current at the fault place is strongly decreasing from 1.8kA to 500A.

Oscillograms from Fig. 55 ...Fig. 60 , for isolated neutral point of the MVN, show that the double phase to ground fault gives a heavier transient for voltages than for currents.

There is not an important difference between the transient produced by the double phase to ground fault in MVN with isolated neutral point compared with the MVN grounded trough compensation inductor, but both cases are much heavier than the transient for MVN grounded trough resistor.

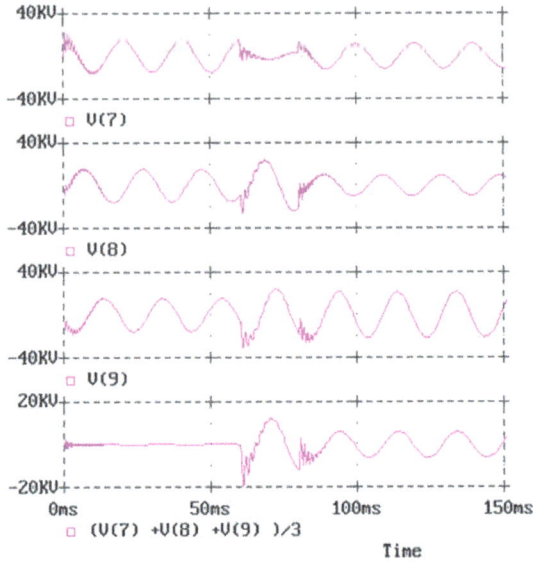

Figure 52. Time variation of the voltages when MVN is grounded trough resistor $R_t = 1\Omega$, $\alpha = 90°$

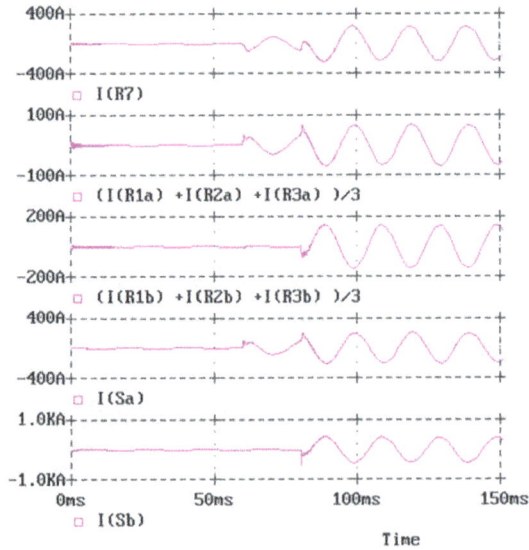

Figure 53. Time variation of the currents when MVN is grounded trough resistor, $R_t = 100\Omega$, $\alpha = 90°$

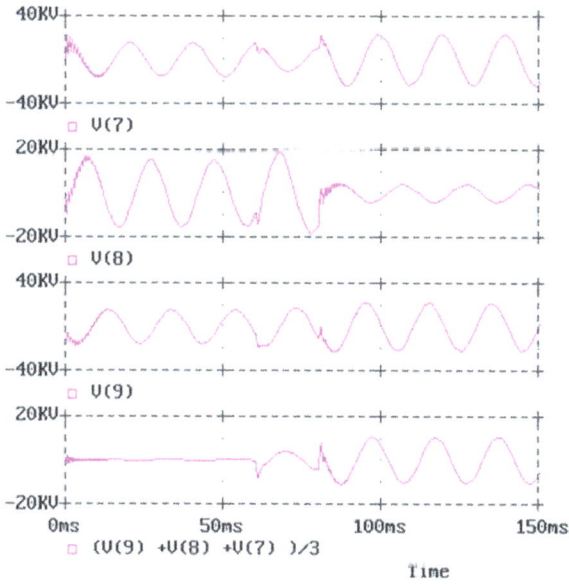

Figure 54. Time variation of the voltages when MVN is grounded trough resistor $R_t = 100\Omega$, $\alpha = 90°$

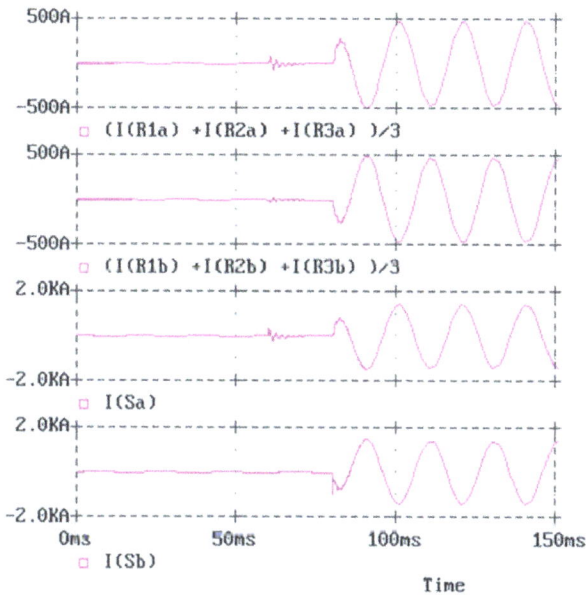

Figure 55. Time variation of the currents when the neutral point of MVN is isolated, $R_t = 1\Omega$, $\alpha = 90°$

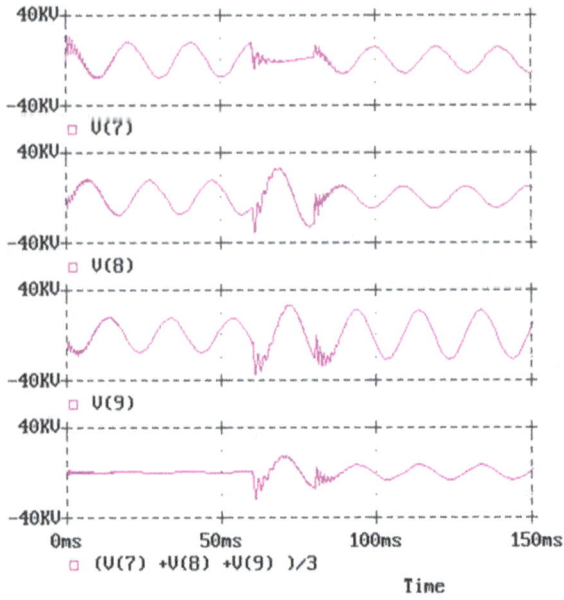

Figure 56. Time variation of the voltages when the neutral point of MVN is isolated, $R_t = 1\Omega$, $\alpha = 90°$

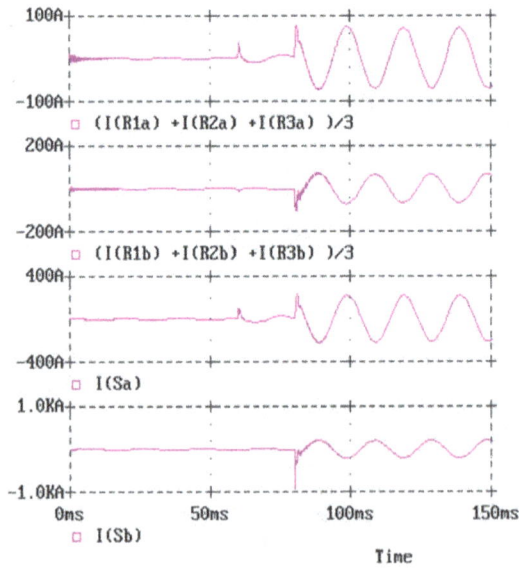

Figure 57. Time variation of the currents when the neutral point of MVN is isolated, $R_t = 100\Omega$, $\alpha = 90°$

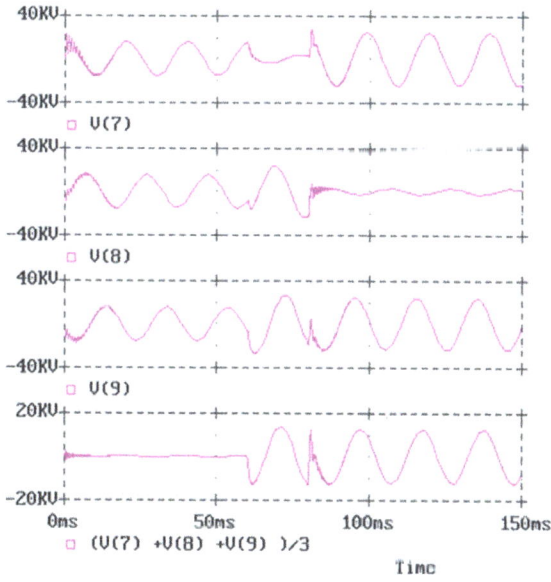

Figure 58. Time variation of the voltages when the neutral point of MVN is isolated, $R_t = 100\Omega$, $\alpha = 90°$

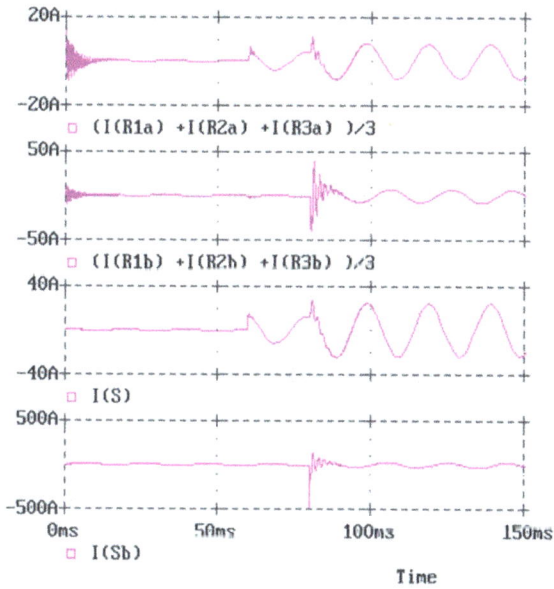

Figure 59. Time variation of the currents when the neutral point of MVN is isolated, $R_t = 1000\Omega$, $\alpha = 90°$

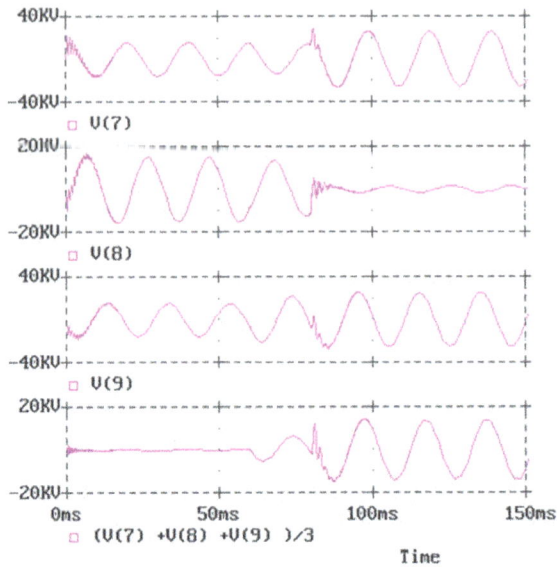

Figure 60. Time variation of the voltages when the neutral point of MVN is isolated, $R_t = 1000\Omega$, $\alpha = 90°$

The greater values of the currents at the two fault places, when the double phase to ground fault is produced, give high thermic solicitation to the instalation, as well as, higher values of the step voltages. This situatiom is dangerous for human being and animals, especially if this events are situated near the earth plate.

The time variation at the two fault places is far from being the same, no matter what grounding methos is used for the neutral point of the MVN.

4.2.3. The interrupted conductor with ground contact towards the customer fault

For this simulation VSGOL is opened , as if the phase conductor would be interrupted, and with a 20ms delay VSa is connected (simulating the ground contact of the broken conductor). The switch VSb remains opened.

The MVN functioning at resonance, with compensation inductor, was simulated with three values of R_t, namely 1Ω, 10Ω, 100Ω. Two values for α were taken into account, $0°$, respectively $90°$.

The equivalent capacitance of the phase conductor from the fault to the consumers is simulated by:

- C_{101a} is simulating the capacitance of phase 1 from the fault place to the bars of the MV transformer station
- C_{102a} represents the model for the capacitance of the phase conductor from the fault place to the consumers supplied by the faulty line.

Two possible situations are considered: $C_{101a} = C_{102a}$, and $C_{102a} = 0,1C_{101a}$.

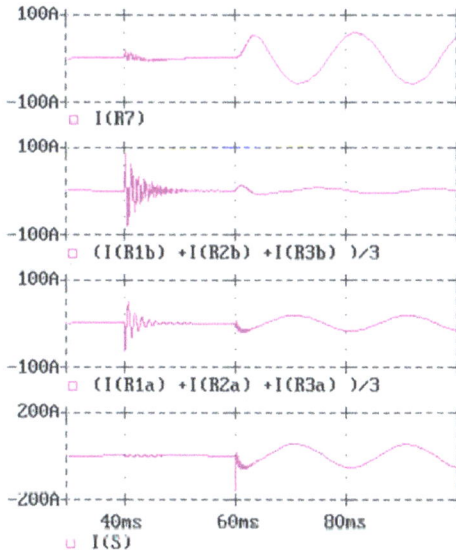

Figure 61. Time variation of the currents, neutral point of MVN is grounded by resistor and $R_t = 1\Omega$, $\alpha = 90°$, $C_{101a} = 9C_{102a}$

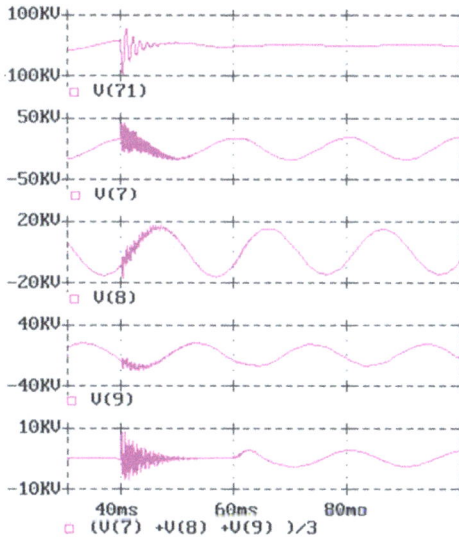

Figure 62. Time variation of the voltages, neutral point of MVN is grounded by resistor and $R_t = 1\Omega$, $\alpha = 90°$, $C_{101a} = 9C_{102a}$

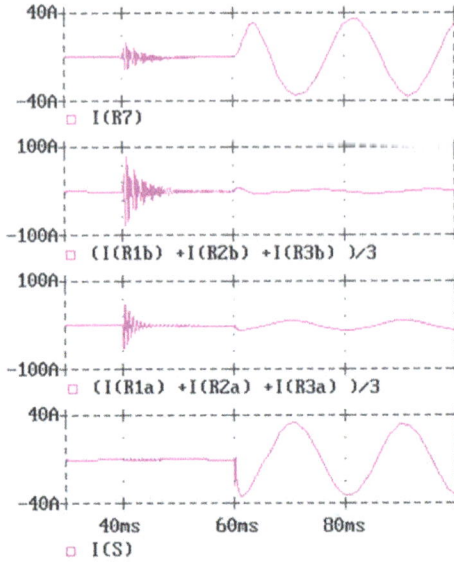

Figure 63. Time variation of the currents, neutral point of MVN is grounded by resistor and $R_t = 100\Omega$, $\alpha = 90°$, $C_{101a} = 9C_{102a}$

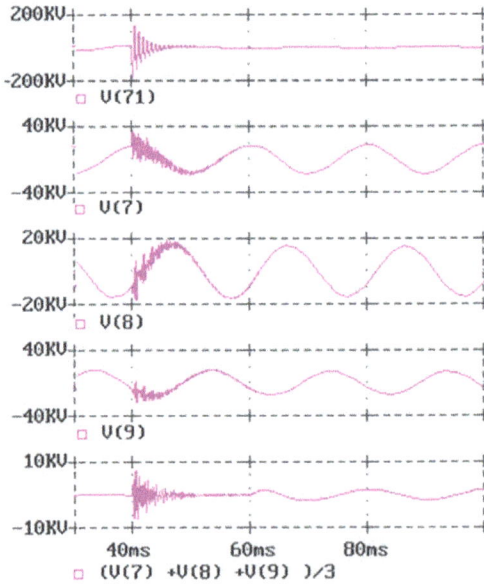

Figure 64. Time variation of the voltages, neutral point of MVN is grounded by resistor and $R_t = 100\Omega$, $\alpha = 90°$, $C_{101a} = 9C_{102a}$

In the oscillograms of the currents I(S) is the current to ground at the place of the fault, I(R7) is the current flowing trough the grounding resistor and the zero sequence current of the faulty line is $(I(R_{1a}) + I(R_{2a}) + I(R_{3a}))/3$. In the oscillogramms $(I(R_{1b}) + I(R_{2b}) + I(R_{3b}))/3$ is the zero sequence current of the healty "b" line.

The phase voltages are V(7), V(8), V(9), the zero sequence voltage on the bars of the MV transform station is $(V(7) + V(8) + V(9))/3$ and V(71) is the voltage of the broken conductor behind the fault place.

In each situation, either the grounding resistance is 1Ω, or 100Ω, the transient regime is stronger in the very moment of broking the phase conductor (t = 40 ms) than in the moment when the broken conductor contacts the ground (t = 60 ms). The voltage on the broken conductor might reach very high, dangerous values, jeopardizing the insulation of the line as well as the insulation of the consumer.

In Figs. 65 to 78 the oscillograms correspond to the MVN at resonance, with compensation inductor for grounding the neutral point. In this cases the current trough the compensation inductor is not represented because the protections used in MVN do not survey this current.

By the point of view of the overvoltages that might appear when the conductor is broken are more dangerous than the overvoltages in the case when the broken conductor is in contact with the ground.

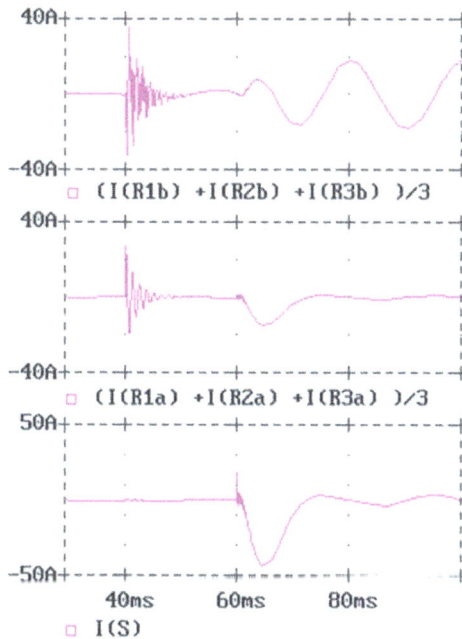

Figure 65. Time variation of the currents when MVN is at resonance, $R_t = 1\Omega$, $\alpha = 0°$, $C_{101a} = C_{102a}$

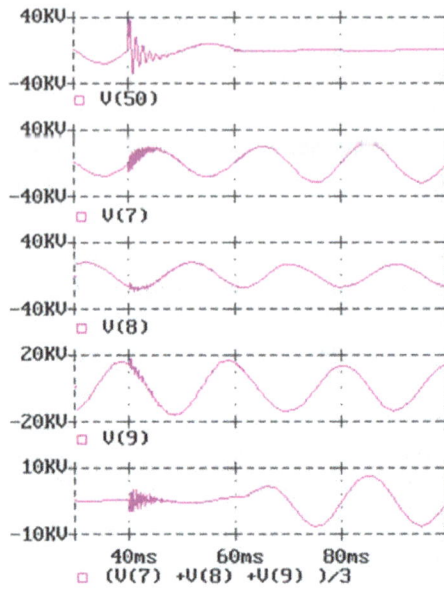

Figure 66. Time variation of the voltages, when MVN is at resonance, $R_t = 1\Omega$, $\alpha = 0°$, $C_{101a} = C_{102a}$

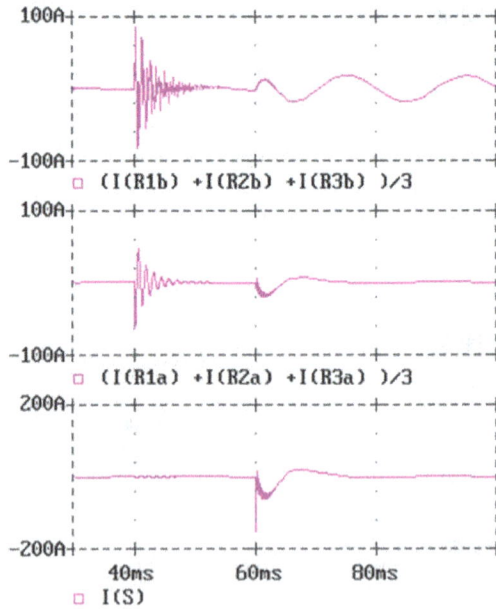

Figure 67. Time variation of the currents when MVN is at resonance, $R_t = 1\Omega$, $\alpha = 90°$, $C_{101a} = C_{102a}$

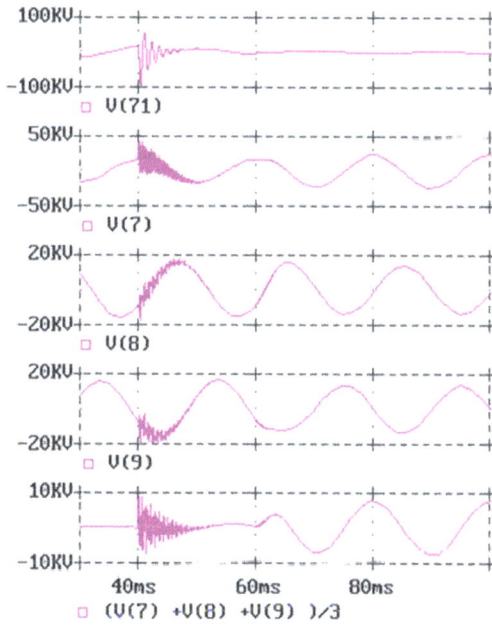

Figure 68. Time variation of the voltages, when MVN is at resonance, $R_t = 1\Omega$, $\alpha = 90°$, $C_{101a} = C_{102a}$

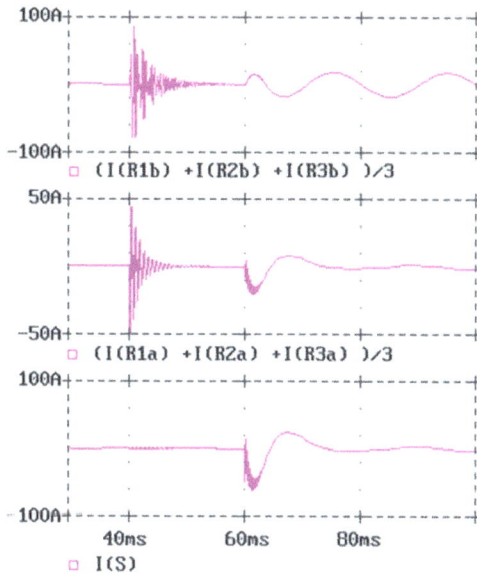

Figure 69. Time variation of the currents when MVN is at resonance, $R_t = 1\Omega$, $\alpha = 90°$, $C_{101a} = 9C_{102a}$

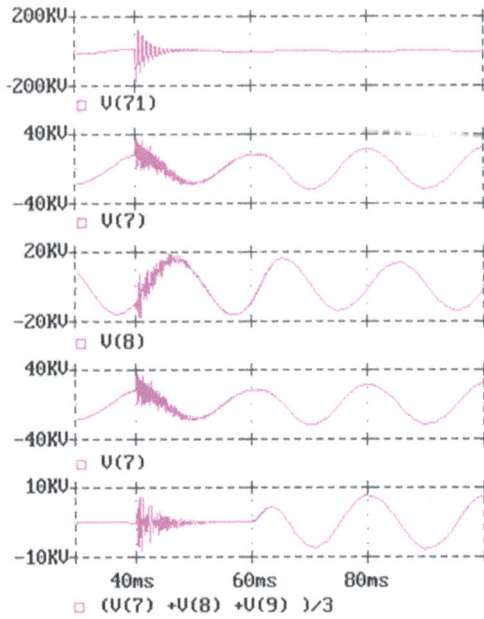

Figure 70. Time variation of the voltages, when MVN is at resonance, $R_t = 1\Omega$, $\alpha = 90°$, $C_{101a} = 9C_{102a}$

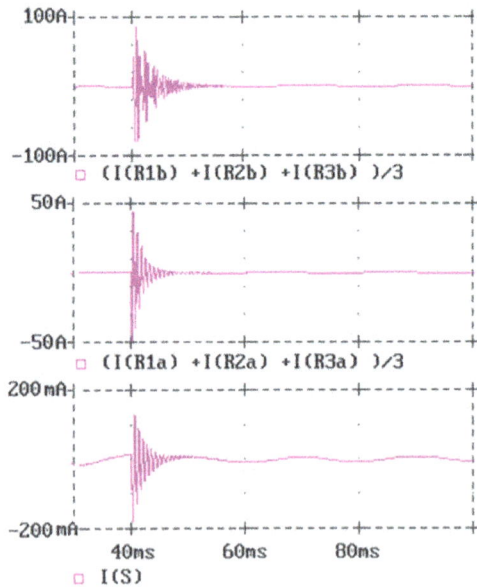

Figure 71. Time variation of the currents when MVN is at resonance, $R_t = 1M\Omega$, $\alpha = 90°$, $C_{101a} = 9C_{102a}$

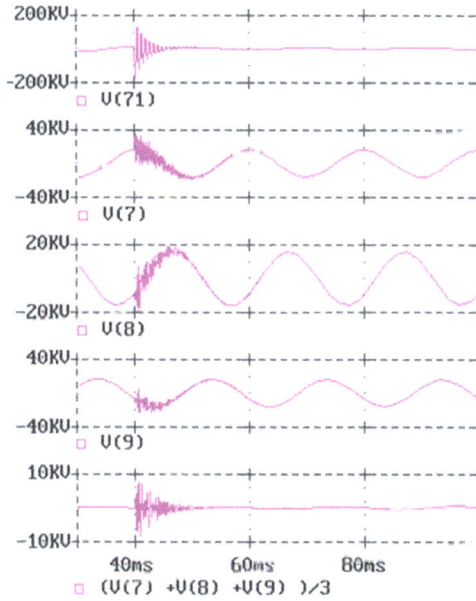

Figure 72. Time variation of the voltages. when MVN is at resonance, $R_t = 1M\Omega$, $\alpha = 90°$, $C_{101a} = 9C_{102a}$

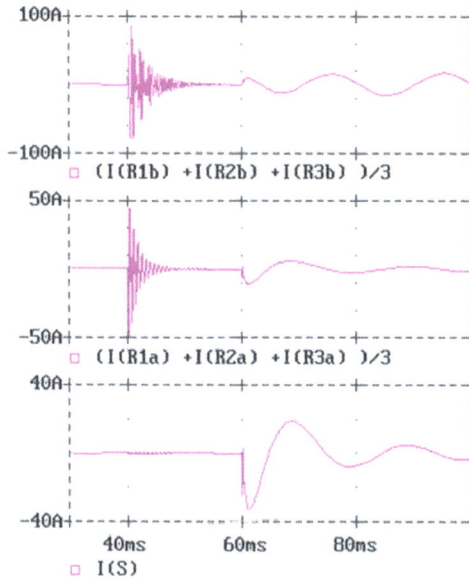

Figure 73. Time variation of the currents when MVN is at resonance, $R_t = 100\Omega$, $\alpha = 90°$, $C_{101a} = 9C_{102a}$

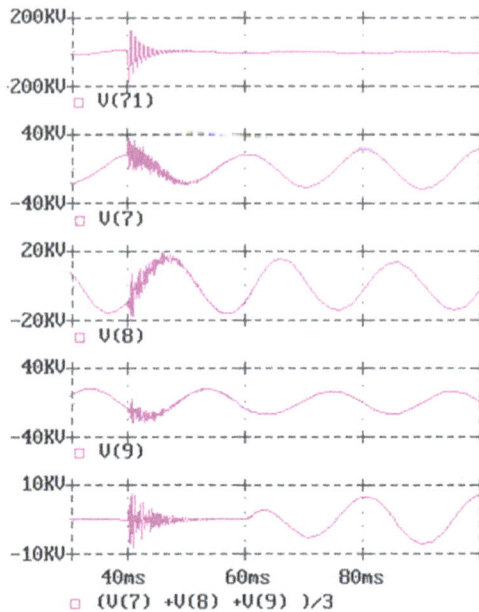

Figure 74. Time variation of the voltages, when MVN is at resonance, $R_t = 100\Omega$, $\alpha = 90°$, $C_{101a} = 9C_{102a}$

When $R_t = 1M\Omega$ the fault becomes of the type broken conductor and at t=60ms does not occur a transient regime.

When the fault is of the type interrupted conductor with ground contact towards the customer the heavyest situation is shown to be when the phase is near 90°.

For higher values of R_t the transient regime is less dangerous.

5. Measurement on real MVN

The results obtained during the measurements in the experiment area, while faults were produced on purpose, are shown in the followings.

The fault produced was of type single phase grounding.

The experimental aria contains 7 electric lines of 6kV, with total value of the capacitive current of 27A From the 7 electric linea 6 are connected at first system of bars and at the second bar system was connected the line on which the faults were produced. The capacitive current of this line is 2.7A.

The transformer station contains a transversal switch in connected position.

For recording the currents and the voltages during the experiment was used a CDR oscilloperturbograph.

Figures 75 to 83 show the following:

- the zero sequence current of the faulty line $I^0_{PT\ Turn}$,
- the current trough the compensation inductor I_B,
- the star voltages, V_a, V_b, V_c; - the zerosequence voltage of MV bars from the transformer station U^0.

The experiment was made in order to verify the concordance with the numeric simulation and to validate the accuracy of the simulator.

The experiment consisted on producing on purpose faults of the type single-phase grounding, namely:

- metallic grounding of a phase $R_t = 0$,
- grounding through the passing resistance $R_t = 250\ \Omega$,
- grounding through the passing resistance $R_t = 500\ \Omega$.

The functioning regimes of the MV network were:

- MV network functioning at resonance; -MV network functions at 10% overcompensation,
- MV network functions at 10% undercompensation.

The three recordings, corresponding to the three functioning regimes for metallic grounding are shown in Figs. 75, 76 and 77.

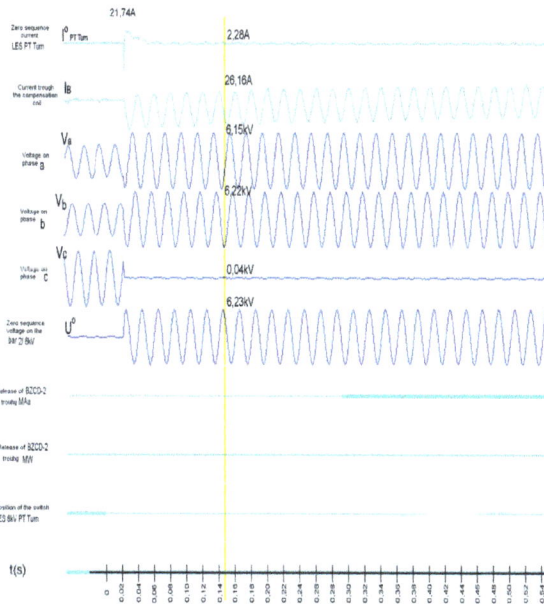

Figure 75. Single phase metallic grounding, $R_t = 5\ \Omega$, network at resonance

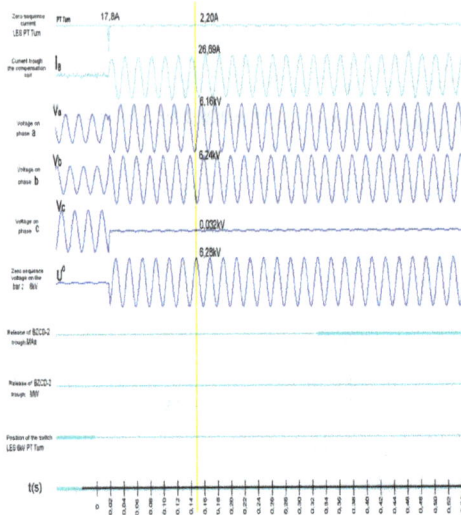

Figure 76. Single phase metallic grounding, Rf= 5 Ω, network at 10% over-compensation network

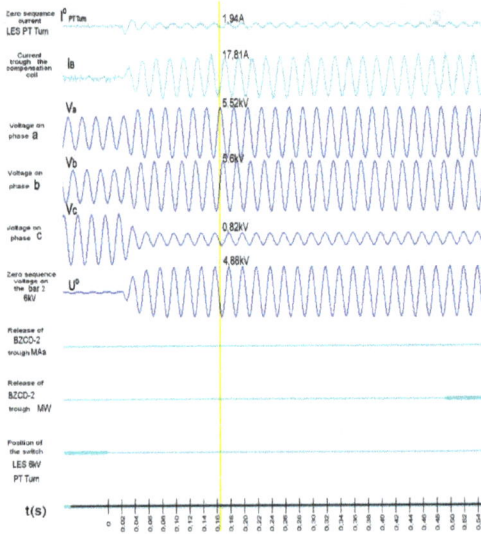

Figure 77. Single phase grounding with Rf=5 Ω, 10% under-compensated network

By comparing the shape of the current of the faulty line during the transient regime (the zero sequence current of the faulty line) obtained by numeric simulation with the one obtained by recording the same current during experiment it can be observed that the simulated current has an oscillating component greater than in recorded experiment.

This difference is due to a greater equivalent resistance of the experimental equivalent circuit than the simulated one.

In the same maner, as for the metallic grounding, but with a 250Ω grounding resistance for the three functioning regimes of the network the, results are presented in Figs. 78, 79 and 80.

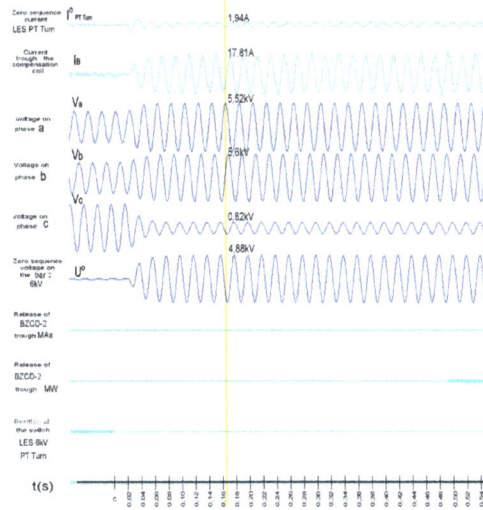

Figure 78. Single phase grounding with R$_t$=250 Ω, network at resonance

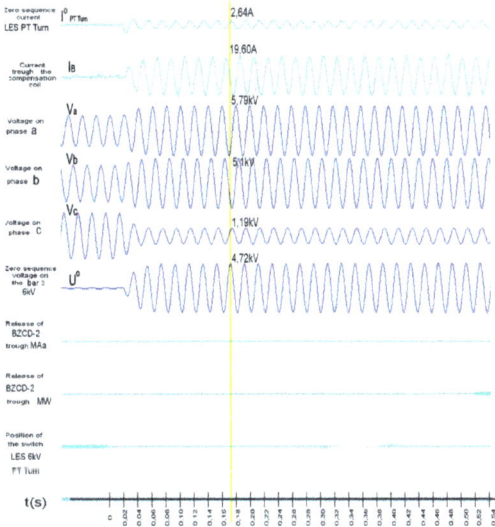

Figure 79. Single phase grounding with R$_t$=250 Ω, 10% over-compensated network

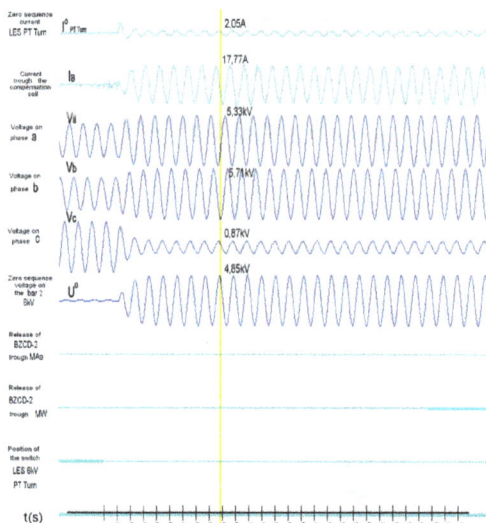

Figure 80. Single phase grounding with Rₜ=250 Ω, 10% under-compensated network

The third value for the grounding resistance used in the experiment is 500Ω and the results for the resonant regime, for 10% overcompensated and for 10% undercompensated are presented, respectively, in the figures 81, 82 and 83.

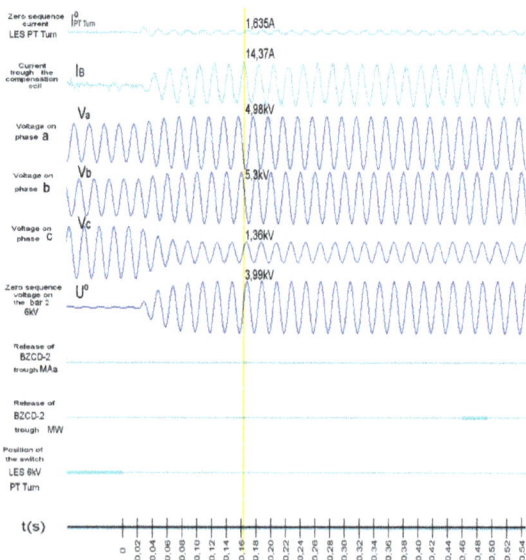

Figure 81. Single phase grounding resistance, Rₜ= 500 Ω, network at resonance

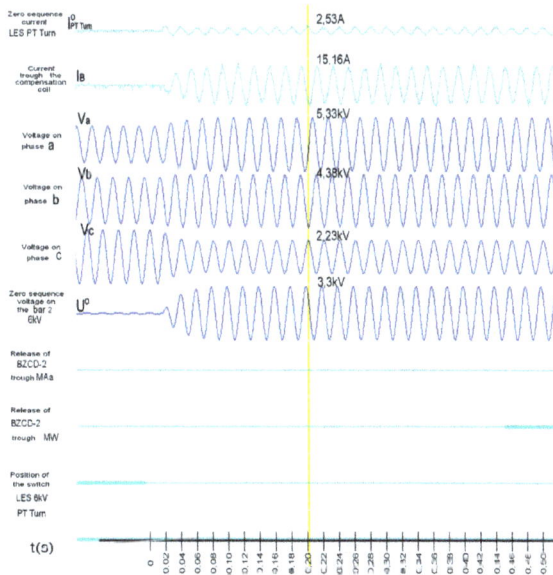

Figure 82. Single phase grounding resistance R_t= 500 Ω, network at 10% overcompensation

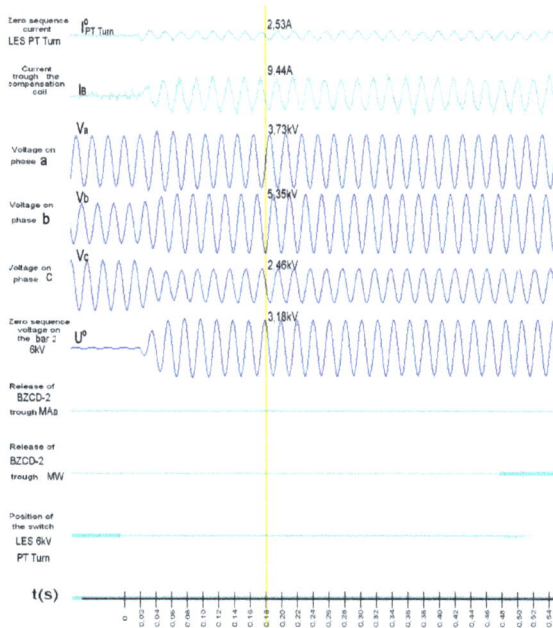

Figure 83. Single phase grounding with R_t=500 Ω network at 10% undercompensation

Table 1, campare some of the values measured during the experiment with the corresponding values obtained by numeric simulation.

The significance of the symbols from the table is as follows:

- I^0_{max} – maximal value of the zero sequence current on the faulty line,
- I_B – current trough the compensation inductor,
- I^0_{stab} – the zero sequence current of the faulty line, after the dumping of the dead-beat component,
- t_{am} – dumping time of the dead-beat component,
- U^0 – the zero sequence voltage at the MV bars of the transform station,
- U_f – voltage of the healty lines.

From Table 1. results that the differences between the values obtained by the numerical simulation and the values measured during the experiment are reasonably small.

Regime of the network	Symbol of quantities	$R_t[\Omega]$								
		5			250			500		
		Values obtained by		Error [%]	Values obtained by		Error [%]	Values obtained by		Error [%]
		Experiment	Simulation		Experiment	Simulation		Experiment	Simulation	
Resonance	$I^0_{max}[A]$	21,7	24,6	13,4	4,8	4,5	6,3	2,8	2,6	7,1
	$I_B [A]$	26,16	24,3	7,1	17,81	16,2	9,1	14,37	12,84	10,6
	$I^0_{stab}[A]$	2,28	2,42	6,1	1,94	1,75	9,8	1,64	1,49	9,1
	t_{am} [ms]	126	130	3,2	38	34	10,5	57	54	5,3
	U^0 [kV]	3,6	3,55	1,4	2,62	2,43	7,3	2,31	2,17	6,1
	U_f [kV]	6,22	6,03	3,1	5,6	5,46	2,5	5,3	4,78	9,8
10% Over-compensated	$I^0_{max}[A]$	17,8	58	226	4,2	4,5	7,1	3,2	2,9	9,4
	$I_B [A]$	26,69	24,82	7,0	19,6	18,3	6,6	15,16	13,86	8,6
	$I^0_{stab}[A]$	2,2	2,06	6,4	2,64	2,42	8,3	2,53	2,27	10,3
	t_{am} ms]	80	86	7,5	35	32	8,6	41	37	9,8
	U^0 [kV]	3,63	3,55	2,2	2,73	2,47	9,5	1,91	1,7	9,9
	U_f [kV]	6,24	5,96	4,5	5,69	5,21	8,4	4,38	4,76	8,7

Regime of the network	Symbol of quantities	$R_t[\Omega]$								
		5			250			500		
		Values obtained by		Error [%]	Values obtained by		Error [%]	Values obtained by		Error [%]
		Experi-ment	Simu-lation		Experi-ment	Simu-lation		Experi-ment	Simu-lation	
10% Under - compensated	$I^0_{max}[A]$	20,99	43	105	4,3	4,5	4,7	3,1	2,8	9,7
	$I_B [A]$	22,82	21,28	6,7	17,77	16,5	7,1	9,44	9,22	2,3
	$I^0_{stab}[A]$	2,5	2,32	7,2	2,05	1,91	6,8	2,53	2,31	8,7
	t_{am} ms]	198	210	6,1	31	28	8,6	23	21	8,7
	U^0 [kV]	3,59	3,56	0,84	2,8	2,53	9,6	1,84	1,67	9,1
	U_f [kV]	6,2	5,97	3,7	5,71	5,39	5,6	5,35	4,81	10,1

Table 1. Comparison of measured and simulated values

6. Conclusions

The numerical simulator designed by us allows the analysis of transients caused by diferent types of faults, such as simple groundings, double groundings, or broken conductor grounded towards the consumer.

We have compared results obtained by simulation with measured values only for simple grounding faults.

In this type of fault the simulator is validated by the measurements, no matter how the neutral point of the medium voltage network is grounded and whatever are the functioning conditions of the electrical network.

The results from Table 1. show that the simplificatory conditions, taken into account for developing the simulator, are correct.

The numerical simulation of the transient regimes caused by simple grounding faults produced in medium voltage networks shows to be an efficient method for analyzing such faults.

The most dangerous transient regimes occur when the initial phase of the voltage of the faulty line is near 90°.

The initial phase of the voltage, that happen to be in the very moment of producing the fault, was taken with the same value in simulation. The differences between the maximal values of zero sequence component, obtained by the two methods, are quite acceptable from a technical point of view and this concordance validates the model.

The resistance of the broken conductor to ground is extremely important, small values of R_t implying long damping periods. If this undesirable condition happens the currents might have high values, as well as important values of the over voltages leading to important supplementary mechanical stress and damages of the insulating devices and, eventually a simple fault can turn to a multiple fault situation.

For small values of the grounding resistance at the fault location ($R_t < 10\Omega$) difference between measured and simulated values is a litle higher than in the case of greater values ($R_t > 100\Omega$).

The model is very useful by giving the values of the voltages on the healthy lines (over voltages that might jeopardize the insulation) and zero sequence currents of the faulty lines (fixing the condition of the fault detection by protection devices).

Using the conclusions of the simulation of a MVN for the values of the currents and voltages during different types of faults it is possible to adjust the prescribed values of the protection.

Simulation of the broken conductor fault, either connected to the ground or not, shows that resonance might occur, over voltages due to this resonances being dangerous to the equipments.

When double phase to ground fault was simulated the values of the voltages showed dangerous voltages for step or touch voltage.

The numerical simulator has the advantage of analyzing rapidly several variants as models of different possible situations.

The simulator is flexible and all kinds of faults can be simulated for different MVN, just by making the proper modification in the parameters of the simulator. These modifications are slightly easier to be performed when PSPICE simulating model is used.

Either using PSPICE or Mathlab-Simulink the accuracy of the analysis is similarly good.

The precision of the results obtained by simulation depends essentially on the accuracy of the MVN parameters.

Author details

Dumitru Toader, Stefan Haragus and Constantin Blaj
"Politehnica" University of Timisoara, Romania

7. References

Blume, W. (1986). *Computer Circuit Simulation*, BYTE, July 1986

Bucatariu, I. & Surianu, F. (2009). Optimal Placement of Fixed Series Capacitor in Distribution Networks, *Proceedings of the 9thWSEA/IASMES International Conference on*

Electric Power Systems High Voltages, Electric Machines Control & Signal Processing, Genova, Italy, October 17-19, 2009, pp.65-70

Chuco, B. (2005). Electrical Software Tools Overview *in* SINATEC-IEEE

Curcanu, G., Toader, D. & Pandia, T. (2006). Determination of overvoltages in high voltage networks at single phase faults by numarical simulation and experiments, *Proceedings 12ᵗʰ International IGTE Symposium,* Gratz

DeCarlo, R & Lin, P. (2001). *Linear Circuit Analysis,* Oxford University Press

Danyek, M., Handl, P.& Raisz, D. (2002). Comparison of Simulation Tools ATP-EMTP and MATLAB-Simulink for Time Domain Power System Transient Studies, *Proceedings of the European EMTP-ATP Conference,* Sopron, Hungary

Dessaint, L.-A., Al-Haddad, K., Le-Huy, H., Sybille, G. & Brunelle, P. (1999). *A power system simulation tool based on Simulink,* IEEE Trans. on Ind. Electronics, vol. 46, no. 6, pp. 1252-1254, Dec. 1999

Dommel, H.W. (1995). *ElectroMagnetic Transient Program. Theory Book,* Bonneville Power Administration, Portland

Drapela, J. (2009). Performance of a Voltage Peak Detection-Based Flichermeter, *Proceedings of the 8thWSEAS International Conference on Circuits, Systems, Electronics, Control & Signal Processing,* Teneriffe, Spain, December 14-16, 2009, pp 296-301

Foltin, M., Ernck, M. & Hnat, J. (2006). SimPowerSystems in Education, *Mezinarodni Konference Technical Computing,* Prague

Hasler, M. & Neirynck, J. (1985). *Circuits non lineares,* Presses Polytechniques Romandes, Lausannes.

Iordache, M. & Mandache, A. (2004). *Computer Aided Analyse of Nonlinear Circuits (in Romanian)* , Politehnica Press Bucharest

Istrate, M., Gavrilas, M., Istrate, C. & Ursuleanu, R. (2009). Algorithms in Transmission Grids Using ATP Simulations, *Proceedings of the 9thWSEA/IASMES International Conference on Electric Power Systems High Voltages, Electric Machines Control & Signal Processing,* Genova, Italy, October 17-19, 2009, pp.109-114

Karlsson, A. (2005). *Evaluation of Simulink/SimPowerSystems and other Commercial Simulation Tools for the Simulation of Machine System Transient,* Swedish Royal Institute of Technology, Stockholm

Mandache, L. & Topan, D. (2009). *Algorithms for Electric Circuit Simulation,* Universitaria Press Craiova

Radoi , C. (1994). *SPICE Simulation and Analysis of Electronic Circuits,* Ameo Press București.

Rashid, M.H. & Rashid, H.M. (2006) *SPICE for Power Electronics and Electric Power,* Taylor & Francis Group, Boca Raton

Ross, C.C. (2004). *Differential Equations,* Springer Science, pp.51-56

Sybille, G., Brunelle, P., Hoang Le-Huy, Dessaint, L.A. & Al-Haddad, K , (2000). *Theory and applications of power system blockset, a MATLAB/Simulink-based simulation tool for power systems,* Power Engineering Society Winter Meeting 2000, IEEE, Vol.1, pp.774-779

Vladimirescu, A. (1999). *SPICE*, Ed.Tehnica, Bucuresti

Vladimirescu, A. (1997) SIMULINK ; *Dynamic System Simulation for MATLAB, User's Guide, version 2.1*, MathWorks Inc.

Chapter 7

Numerical Simulation of Combustion in Porous Media

Masoud Ziabasharhagh and Arash Mohammadi

Additional information is available at the end of the chapter

1. Introduction

The demand of fossil fuel resources and air pollution are the major problems that caused by using of fuels. In recent years, many researchers have developed new methods for efficient combustion of fuels. Porous media combustion, also known as filtration combustion in a packed bed, due to the interaction between two different phases, solid and gas or liquid. The theory of filtration combustion involves a new type of flame with exothermic chemical reactions during fluid flow in a porous medium. The term 'filtration combustion' was introduced by Russian scientists for combustion of gas flow through porous media. This term does not correspond to western scientific terminology, still it can be found in special literature as a synonym to combustion within porous media (PM). This process facilitates a combustion process with stability in a wide range of reactant fluid velocities, air-fuel ratios, and power density. PM combustion has some unique characteristics. It gives rise to high radiant output, low NO_x (Oxides of Nitrogen) and CO (Carbon Monoxide) emissions, high flame speed and higher power density. A porous material means a material with connected voids that flow can easily penetrate through its structure. This technology is different from conventional combustion, with free flame, thin reaction zone and high temperature gradients. Compared to conventional combustion devices, the combustion efficiency of PM burner is reasonably high and better heat transfer from burned gases to unburned mixture, takes place. On the other hand, in PM combustion three modes of heat transfer conduction, convection, and radiation are significant. In addition, there is a better homogenization of temperature across the PM and the significant amount of radiation helps to preheat the incoming air-fuel mixture at upstream. The technique of premixed combustion within PM has been studied and applied to steady combustion with great success. The porous media combustion has proved to be one of the applicable options to solve the problems to a remarkable extent in both technical and economic perspectives. This technique has been used for both gaseous and liquid fuels in steady or unsteady combustion. Flame stability in

PM with lean and rich mixtures, significant reduction in pollutants and increasing combustion efficiency, was proven.

In recent years many researchers investigated the PM combustion technology both experimentally and theoretically. The most of researches are in field of steady combustion in PM and few of them are about transient flame propagation, both approaches are employed in PM combustion. Steady combustion is widely used in radiant burners and surface combustor-heaters due to its high radiant emissivity of the solid. The combustion zone is stabilized by its solid. The other, transient leads to an unsteady reaction zone freely propagate as a filtration combustion wave in the downstream direction. Combustion in PM differs considerably from the homogeneous flames flame front. Considerable features of PM for application of combustion technology are: large specific surface area, excellent heat transfer properties, heat capacity, transparency for fluid flow, thermal resistance, mechanical resistance, recuperation of energy and electrical properties.

2. Background of combustion in porous media

2.1. Stability of flame

The flame stabilization and propagation in a PM are governed by the modified Peclet number:

$$Pe = \frac{(S_L \ d_m \ C_p \ \rho)}{k},$$

(1)

where S_L is the laminar flame speed, d_m is the equivalent diameter of the average hollow space of the porous material, C_p is the specific heat of the gas mixture, ρ is the density of the gas mixture and k is the thermal conductivity of the gas mixture. For flame propagation through a porous material, the critical Peclet number of 65 has been found. Thus, $Pe < 65$ for quenching, and $Pe > 65$ for flame propagation.

2.2. Premixed and non-premixed mixture

PM may work with a premixed flow or with a non-premixed fuel flow. Premixed porous burners consist of two zones: the premixed fuel–air mixture first enters a hot solid matrix, where it is heated until it enters to the second hot solid matrix. Depending on its application, a third section, a compact heat exchanger may be added to the burner. The schematic of a two-layer premixed combustor is shown in Fig. 1.

A premixed mixture of methane-air and hydrogen-air were used in different burners with two section PM. Measurements show considerable reduction in the concentrations of NO_x, CO in the fluid gases. Also the effects of hydrogen addition to methane, were investigated. For the porous burners hydrogen was found to lower the NO_x emissions slightly, while for the other burners an increase, or no obvious effect, was found. The enhancement of the radiation flux from PM burners operating with non-premixed flames by using a vane-rotary burner, in which the swirling fuel flow was confined by an air duct, is necessary. They also studied emission characteristics of ceramic foam burners operating with non-premixed flame.

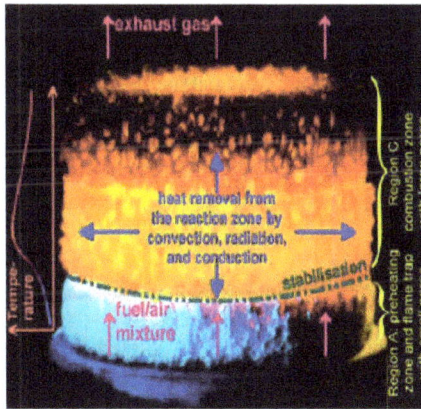

Figure 1. Schematic of a premixed combustor (M. A. Mujeebu et al. , 2009)

2.3. Reciprocating flow

For the excess enthalpy combustion, a combustion system using reciprocating flow in PM was introduced. By the reciprocating flow, the combustion gas enthalpy is regenerated into increase in enthalpy of the combustible gas through the PM, which store heat. For this technique a new arrangement of the PM that stabilized flame for a wide operating condition, was used. The mixture first flow in, and the gas and solid temperatures reaches to a maximum at the exit side. Then the flow direction is reversed by means of valves. On the reverse flow half-cycle, the fresh mixture encounters much higher solid temperatures at the entering side. Therefore, the amount of heat recycled becomes larger than that with the single flow direction. Hence in the reciprocating flow system, the heat transfer from the combustion gases raises the solid temperatures from both directions.

2.4. Hydrogen production

Production of hydrogen from gases such as methane and hydrogen-sulphide is another potential application of PM combustion. These reactants convert into products such as hydrogen, syngas (H_2 and CO), and sulfur. For both methane and hydrogen-sulphide combustion, upstream propagation corresponded to the range of equivalence ratios from stoichiometry to 1.7, and downstream wave propagation was observed for ultra-rich (1.7- 4) mixtures.

2.5. Materials for porous media combustion

Aluminum oxide (Al_2O_3), silicon carbide (SiC), and zirconium dioxide (ZrO_2) proposed as suitable materials for application. Al_2O_3 and ZrO_2 were recognized as high temperature resistant materials. SiC shows good thermal shock resistance, mechanical strength, and conductive heat transport. SiC also has high melting point (3260 K), against cyclic thermal stress and strength retention at the peak regenerator temperature (1673 K), and excellent oxidation resistance. Metallic materials were found less suitable for PM because of their inadequate thermal stability and high thermal inertia. Fe–Cr–Al-alloys and nickel-base alloys were found suitable for some applications but they were said to be comparatively less heat resistant. Structures of ceramic foams with different base materials were observed to possess high porosity, good conduction heat transport, low thermal inertia, low radiation heat transport properties and relatively high pressure drop. The effective thermal conductivity of anisotropic porous composite medium could vary largely with the component fractions.

3. Numerical modeling

The thermophysical properties of the air such as density, thermal conductivity and specific heat are assumed to be functions of the temperature. Usually, the pressure drop through the porous burner is not that high (with high porosity of PM) and its effect on the thermophysical properties can be neglected. In general, the properties of the solid phase may be assumed to be constant and assumed that there is thermal non-equilibrium between the gas and solid phases. Therefore, there are two energy equations to model the energy transport in the system. The porous material can be assumed as a scattering, emitting and absorbing medium. Gaseous radiation is assumed to be negligible compared to the solid radiation.

3.1. One-dimensional modeling

Many researchers studied the combustion and heat transfer in porous radiant burners. The PM was assumed to emit, absorb, and scatter radiant energy. Non-local thermal equilibrium between the solid and the gas is assumed for and combustion was described by a one-step, multi-steps or kinetic reactions. The effect of the optical depth, scattering albedo, solid thermal conductivity, upstream environment reflectivity, and interphase heat transfer coupling on the burner performance can be considered. Also, low solid thermal conductivity, low scattering albedo, and high inlet environment reflectivity produced a high radiant efficiency. The system consisted of a packed bed or foams in which a natural gas–air mixture combusts inside it. Radiative heat transfer in the packed bed or foams was modeled as a diffusion process, and the flow and temperature distribution in the packed bed or foams can be determined. The numerically results usually were compared with available experimental data for a similar system. Subsequently, the one-dimensional predictions of methane/air combustion in inert PM can include full mechanism (49 species and 227 elemental reactions), skeletal mechanism (26 species and 77 elemental reactions), 4-step reduced mechanism (9 species and 1-step global mechanism). The effects of these models on temperature, species, burning speeds and pollutant emissions were examined by researchers. Experimental and numerical investigations found that the flammability limits of the gaseous mixture in PM were more sensitive to their geometric properties than the physical properties.

3.1.1. One-dimensional governing equations

The following assumptions are made to simplify the problem:

1. Gas radiation is neglected, gas flow and heat transfer are one-dimensional.
2. PM is considered to be non-catalytic, homogeneous and optically thick.
3. The radiation of solid phase is treated using Rosseland approximation.
4. The PM consists of solid dispersed homogeneously, and the porosity variation near the tube wall is neglected. Under the above assumptions, a set of differential equations can be obtained in the following form.

Continuity equation:

$$\frac{\partial}{\partial t}\left(\rho_g \varepsilon\right) + \frac{\partial}{\partial x}\left(\rho_g u \varepsilon\right) = 0. \tag{2}$$

Gas phase energy equation:

$$\rho_g c_g \varepsilon \frac{\partial T_g}{\partial t} + \rho_g c_g \varepsilon u \frac{\partial T_g}{\partial x} + \sum_i \rho_g \varepsilon Y_i V_i c_{gi} \frac{\partial T_g}{\partial x} + \varepsilon \sum_i \dot{\omega}_i h_i W_i = \varepsilon \frac{\partial}{\partial x}\left[\left(k_g + \rho_g c_g D_{\parallel}{}^D\right)\frac{\partial T_g}{\partial x}\right] - h_v(T_g - T_s). \tag{3}$$

Solid phase energy equation:

$$\rho_S c_S(1-\varepsilon)\frac{\partial T_s}{\partial t} = k_s(1-\varepsilon)\frac{\partial^2 T_s}{\partial x^2} + h_v(T_g - T_s) - \frac{dq_r}{dx} = 0. \tag{4}$$

Species transport equation:

$$\rho_g \varepsilon \frac{\partial Y_i}{\partial t} + \rho_g \varepsilon u \frac{\partial Y_i}{\partial x} + \frac{\partial}{\partial x}\left(\rho_g \varepsilon Y_i V_i\right) - \varepsilon \dot{\omega}_i W_i = 0. \tag{5}$$

$$V_i = -\left(D + D_{im}^d\right)\frac{1}{X_i}\frac{\partial X_i}{\partial x}. \tag{6}$$

density:

$$\rho = \text{constant density.} \tag{7}$$

Radiation:

$$q_r(x) = -\frac{16}{3}\frac{\sigma T_s^3}{\beta}\frac{dT_s}{dx}. \tag{8}$$

3.1.2 Boundary conditions for one-dimensional model

The following boundary conditions are considered in the computations:

| Inlet | $T_g = T_{f,inlet}$ | $\dfrac{\partial T_s}{\partial x} = 0$ | $\dfrac{\partial T_s}{\partial x} = 0$ |
| Outlet | $\dfrac{\partial T_g}{\partial x} = 0$ | $\dfrac{\partial T_s}{\partial x} = 0$ | $\dfrac{\partial Y_i}{\partial x} = 0$ |

3.1.3. Solution procedure

The time integration is performed using combination of implicit finite difference and finite volume method on a uniform- adaptive or non uniform mesh. The convective terms are upwinded and the diffusive terms are discretized using a second-order technique. The initial condition was also given to the system of discretized equations. For more information interested reader can see (Henneke et al , 1999)

3.2. Two-dimensional modeling

In the two dimensional model a comparison is made between the local thermal equilibrium and thermal non-equilibrium approaches by different researchers. The volume-averaged treatments unable to predict the pore-level, local high temperature region in the gas phase and the pore-level variation in the flame speed with respect to the flame location in the pore. The gas and the solid phases consider in non-local thermal equilibrium, and separate energy equations use for the two phases. The solid phase can be assumed absorbing, emitting and scattering, while the gas phase was considered transparent to radiation. The alternating direction implicit (ADI) scheme can be used to solve the transient two dimensional energy equations. Methane–air combustion with one-step, multi step and detailed chemical kinetics are used to model the combustion, like one dimensional combustion modeling. The radiative part of the energy equation can be modeled using simple Rosseland method until complex discrete ordinate method. In the flat plate burner, air flows axially through a constant area duct filled with a porous layer of thickness L. In the cylindrical and spherical burners, the air flows radially through an annular porous matrix.

3.2.1. Two dimensional governing equations

The energy equation for the gas phase is as follows:

$$\frac{\partial}{\partial t}\left(\varphi \rho_g C_{pg} T_g\right) + \frac{1}{r^n}\frac{\partial}{\partial r}\left(\varphi \rho_g C_{pg} r^n v T_g\right) = \frac{1}{r^n}\frac{\partial}{\partial r}\left(\varphi k_g r^n \frac{\partial T_g}{\partial r}\right) - (1-\varphi)h_v\left(T_g - T_s\right) + \varphi \Delta H_c S_{fg} \quad (9)$$

Where φ, ρ_g, r, C_{pg}, T_g, v, k_g, h_v, H_c and S_{fg} are the porosity, density, radial position, specific heat, temperature, velocity, thermal conductivity, volumetric heat transfer coefficient, enthalpy of combustion and rate of fuel consumption per unit volume, respectively. Subscripts g and s refer to gaseous and solid phases, respectively.

The energy equation for the solid phase is:

$$\frac{\partial}{\partial t}\left(\rho_s C_s T_s\right) + \frac{1}{r^n}\frac{\partial}{\partial r}\left(r^n k_s \frac{\partial T_s}{\partial r}\right) - h_v\left(T_s - T_g\right) - \nabla. F. \quad (10)$$

The term F represents the radiative transport equation and is given by:

$$\nabla.F = -(1 - \omega)(G - 4E_b),\tag{11}$$

where ω is the single scattering albedo and the irradiance G is governed by

$$\nabla^2 G = \eta^2(G - 4E_b),$$

$$\eta^2 = 3\beta^2(1 - \omega)(1 - g\omega),\tag{12}$$

where E_b is the Planck black body emitted flux and G is the radiative flux.

The conservation equation for the mass fraction of the fuel is given as follows:

$$\frac{\partial}{\partial t}(\rho_g m_f) + \frac{1}{r^n}\frac{\partial}{\partial r}(\rho_g r^n v m_f) = \frac{1}{r^n}\frac{\partial}{\partial r}\left(r^n D_{AB}\rho_g \frac{\partial m_f}{\partial r}\right) - S_{fg},\tag{13}$$

where m_f is the fuel mass fraction and D_{AB} is the diffusion coefficient. The n value is set to 0, 1 and 2 for spherical, radial and axial flow burners, respectively.

A single-step Arrhenius type chemical kinetic equation as given below is normally adopted in modeling the combustion:

$$S_{fg} = f\rho_g^2 m_f m_{o_2} \exp\left(-\frac{E}{RT_g}\right),\tag{14}$$

where f, m_{o2}, E and R refer to pre-exponential factor, oxygen mass fraction, activation energy and gas constant, respectively.

3.2.2. Boundary conditions for two-dimensional model

The following boundary conditions are adopted for the gas, solid and species:

Gas:

$$T_{g|r=r_{in}} = 0 \qquad \text{at} \qquad r = r_{in}.$$

$$\left.\frac{\partial T_g}{\partial r}\right|_{r=r_{out}} = T_{in} \qquad \text{at} \qquad r = r_{out}.\tag{15}$$

Solid:

$$h_{in}\left[(T_{g,in} - T_{s|r=r_{in}})\right] + \sigma\epsilon_{in}\left[(T^4_{in,amb} - T^4_s|_{s|r=r_{in}})\right] = -k_s\left.\frac{\partial T_s}{\partial r}\right|_{r=r_{in}} \qquad \text{at} \qquad r = r_{in}.$$

$$h_{out}\left[(T_{out,in} - T_{s|r=r_{out}})\right] + \sigma\epsilon_{out}\left[(T^4_{out,amb} - T^4_s|_{s|r=out})\right] = -k_s\left.\frac{\partial T_s}{\partial r}\right|_{r-r_{out}} \qquad \text{at} \quad r = r_{out}.\tag{16}$$

Species:

$$m_f = m_{f,in} \qquad \text{at} \qquad r = r_{in}.$$

$$\frac{\partial m_f}{\partial r} = 0 \quad \text{at} \quad r = r_{out}.$$ (17)

The control volume approach, or the finite-difference method, can be used to solve the governing equations. The solution is advanced in time by using a fully implicit technique and this was necessary due to the stiffness of the governing matrix of the problem. Also, it is necessary to use an adaptive grid, or a very fine grid, to insure the accuracy of the solution.

3.2.3. Solution procedure

The based code can be solved using alternative direction implicit (ADI) finite volume formulation. The pressure field is solved using the SIMPLE method in steady state and PISO in transient state. The gridding system should prove to be sufficient by testing several grids sizes. For more information interested reader can see (Mohammadi, 2010, Hackert et al, 1998).

4. Three-dimensional modeling

Although there exist many simulation of one-dimensional and two dimensional model in literature but only several papers discussed three-dimensional simulation in PM. Three-dimensional simulations of combustion inside the PM burner, are complex and time consuming. A finite-volume calculation for three-dimensional reacting flow in a porous burner can be used. The Navier–Stokes equations, energy for both phase of PM and species transport equations were solved, and radiative heat transfer under local thermal non equilibrium between the solid and gas phases was considered.

4.1. Three-dimensional governing equations

Following assumptions were used in modeling and simulation of the PM in modified KIVA-3V code:

1. There is thermal non-equilibrium between gas and solid phases.
2. Solid is homogeneous, isotropic, variable property with temperature and has no catalyst effects.
3. Only radiation heat transfer from the solid phase is considered and the gas phase is transparent.

Based on the above assumptions, the general governing equations are simplified as follows: Continuity equation for species i:

$$\frac{\partial(\varphi \rho_i)}{\partial t} + \nabla. (\varphi \rho_i u) = \nabla. \left[\rho \, \varphi \, D_{im} \, \nabla \left(\frac{\rho_i}{\rho} \right) \right] + \varphi \, \dot{\rho}_i^c + \dot{\rho}^s \, \delta_i ,$$ (18)

where diffusion coefficient modified according to kinetic theory of gases. ρ is mixture density, ϕ is porosity (void fraction) of PM, ρ_i is density of species i, D_{im} is diffusion coefficient of species i in the mixture and u is the velocity vector.

Gas phase momentum equation:

$$\frac{\partial(\rho u)}{\partial t} + \nabla \cdot (\rho u u) = -\frac{1}{a^2}\nabla P - \nabla\left(\frac{2}{3}\rho k\right) + \nabla \cdot \sigma + F^s - \left(\frac{\Delta P}{\Delta L}\right). \tag{19}$$

The last term on the right-hand side of Eq. (19) is due to pressure drop caused by PM according to Ergan equation and σ is stress tensor. The dimensionless quantity a is used in conjunction with the Pressure Gradient Scaling (PGS) method. This is a method for enhancing computational efficiency in low Mach number flows, where the pressure is nearly uniform.

$$\left(\frac{\Delta P}{\Delta L}\right) = \left(\frac{\mu}{\alpha} u + c_2 \frac{1}{2}\rho |u| u\right). \tag{20}$$

where α is the permeability and c_2 is the inertial resistance factor of PM, which are determined according to Eq. 21.

$$\alpha = \frac{d^2}{150}\frac{\epsilon^3}{(1-\epsilon)^2} \quad , \quad c_2 = \frac{3.5}{d}\frac{(1-\epsilon)}{\epsilon^3}. \tag{21}$$

Gas phase energy equation:

$$\frac{\partial}{\partial t}\left(\varphi\rho c_p T_g\right) + \nabla \cdot \left(\varphi\rho c_p T_g u\right) + \varphi\sum_i \dot\omega_i H_i W_i = -\varphi P \nabla \cdot u + \varphi A_0 \rho\epsilon + (1 - A_0)\sigma: \nabla u$$

$$\psi \nabla \cdot \left((k_g + \rho_g c_g D_\parallel^d)\nabla T_g\right) - h_v\left(T_g - T_s\right) + \dot Q^s . \tag{22}$$

where the fourth term on the right hand side is the conduction heat transfer due to thermal conductivity of fluid and dispersion term due to existence of PM and the fifth term represent the convective heat transfer between gas and solid phase of PM which are determined according to Eqs. (23 – 25):

$$D_\parallel^d = 0.5\ \alpha_g Pe \tag{23}$$

$$\mathrm{Nu}_v = 2 + 1.1\ \mathrm{Re}^{0.6}\ \mathrm{Pr}^{0.33}, \tag{24}$$

$$h_v = \frac{6\varphi}{d^2}\ k_g Nu_v \tag{25}$$

c_p is specific heat of the mixture, T_g is gas temperature, Y_i is mass fraction species i, is rate of reaction i, H_i is enthalpy of species i, W_i is molecular weight of species i, k_g is thermal conductivity of the fluid, is thermal dispersion coefficient along the length of the porous medium, h_v is volumetric heat transfer coefficient between solid and gas phase of PM. Correlation (23 - 25) was estimated from experimental data by Wakao and Kaguei for heat transfer between packed beds and fluid.

Solid phase energy equation:

$$\frac{\partial}{\partial t}\left((1 - \varphi)\rho_s c_s T_s\right) = \nabla \cdot [k_s (1 - \varphi)\nabla T_s] + h_v\left(T_g - T_s\right) - \nabla \cdot q_r, \tag{26}$$

T_s is solid temperature, ρ_s is solid density, c_s is specific heat of solid phase, k_s is thermal conductivity of solid phase, q_r is the radiation heat transfer in solid in r direction.

Chemical species continuity equation:

$$\frac{\partial}{\partial t}(\varphi\rho Y_i) + \nabla.(\varphi\rho Y_i u) + \nabla.(\varphi\rho Y_i v_i) - \varphi \dot{\omega}_i W_i = 0, \tag{27}$$

$$v_i = -\left(D + D_{m\parallel}^d\right)\frac{1}{X_i} \nabla X_i, \tag{28}$$

$$Pe = \frac{\rho\, c_{p\,|u|\,d}}{k_g}, \tag{29}$$

$$D_{m\parallel}^d = 0.5\, D_{im}Pe, \tag{30}$$

X_i is molar fraction of species i, Pe is Peclet number, d is diameter of sphere in packed bed, is species dispersion coefficient, Y_i is mass fraction of species i and v_i is diffusion velocity of species i in the mixture.

Turbulence model

Since there is no model presented for simulation of turbulent compressible-flow in PM by any researcher. Hence the basic κ- ε equations were used without any modification.

The transport equation for κ turbulent kinetic energy:

$$\frac{\partial(\rho k)}{\partial t} + \nabla.(\rho u\, k) = -\frac{2}{3}\rho \kappa \nabla.u + \sigma : \nabla u + \nabla.\left[\left(\frac{\mu}{Pr_k}\right)\nabla k\right] - \rho\varepsilon + \dot{W}^{\,s}, \tag{31}$$

where σ is stress tensor. With a similar consideration dissipation rate, ε:

$$\frac{\partial(\rho\varepsilon)}{\partial t} + \nabla.(\rho u\varepsilon) - \left(\frac{2}{3}C\epsilon_1 - C\epsilon_3\right)\rho\varepsilon\nabla.u + \nabla.\left[\left(\frac{\mu}{Pr_\varepsilon}\right)\nabla\epsilon\right] + \frac{\epsilon}{k}\left[C\epsilon_1\,\sigma:\nabla u - C\epsilon_2\,\rho\varepsilon + c_s\dot{W}^s\right], \tag{32}$$

The parameters, ,, and are constant whose values are determined from experiments and some theoretical considerations and is viscous stress tensor.

Equation of state:

$$P = \rho R\, T_g\, /\, \overline{W}, \tag{33}$$

R is universal gas constant, average molecular weight of mixture, P is pressure inside the combustion chamber and the PM.

Radiation model

Due to extreme temperature of combustion zone and solid phase, radiation heat transfer is very important. Gas phase radiation in comparison with solid phase radiation that has a high absorption coefficient, is negligible. Several relations for modeling of radiation heat transfer and derived radiation intensity are presented. The heat source term, due to radiation in solid phase that appears in Eq. 26, can be calculated from Rosseland model.

$$q_r = -\frac{16}{3}\frac{\sigma\, T_s^{\,3}}{\beta} \nabla T_s, \tag{34}$$

where σ is Boltzmann constant and β is extinction coefficient.

Combustion model

Chemical mechanism for oxidation of methane fuel is considered chemical production rate:

$$\omega_i = \sum_{i=1}^{NR} (v''_{k,i} - v'_{k,i}) R_i, \tag{35}$$

and are stoichiometric coefficients. Combustion process includes ten equations and twelve species. These equations are presented in Table .1, which includes one-step reaction for methane fuel and equations 2-4, are Zeldovich mechanism for NO formation. Rate of reactions are computed with Arrhenius method. But for six other equations that reaction rate is very quick relative to last four equations hence, equilibrium reactions are considered. In order to consider effects of turbulence on combustion the common Eddy-Dissipation model can be used.

Number	Equation
1	$CH_4 + 2\,O_2 \rightarrow CO_2 + 2\,H_2O$
2	$O_2 + 2\,N_2 \rightarrow 2\,N + 2\,NO$
3	$2\,O_2 + N_2 \rightarrow 2\,O + 2\,NO$
4	$N_2 + 2\,OH \rightarrow 2\,H + 2\,NO$
5	$H_2 \leftrightarrows 2\,H$
6	$O_2 \leftrightarrows 2\,O$
7	$N_2 \leftrightarrows 2N$
8	$O + H \leftrightarrows OH$
9	$O_2 + 2\,H_2O \leftrightarrows 4\,OH$
10	$O_2 + 2\,CO \leftrightarrows 2\,CO_2$

Table 1. Kinetic and equilibrium reactions

Spalding suggested that combustion processes are best described by focusing attention on coherent bodies of gas, which squeezed and stretched during their travel through the flame. This model relates the local and instantaneous turbulent combustion rate to the fuel mass fraction and the characteristic time scale of turbulence. The application of this model requires adjustment of a specific coefficient and ignition time to match the experimental combustion rate with the computational combustion rate. In combustion chamber of engine, high turbulence intensity exists and hence combustion for such device lies in the flamelets-in-eddies regime. The intrinsic idea behind the model is that the rate of combustion is determined by the rate at which parcel of unburned gas are broken down into the smaller ones, such that there is sufficient interfacial area between the unburned mixture and hot gases to permit reaction and also the turbulence length scale which is quite important can determine the turbulent burning rates. In this case, the coefficients of model were determined from experimental analysis of conventional engine.

5. Foundation of reaction

5.1. Chemical equilibrium

A general chemical equilibrium reaction with $v'_{i,s}$ and $v''_{i,s}$ representing the stoichiometric coefficient of reaction and products for the chemical species M_i

$$\sum_{i=1}^{N} v'_{i,s} \, M_i \; \rightleftharpoons \; \sum_{i=1}^{N} v''_{i,s} \, M_i \, . \tag{36}$$

State of equilibrium can be interpreted as a situation, in which both the forward as well as reverse reactions progresses with identical speed.

$$K_p(T) = \prod_i \left(\frac{pi}{p0}\right)^{(v''_{i,s} - v'_{i,s})} . \tag{37}$$

The equilibrium constant K_p now contains the information about the equilibrium material composition in term of partial pressure p_i of the various species i . For more information interested reader can see (Mohammadi 2010).

5.2 Reaction kinetics

A one step chemical reaction of arbitrary complexity can be represented by the following stoichiometric equation:

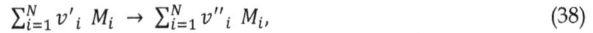

$$\sum_{i=1}^{N} v'_i \, M_i \; \rightarrow \; \sum_{i=1}^{N} v''_i \, M_i, \tag{38}$$

where v'_s are the stoichiometric coefficient of reactions and v''_i representing the stoichiometric coefficient of products, M_i the specification of molecule of ith chemical species, and N total number of component involved. Usually, they are represented with an Arrhenious formulation form:

$$K = AT^b \exp\left(-\frac{E_A}{RT}\right). \tag{39}$$

The constant A and the exponent b as well as the so-called activation energy E_A are summarized for many chemical reactions in extensive table.

6. The finite volume method

Customarily, CFD codes work with the finite volume method. This approach guaranties the numerical preservation of conservative quantities for the incompressible flows. The finite volume (FV) method uses the integral form of the conservation of equations. The solution domain is subdivided into a finite number of control volumes, and the conservation equations are applied to each control volume. As result, an algebraic equation for each CV is obtained. The FV method accommodates any type of grid, so it is suitable for complex geometries. However, the computational mesh ideally, be built hexahedratically. The conservation law for transport of a scalar in an unsteady flow has the general form:

$$\frac{\partial}{\partial t}(\rho \, \Phi) + \nabla.(\rho u \Phi) = \nabla.(\Gamma \nabla \Phi) + S_\Phi \, , \tag{40}$$

$(\rho u \Phi)$ designates convection, $(\Gamma \nabla \Phi)$ diffusion flows of and S_Φ the corresponding local source. For more information interested reader can see (Mohammadi 2010).

6.1. The finite volume equations formulation

Finite volume equations are derived by the integration of above differential equations over finite control volumes that taken together fully cover the entire domain of interest (Fig. 2). These control volumes are called "cells" P, for which the fluid-property value, are regarded as representative of the whole cell. It is surrounded by neighboring nodes which we shall denote by N, S, E, W, B and T. Cells and nodes for velocity components are "staggered" relative to those for all other variables.

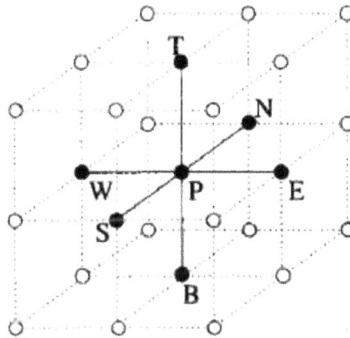

Figure 2. Computational molecule in 3D domain (Patankar, 2002)

For more information interested reader can see (Mohammadi, 2010).

6.2. Discretisation and numerical solution of the momentum equation

Finally, the momentum equation for the calculation of velocity and pressure by use of continuity equation should be considered. For numerical reasons, it is recommendable to resort to so called staggered grid, i.e. pressure and velocity are calculated on computational grids shifted to each other, the pressure for example in the cells and the velocity on the nodes. The calculations of velocity commonly take place iteratively, for which several algorithms are known (e.g. SIMPLE, PISO, SIMPLER…). In final analysis, all have the fact in common that is first step the momentum equation is solved for the velocities of momentums kept constant. In the second step, pressure corrections are then calculated with the help of a Poisson equation. For pressure with these pressure corrections, new velocities are then calculated again, and that again, until a pre-given break off threshold for the convergence is reached.

6.2.1. Discretisation of transient convection diffusion equation

Transient three dimensional convection diffusion of a general property Φ in a velocity field that govern by equation (40). The fully implicit discretisation equation is:

$$a_P\,\Phi_P = a_W\,\Phi_W + a_E\,\Phi_E + a_S\,\Phi_S + a_N\,\Phi_N + a_B\,\Phi_B + a_T\,\Phi_T + a^{\circ}_P\Phi^{\circ}_P + S_u \,, \tag{41}$$

where:

$$a_p = a_w + a_E + a_s + a_N + a_B + a_T + a^o{}_p + \Delta F - S_p \qquad (42)$$

with $a_p^o = \frac{\rho_p^o \Delta V}{\Delta}$ and $\bar{S}\Delta V = S_u + S_p \Psi_p$

For more information interested reader can see (Mohammadi, 2010).

7. Simple algorithm

As discussed in the preceding section, the governing equation for the flow may be solved in terms of derived variables, or in term of primitive variables consisting of the velocity components and the pressure.

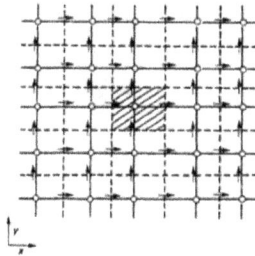

Figure 3. Staggered location for the velocity components in a two dimensional flow (Patankar, 1980)

However, in the advent of Simple (Semi Implicit Method for Pressure Linked Equations) algorithm, along with its revised version Simpler and the enhancement such as Simplec, the solution of the equations using primitive variable approach has become very attractive. In fact, Simple and Simple like algorithms are extremely popular for the solution of problems involving convective flow and transport. The basic approach involves the control volume formulation, with the staggered grid, as outlined in the proceeding section. This avoids the appearance of physically unrealistic wavy velocity fields in the solution to equations. The pressure at a chosen point is taken at arbitrary value and the pressures at other points are calculated as differences from the chosen pressure value.

Following (Patankar, 1980), if a guessed pressure field p^* is taken, the corresponding velocity field can be calculated from the discretised equations for the control volume shown in Fig. 4 These equations are of the form:

$$a_e u_e = \sum a_{nb} u_{nb}^* + b + \left(p_p^* - p_E^*\right) A_e, \qquad (43)$$

where the asterisk on the velocity indicates the erroneous velocity field based on guessed pressure field. Here, a_{nb} is a coefficient that accounts for the combined convection-diffusion at the faces of the control volume, with nb referring to the neighbors e to the control volume, b includes the source terms except the pressure gradient, and A_e is the area on which pressure acts, being $\Delta y^* \Delta z$ for 3D. The numbers of neighbor terms are 6 for three dimensional ones. Similar equations can be written for $v_n{}^*$ and $w_t{}^*$, where t lies on the z-direction grid line between grid points P and T. if p is the correct pressure and p is the correct pressure and p' the pressure correction, we may write:

$$p = p^* + p', \; u = u^* + u', \; v = v^* + v', \; w = w^* + w', \tag{44}$$

where the prime indicate corrections needed to reach the correct values that satisfy the continuity equation. Omitting the correction terms due to the neighbors, an iterative solution may be developed to solve for the pressure and the velocity field. Then, the velocity correction formula becomes:

$$u_e = u_e^* + \frac{A_e}{a_e}(p_p' - p_E').$$

$$v_n = v_e^* + \frac{A_n}{a_e}(p_p' - p_N'), \tag{45}$$

And similarly for w_t. From the time dependent continuity equation, the pressure correction equation in then developed as:

$$a_P\,p'_P = a_E\,p'_e + a_W\,p'_W + a_N\,p'_N + a_S\,p'_S + a_T\,p'_T + a_B\,p'_B + b, \tag{46}$$

where b is a mass source which must be eliminated through pressure correction so that continuity is satisfied. Here, T and B are neighboring grid points on the z direction grid line.

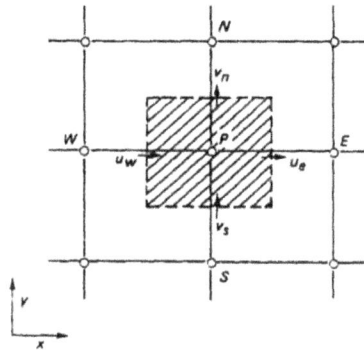

Figure 4. Control volume for driving the pressure correction equation (Patankar, 1980)

The simple algorithm has the following main steps:

1. Guess the pressure field p^*.
2. Solve the momentum equation to obtain $u^*, v^*,$ and w^*.
3. Solve the pressure correction equation to obtain p'.
4. Add p' to p^* to obtain the corrected pressure p.
5. Calculate u, v and w from u^*, v^* and w^* using velocity correction equations.
6. Treat the corrected pressure p as the new guess p^* and iterate the preceding procedure to convergence.

The revised version Simpler is quite similar to preceding algorithm and was developed mainly to improve the rate of convergence. In this case, the mail steps are:

1   Guess the velocity field
2. Solve the pressure equation, which is similar to pressure correction equation, Eq. 46, to obtain the pressure distribution. In this equation p' is replaced by p and a different expression arise for b.
3. Treating the pressure field as p^*, solve the momentum equations to obtain u^*, v^* and w^*.
4. Solve the pressure correction equation to obtain p'.
5. Correct the velocity field but not the pressure.
6. Use the velocity field as the guessed distribution and iterate the preceding procedure to convergence.

The pressure at any arbitrary point in the computational domain is specified and pressure differentials from this value are computed. The boundary condition may be a given pressure, which makes $p' = 0$, or a given normal velocity which makes the velocity a known quantity at the boundary and not a quality to be corrected so that p' at the boundary is not needed. For further details, (Patankar, 1980) may be consulted.

8. Multi-step reaction models

Models for premixed PM combustion are complicated by the highly nonlinear radiative exchange terms in the energy equation for the solid matrix in addition to the stiffness of the set of gas phase equations. Therefore researchers have simulated the gas phase reactions using single-step chemistry. However, few researchers had taken up this issue and presented multi-step reaction models. It was concluded that use multi-step kinetics is essential for accurate predictions of the temperature distributions, energy release rates, and emissions. Single-step kinetics was shown to be adequate for predicting all the flame characteristics except the emissions for the very lean conditions under which equilibrium favors the more complete combustion process dictated by global chemistry. full mechanism (49 species and 227 elemental reactions), skeletal mechanism (26 species and 77 elemental reactions), 4-step reduced mechanism 9 species and 1-step global mechanism.

In the open literature, there are few articles concerning the interaction between a fuel spray and a PM. The PM under study was of high porosity with uniformly distributed spheres and with uniform distribution of cavities with equal mean pore size in the porous medium. The interaction between droplets during evaporation was neglected. Physical properties of fuel such as latent heat of evaporation and healing value were constant. The temperature distribution inside of the fuel droplet was uniform, but time-varying. Laminar isobaric flow consits of air, fuel droplets, gaseous combustible mixture and hot products of combustion. After evaporation the fuel mixes immediately with air to form a homogeneous combustible mixture. Mass fraction of the liquid was negligible, and the gas was optically transparent. Also, radiative heat transfer between the skeleton surfaces of the porous medium can be considered.

9. Examples

9.1. Mesh preparation

Prior to CFD simulation, computational mesh of cylinder was generated with Kiva-Prep (preprocessor for mesh generation for KIVA-3V main code). The geometry of a mesh is composed of one block. Fig. 5 shows the grid configuration of porous tube, About 300000 grids were generated for computational studies.

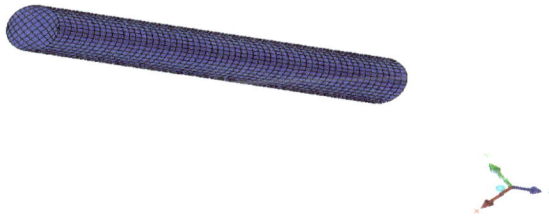

Figure 5. Computational mesh for the CFD calculation

9.2. Initial and boundary conditions

The test section was a vertical quartz glass-tube with 1.3 m in length and 0.076 m in diameter, that was isolated from the environment. The test section was filled with a packed bed of 0.0056 m solid alumina spheres. For simulation a cylinder with 0.076 m in diameter and 0.60 m in length that filled with PM, was considered (Fig. 5). The boundary condition applied to the momentum and energy equation with the assumption of zero gradients for temperature of both phase of PM and for species transport through the downstream boundary. At the upstream boundary, the gas temperature is 300K, composition is premixed methane-air with equivalence ratio 0.15, and velocity is 0.43 m/s of the premixed reactants and zero gradient for solid phase, were specified. For initial temperature for both phase of PM experimental measured data was used. Fuel is methane, porosity of PM is 0.4. The laminar flow considered for simulation. For validation of numerical simulation, modified KIVA code was used for simulation of unsteady combustion is a cylindrical tube with the experiments of Zhdanok et al. Fig. 6 plots a comparison of computation results to the experimental results of Zhdanok at the time of 147 s, which shows that the computed speed of combustion wave agrees well with the same condition of the experimental results. It is seen that methane is completely consumed in flame front that has maximum temperature.

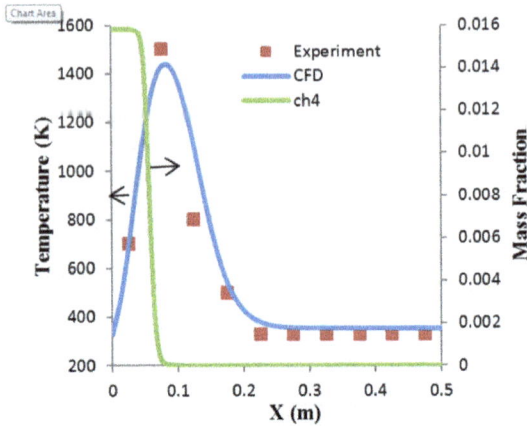

Figure 6. Comparison of combustion wave propagation between CFD and experiment in time 147 s

9.3. Discussion

In Figs. 7-10, contours of methane, carbon dioxide, temperature in gas and solid phases of PM, in cross section of tube for several times (10, 50, and 100 second) are shown. In Fig. 7 mass fraction of methane is shown. With entering of methane-air to porous tube that has high initial temperature, approximately 1800 K, in a narrow zone near inlet location, the temperature of gas increases until it reaches to self-ignition temperature. Methane consumes in a narrow region that is thicker relative to conventional flame front. In Figs. 8a, b, c, it is seen that after 10, 50 and 100 second combustion is started in respectively at x = 0.8, 2, 6 cm from entrance of mixture. The value of mass fraction of methane in PM tube is between 0.002 and this Fig indicates that the flame front has arc-shape.

Fig. 8 shows mass fraction of carbon dioxide in different time after start of simulation. Mixture flows through the porous tube that has initially heated and combustion in narrow zone of high temperature takes place. It is seen that reactions occur around the flame front in PM and its thickness is about 0.4 cm, which is very thick in compare with flame front in normal combustion. Carbon-dioxide disperses in pre heat region of entering mixture by diffusion of CO_2 and disperses in post flame by flow motion. In Fig. 9 temperature distribution in gas phase of PM in different time is shown. Flame front is recognizable from its high temperature region. Maximum fluid temperature is about 1600 K. Also, because effect of solid phase of PM, part of heat release of combustion is absorbed by it and prevents from high temperature gradient in fluid. Energy is re-circulated to the unburned gas mixture through the heat combustion and radiation of the solid. Fig. 10 shows temperature distribution in solid phase. At initial condition, mixture temperature is 300 K. The inlet temperature of solid phase is higher than gas temperature and heat is transferred from solid to gas, so allow it to reach the ignition temperature. Maximum temperature in solid phase is about 1600 K. Then the gas delivers its energy to the solid. Also, due to high heat capacity of solid phase of PM, low-temperature gradient occurs in it.

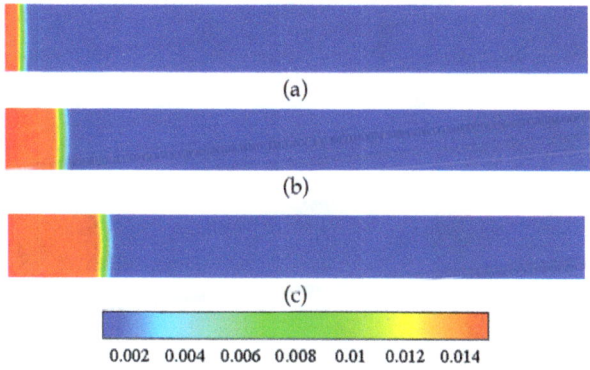

Figure 7. Mass fraction of methane in cross section x = 0 after a) t = 10 s b) t = 50 s c) t = 100 s

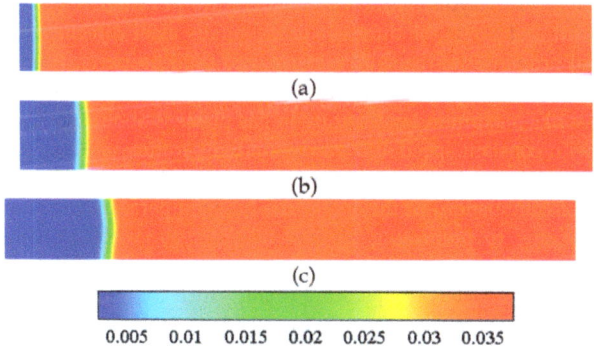

Figure 8. Mass fraction of Carbon-dioxide in cross section x = 0 after a) t = 10 s b) t = 50 s c) t = 100 s

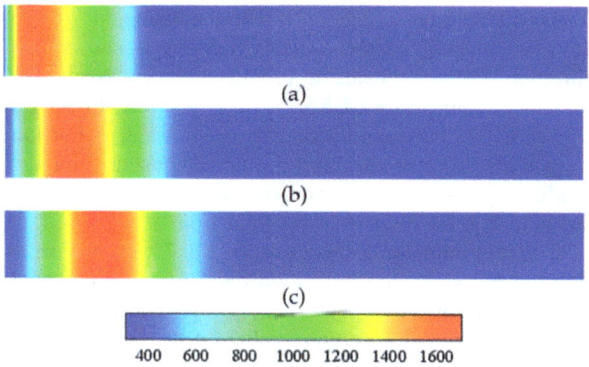

Figure 9. Gas phase temperature in cross section x = 0 after a) t = 10 s b) t = 50 s c) t = 100 s

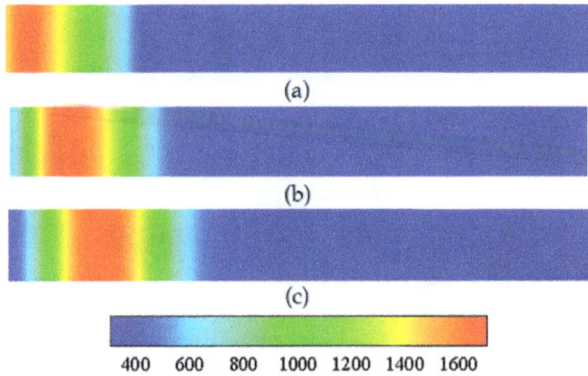

Figure 10. Temperature of solid phase in cross section x = 0 after a) t = 10 s b) t = 50s c) t= 100 s

Figs. 11, 12 show results for distribution of temperature in center line of PM tube in both phase of PM (solid and fluid) for different times versus axial direction. Heat transport is related to thermal properties of the solid material and fluid property. Flame core transports heat to incoming methane-air mixture with conduction and radiation, that its temperature is 300 K. At time t = 10 s maximum temperature in gas phase is about 1700 K. After 10 s flame moves to right with constant speed and maximum temperature of about 1600 K. After this time equilibrium between conduction, convection and radiation, causes to no change in maximum value of combustion. At the end of the tube temperature in all cases is about 325 K. In solid phase due to preheating of inlet mixture, at t = 10 s in inlet solid phase temperature is about 1450 K, in t = 50 s and t = 100 s, upstream temperature is about 850 K and 580 K, respectively and this temperature finally reach to 300 K approximately. Fig. 13 shows distribution of methane mass fraction from 10 to 100 s. Inlet mass fraction of methane is 0.016. Decrease in mass fraction of methane shows location of flame front. Flame location after 10 s is in location 3.8 cm and its thickness is thickness of 4 cm, after 50 s, is about 1.6 cm with the thickness 4.3 cm, and after 100 s is about 3.3 cm with the thickness 2.4 cm. From this Fig. inferred in 50 s and 100 s after simulation variation in flame thickness is very low value and the CH_4 is almost completely consumed in this zone. Fig. 14 shows mass fraction of CO_2 value in axial direction. Carbon dioxide is produced during combustion and its mass fraction reaches to highest value in x = 2.4, 5.5, 7.1 cm respectively to 0.036, 0.035, 0.024 mass fraction after 10, 50 and 100 s from simulation. Fig. 15 shows mass fraction of CO value in axial direction. Carbon monoxide is produced during combustion with entering of mixture and as an intermediate species produced and consumes gradually in axial direction. CO concentration reaches its highest value near the flame front at x = 1.8, 4.1, 6 cm respectively after 10, 50 and 100 s and gradually decreases at x = 4.8, 7.2 and 9.1 cm to approximately zero. With completeness of combustion and it oxidizes slowly and converted to CO_2, because of enough accessible oxygen for converting CO to CO_2. Fig. 16 shows mass fraction of CH_3 value in axial direction. Methyl concentration reaches its highest value in x = 1.4, 3.6, 9.2 cm respectively 10, 50 and 100 s after simulation. But after 100s due to fluid flow, it disperses in porous tube and oscillation occurs in value of it.

10. Two applications of PM technology

10.1. Internal combustion engines

The major target for further development of the current IC engines is to reduce their harmful emissions to environment. The most important difficulty with existing IC engines that currently exists is non-homogeneity of mixture formation within the combustion chamber which is the cause heterogeneous heat release and high temperature gradient in combustion chamber which is the main source of excess emissions such as NOx, unburned hydrocarbons (HC), carbon monoxide (CO), soot and suspended particles. At present, the IC engine exhaust gas emission could be reduced by catalyst, but these are costly, sensitive to fuel and with low efficiency. Another strategy has been initiated to avoid the temperature gradient in IC engines that is using homogeneous charge compression ignition (HCCI)

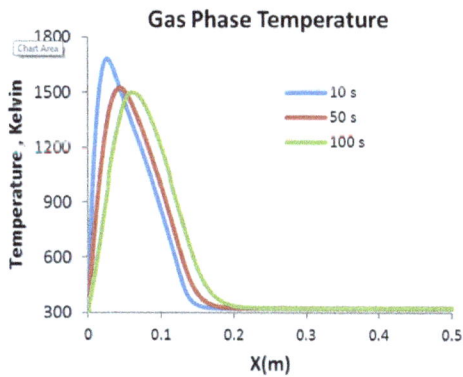

Figure 11. Mean temperature distribution in gas phase of PM versus axial direction

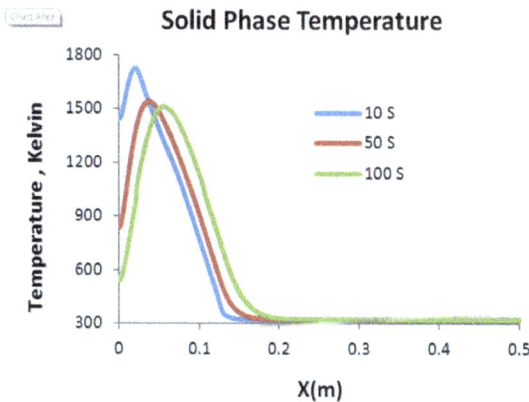

Figure 12. Mean mass fraction of methane versus axial direction

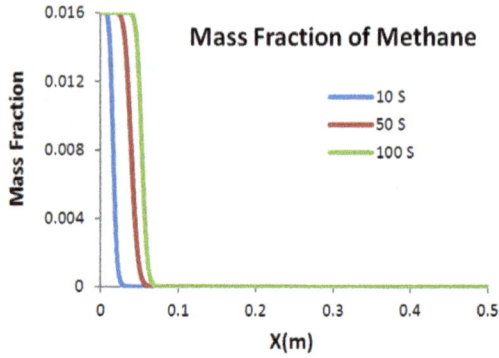

Figure 13. Mass fraction of methane versus axial direction

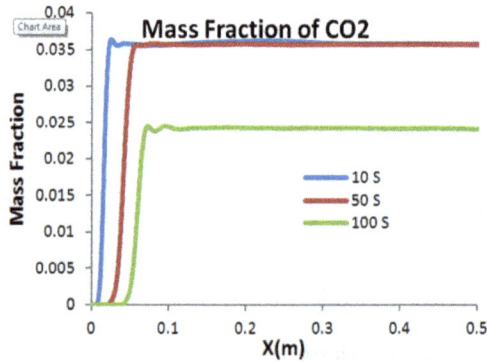

Figure 14. Mass fraction of Carbone dioxide versus axial direction

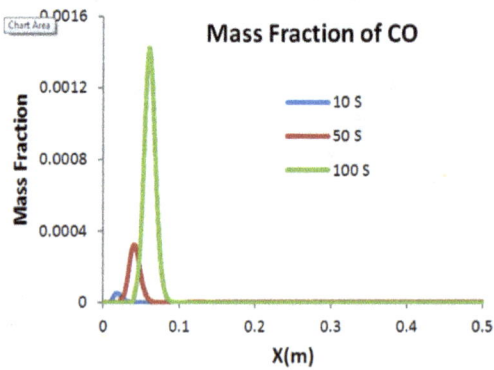

Figure 15. Mass fraction of CO versus axial direction

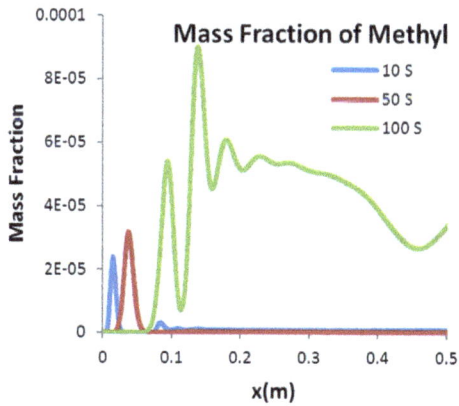

Figure 16. Mass fraction of CH₃ value versus axial direction

engines but there still exist some challenges including, higher HC and CO emissions and control of ignition time and rate of heat release under variable engine operating condition. In such engines, lean mixture with high amount of exhaust gas recycled (EGR), could be used that increases amount of CO and soot. Also by increasing the load of HCCI engine, NO_x formation and fuel consumption increases. It means that low compression ratio should be used for these engines, while in reality, the compression ratio must be high enough that temperature near the end of compression process, lead to self-ignition of mixture with reasonable time delay. Therefore, these engines are suitable for low and medium loads and it better for high loads the engine can change mode of operation to compression ignition. Also direct fuel injection engines generally have some unresolved problems due to lack of homogeneity of mixture formation and combustion. Several other technologies have been used to reduce emissions in engines, such as electronically controlled high pressure fuel injection systems, variable valve timing, EGR but still in these methods still could not solve the problem completely under all engine operating conditions. Could there be any other homogeneous combustion in IC engines to meet all operational conditions (various load and speed)? The demand target may be possible with homogeneous mixture formation and a 3D-ignition of a homogeneous charge to prevent formation of flame front that lead to temperature gradient in the entire combustion chamber which is ensuring a homogeneous temperature field. In conventional direct injection engines mechanisms also there is a lack of homogenization of combustion process. PM-engine is defined as an engine with homogeneous combustion process. The following distinct process of PM-engine is realized in PM volume: energy recirculation in cycle, fuel injection in PM, fuel vaporization for liquid fuels, perfect mixing with air, homogeneity of charge, 3D-thermal self-ignition, and homogeneous combustion. PM-engine may be classified as heat recuperation timing in an engine as: Engine with periodic contact between PM and cylinder is called closed PM-chamber and Engine with permanent contact between PM and cylinder which is called open

PM-chamber. In this paper an open PM-chamber is studied. Permanent contact between working gas and PM-volume is shown schematically in Fig. 17. The PM is placed in cylinder head. During the intake stroke there is a not much influence from PM-heat capacitor with in-cylinder air thermodynamic conditions. Also during early stages of compression stroke only a small amount of air is in contact with hot porous medium. The heat transfer process (non-isentropic compression) increases during compression, and at TDC the air penetration is cut to the PM volume. At final stages of compression stroke the fuel is injected into PM volume and with liquid fuels rapid fuel vaporization and mixing with air occurs in 3D-structure of PM-volume. A 3D-thermal self-ignition in PM-volume together with a volumetric combustion is characterized by a homogeneous temperature distribution. Therefore, all essential conditions exist for having homogeneous combustion in the PM engine. The initial idea to use PM in IC engines was proposed by Weclas. Their investigation was performed in a single-cylinder air-cooled PM Diesel engine without any catalyst. A Silicon Carbide (SiC) PM was mounted in the cylinder head between the intake and exhaust valves and fuel was injected through the PM volume. The implementation has improved engine thermal efficiency, reduced emissions and noise in comparison to the base engine. The mean cylinder temperature was about 2200 K for base engine without having any PM. The temperature reduces to about 1500 K when PM is used which is significantly even lower during combustion.

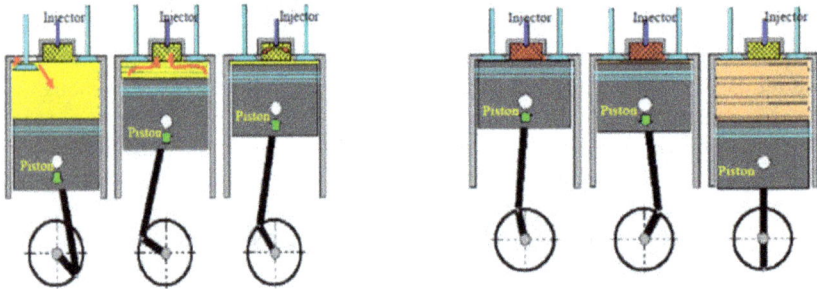

Figure 17. A permanent contact PM-engine in operation (Weclas, 2001)

The effect of SiC PM as a regenerator was simulated by Park and Kaviany. In their study a PM disk like shape was connected to a rod and was moving near piston within the cylinder of diesel engine. A two-zone thermodynamic model with single-step reaction for methane-air combustion is carried out. It is shown that the maximum cylinder pressure during combustion increases and more work is done during a full cycle, also engine efficiency increases, but due to high temperature of PM that its temperature is higher than adiabatic flame temperature of methane-air, the production of NO_x is rather higher while as its soot decreases. Macek and Polasek simulated and studied a PM engine with methane and hydrogen respectively and its potential for practical application was shown. Weclas and Faltermeier investigated penetration of liquid-fuel injection into a PM (as arrangement of cylinders which were mounted on a flat plate with different diameters). The arrangement was changed to obtain optimum geometry which produces the best mixture formation.

Porous regenerator as shown in Fig. 18 has the potential to improve fuel–air mixing and combustion. The porous insert is attached to a rod and moves in the cylinder, synchronized, but out of phase with the piston. During the regenerative heating stroke, the regenerator remains just beneath the cylinder head for most of the period and moves down to the piston (as it approaches the TDC position). During the regenerative cooling stroke, the regenerator moves up and remains in the original position until the next regenerative heating stroke. Following the combustion and expansion, the products of combustion (exhaust gases) retain an appreciable sensible heat. During the regenerative cooling stroke, the hot exhaust gas flows through the insert and stores part of this sensible heat by surface-convection heat transfer in the porous insert (with large surface area). For the proposed engine, a thermal efficiency of 53% was claimed, compared to 43% of the conventional Diesel engines. Macek and Polasek presented a finite volume based simulation of porous medium combustion for reducing emissions from reciprocating internal combustion engines.

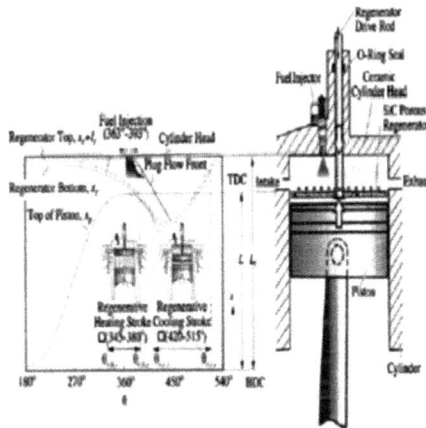

Figure 18. Sequence of motion of the regenerator and physical of fuel injection and air blowing during the regenerative heating stroke (Park and Kaviany, 2002)

The application of a highly porous open cell structures to internal combustion engines for supporting mixture formation and combustion processes was introduced by Weclas. Novel concepts for internal combustion engines based on the application of PMC technology were presented and discussed. His study proved that gas flow, fuel injection and its spatial distribution, vaporization, mixture homogenization; ignition and combustion could be controlled or positively influenced with the use of porous media reactors. The key features of the highly porous medium for supporting the mixture formation, ignition and combustion in IC engines are illustrated in Fig. 19. A study on the use of PMC in direct injection (diesel or gasoline) IC engines was performed by Durst and Weclas. Polasek and Macek presented the simulation of properties of IC engine equipped with a PM to homogenize and stabilize the combustion of CI engines. The purpose of the PIM matrix use was to ensure reliable ignition of lean mixture and to limit maximum in-cylinder temperature during combustion.

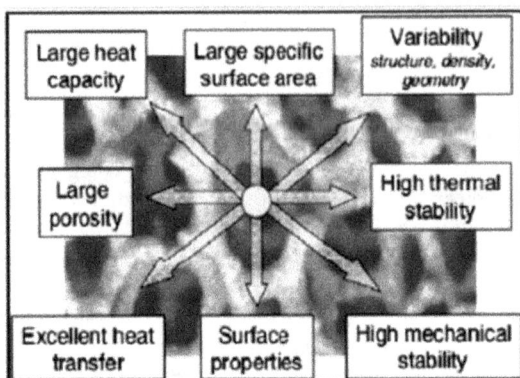

Figure 19. Main feature of porous structure to be utilized to support engine process (Durst and Weclas, 2001)

10.2. Gas turbines and propulsion

A porous burner with matrix stabilized combustion for gas turbine applications. A numerical model for analyzing the evaporation processes in PM for gas turbine applications had been developed. Evaporation of a point wise-injected kerosene spray in a carbon-carbon porous medium was considered. The effects of porous medium temperature, fuel flow rate, air inlet temperature and porous medium geometry on the evaporation of spray can be analyzed. Evaporation characteristics were not found to vary much with porous medium geometry, as the porous medium was modeled as a momentum sink. But thermal effects of PM were found to be more dominant. The characteristics of combustion within porous media which are attractive in a propulsion context are the ability to burn leaner and hotter than a free flame with low emissions, there no cooling requirement for the combustor itself and the potential to operate free from combustion-induced noise. The performance of a PM combustor is applicable for gas turbines, at elevated pressures and inlet temperatures. The combustor was formed of reticulated porous ceramics, untreated to augment or sustain chemical reaction. The results showed that the combustor could operate in a "super-adiabatic" mode, with low emissions.

Author details

Masoud Ziabasharhagh and Arash Mohammadi
K. N. Toosi University of Technology, Iran

11. References

Stanglmaier, R. H., Roberts, C. E., (1999). Homogeneous charge compression ignition: benefits, compromises and future engine applications, *SAE Paper*, 1999-01-3682.

Weclas, M., (2001). Potential of porous medium combustion technology as applied to internal combustion engine, *MECA/AECC*, Nurnberg, Germany.

Trim, D., Durst, F., (1996). Combustion in porous medium – advances and application, *Combustion Sci. and Tech.*, Vol 121, pp. 153-168.

Howell, J. R., Hall, M. J., Ellzey, J. L., (1996). Combustion of hydrocarbon fuels within porous inert media, *Progress in Energy and Combustion Science*, Vol. 22, Issue. 2, pp. 121–145.

Kamal, M. M., Mohamad, A. A., (2006). Combustion in porous media, a review, *Journal of Power and Energy*, Vol. 5,pp. 487–508.

Innocentini, M. D. M., Tanabe, E. H., Aguiar, M. L., Courly, J. R., (2012). Filteration of gases in high pressure : Permeation behavior of fiber-based media used for natural gas cleaning, *J. Chemical Engineering Science*, Vol. 74, pp. 38-48.

Monmont, F. B. J., Van-Odyck, D. E. A., Nikiforakis, N. (2012). Experimental and theoretical of the combustion of n-tridecane in porous media, *J. Fuel*, Vol. 93, pp. 28-36.

Loukou, A., Frenzel, I., Klein, J., Trimis, D., (2012). Experimental study of hydrogen production and soot particulate matter emissions from methane rich-combustion in inert porous media, *Int. J. Hydrogen Energy*, Article in press.

Wood, S., Harris, A. T., (2008). Porous burners for lean-burn applications. *Progress in Energy and Combustion Science*, Vol. 34, pp. 667-684.

Mujeebu, A. A., Abdullah, M. Z., Mohammad, A. A., Bakar, M. Z. A., (2010). Trend in Modeling of Porous Media Combustion, *J. Progress in Energy and Combustion Science*, Vol. 2, pp. 1-24.

Kamal, M. M., Mohamad, A. A., (2006). Development of a cylindrical porous-medium burner. *Journal of Porous Media*, Vol. 9, Issue. 5, pp. 469–481.

Kamal, M. M., Mohamad, A. A., (2007). Investigation of liquid fuel combustion in a cross flow burner. Proceedings of the Institution of Mechanical Engineers: Part A – *Journal of Power and Energy*, 221, pp. 371–385.

Rortveit, G. J., Stromman, A. H., Ditaranto, M., Hustad, J. E., (2005). Emissions from combustion of H2 and CH4 mixtures in catalytic burners for small-scale heat and power applications, *Clean Air: International Journal on Energy for a Clean Environment*, Vol. 6, Issue. 2, pp. 187–194.

Contarin, F., Saveliev, A.V., Fridman, A. A., Kennedy, L. A., (2002). A reciprocal flow filtration combustor with embedded heat exchangers: numerical study, *International Journal of Heat and Mass Transfer*, Vol. 46, pp. 949–961.

Bingue, J. P., Saveliev, A.V., Fridman, A. A., Kennedy, L. A., (2002). Hydrogen production in ultra-rich filtration combustion of methane and hydrogen sulfide. *International Journal of Hydrogen Energy*, Vol. 27, Issue. 6, pp. 643–649.

Bingue, J. P., Saveliev, A.V., Kennedy, L. A., (2004). Optimization of hydrogen production by filtration combustion of methane by oxygen enrichment and depletion. *International Journal of Hydrogen Energy*, Vol. 29, pp. 1365–1370.

Slimane, R. B., Lau, F. S., Khinkis, M., Bingue, J. P., Saveliev, A. V., Kennedy, L. A., (2004). Conversion of hydrogen sulfide to hydrogen by super adiabatic partial oxidation:

thermodynamic consideration, *International Journal of Hydrogen Energy*, Vol. 29, pp. 1471–1477.

Raviraj, S. D., Ellzey, J. L., (2006). Numerical and experimental study of the conversion of methane to hydrogen in a porous media reactor. *Combustion and Flame*, 144, pp. 698–709.

Hendricks, T. J., Howell, J. R., (1994). Absorption/scattering coefficients and scattering phase functions in reticulated porous ceramics. Radiative Heat Transfer: Current Research ASME HTD-276, pp. 105–113.

Barra, A. J., Diepvens, G., Ellzey, J. L., Henneke, M. R., (2003). Numerical study of the effects of material properties on flame stabilization in a porous burner. *Combustion and Flame*, 134, pp. 369–379.

M. Weclas, *porous media in internal combustion engine*, M. Scheffle, P. Colombo, editors, Cellular ceramics-structure , manufacturing, properties and application, Wiley, 2005.

Durst, F., Weclas, M., (2001). A new type of internal combustion engine based on the porous medium technology, *Proc Inst Mech Eng*, Vol. 215, pp. 63-81.

Park, C. W., Kaviany, M., (2002). Evaporating combustion affected by in cylinder reciprocating porous regenerator, *ASME J. Heat Transfer*, Vol. 124, pp. 184-194.

Polasek, M., Macek, J., (2003). Homogenization of combustion in cylinder of CI engine using porous medium, *SAE* Paper, 2003-01-1085.

Mohammadi, A., (2010). *Numerical Simulation of Spark Ignition Engines, Numerical Simulations – Examples and Applications in Computational Fluid Dynamics*, Lutz Angermann (Ed.), ISBN: 978-953-307-153-4, InTech, Austria.

Amsden., A. A., O'Rourke, P. J., and Butler, T. D., (1989). KIVA-II: A Computer Program for Chemically Reactive Flows with Sprays, Los Alamos National Laboratory Report LA-11560-MS, Los Alamos.

Amsden., A. A., (1997). KIVA-3V: A Block-Structured KIVA Program for Engines with Vertical or Canted Valves, Los Alamos National Laboratory Report LA-13313-MS, Los Alamos.

Wakao, N. and Kaguei, S., (1982). *Heat and Mass Transfer in Packed Beds*, Gordon and Breach Science Publications, New Yourk, USA.

S. R. Turn, (1996). *An introduction to combustion*, McGra-Hill, New York, pp. 399-400.

Modest, M., F., (2003). *Radiative Heat Transfer*, Academic Press, California, USA.

Zhdanok, S., (1995). Super Adiabatic Combustion of methane air mixture under filtration in a Packed Bed, *Combustion and Flame*, Vol. 100, pp. 221-231.

Henneke, M. R., Jellzey, J. L., Modeling of Filtration Combustion in a Packed Bed, *Combustion and Flame*, Vol. 117, pp. 832-840.

Mujeebu, M. A., Abdullah, M. Z., Abu Bakar, M. Z., Mohamad, A. A., Muhad, R. M. N., Abdullah, M. K., (2009). Combustion in porous media and its applications – A comprehensive survey, *Journal of Environmental Management*, Vol. 90, pp. 2287–2312.

Patankar. S. V. (1980). *Numerical Heat Transfer and Fluid flow*, McGraw-Hill, ISNB 0-07-048740-5, USA.

Chapter 8

Numerical Simulation of Slab Broadening in Continuous Casting of Steel

Jian-Xun Fu and Weng-Sing Hwang

Additional information is available at the end of the chapter

1. Introduction

Broadening is the deformation of materials in the vertical direction of force in the steel rolling process. Slab broadening in continuous casting increases the slab width in the secondary cooling zone. Continuous casting is a process in which the temperature drops sharply. The drop in temperature leads to slab shrinkage; the linear shrinkage of carbon steel is about 2.5% in the width direction. The decrease in slab width from the initial shell to the cooling slab is considered to be almost negligible and the width may even increase under some conditions.

Although the slab shrinks in the secondary cooling zone, the width of the slab is sometimes greater than that of the entrance of the corresponding mold. The change in slab width is due to broadening being greater than shrinkage. It is rarely well-know that this phenomenon often occurs for slab in continuous casting. Slab broadening makes it difficult to accurately control the size of the slab and has adverse effects on the subsequent rolling processes. Slab broadening becomes increasingly obvious with increasing casting speed. If no vertical miller is used in the rolling process, the broad part of the slab is cut off, wasting material. With a vertical miller, the broad part of the slab is rolled in the width direction, which leads to fluctuation in the slab thickness. The study of slab broadening in the continuous casting process is thus necessary. The present work (FU JianXun et al. 2010(a-c),2011(a-b)) investigates slab broadening in continuous casting using mathematical simulation, industrial measurements, and experiments. Assessments of slab width in several continuous casting factories indicate that slab broadening is common in the continuous casting process. Slab broadening occurs in the secondary cooling zone, as confirmed by experiments. The effects of the productive factors on the slab broadening were also derived. The mechanism of slab broadening is investigated and discussed.

2. Research method

2.1. Index definition

In order to measure slab broadening, the ratio of apparent shrinkage (RAS) and the ratio of ultimate broadening (RUB) were defined respectively as:

$$RAS = \left(U / W - 1 \right) \times 100\%, \tag{1}$$

$$RUB = \left(S / T - 1 \right) \times 100\%, \tag{2}$$

where:
U is size of mold on the top entrance (mm);
W is the measured width of the slab (mm);
T is the ultimate width of the slab (in mm);
S is the width of the slab (in mm).

The value of the RAS, which denotes the degree of mold shrinkage, is positive when the top width exceeds the slab width. This index can be used to set the mold size. The value of the RUB, which denotes the degree of broadening, is positive when the slab width exceeds the ultimate width.

2.2. Online measurements

An online measurement system was designed to measure slab broadening at the exit of the caster.(FU JianXun et al. 2011(b)) The system comprises an optical lens, a digital camera, a data cable (IEEE 1394), and a computer. The system is controlled using the computer. The digital camera is used to take infrared images of the hot slab. Data is read from the camera at a preset frequency.

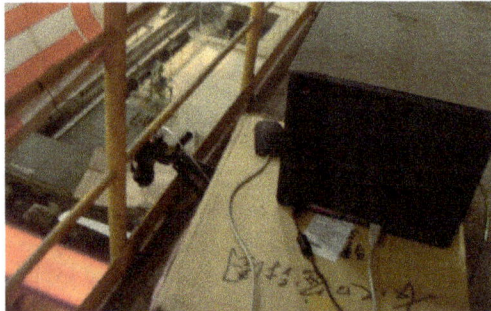

Figure 1. Online width measurement system for hot slabs.(FU JianXun et al. 2011(b))

Then, the graphics module creates images of the hot slab according to color aberration, and saves the images as files. The accuracy of the online system is 1 mm. A photograph of the system is shown in Figure 1.

2.3. RAS and RUB investigation

Table 1 lists the RAS and RUB values obtained for five companies. The data in the right column is from a handbook (XIONG Yi-gang, 1994). The RAS values range from 0.47% to 2.16%, with most being higher than the handbook data. The linear shrinkage of carbon steel from the initial shell to the cooled slab is 2.5%. The RUB values range from 0.34% to 2.03%, which is the result that the linear shrinkage subtracting the RAS. In general, the width of a slab is smaller than the top width of a mold. Broadening may overcome shrinkage under certain operating conditions for some particular grades of steel.

The width of a cooled slab is larger than the ultimate width when broadening exists. To obtain a slab with the desired dimensions, the top and bottom widths of the mold must be reset. Therefore, the measured width (W) could replace the ultimate width (T). The RAS of a slab can thus be set by changing the values of the top and bottom widths of the mold. Then compare these values with the linear shrinkage of each steel grade, and it could consequently be found out whether the slab broadening exists and the approximate range of it could also be derived.

	CompanyA	Company B	Company C	Company D	Company E	Handbook
RAS (%)	0.47~0.54	1.81	1.93~2.16	1.70~1.90	1.10~2.11	2.1
RUB (%)	1.96~2.03	0.69	0.34~0.57	0.60~0.80	0.39~1.40	0.4

Table 1. RAS and RUB values at various manufactures (FU JianXun et al. 2010(a))

The data show that slab broadening is common in continuous casting. The slab width is the result of shrinkage and broadening. The linear shrinkage of a carbon steel is about 2.5%, which is slightly larger than the slab broadening.

2.4. Mechanics calculations (FU JianXun et al.2011(a))

In the secondary cooling zone, the slab has to release sensible heat and latent heat to avoid complete solidification and to maintain the surface temperature according to the technical requirements of the metallurgy process. In this zone, the stress and strain of the slab are the result of mechanical action and thermal effects (S. Kobayashi et al,1988). Some parts of the slab may have a low temperature, which causes thermal stress in the secondary cooling zone. The thermal stress of the slab in the secondary cooling zone is small enough to be ignored compared to the stress caused by the bulging and the roller disalignment. Thus, mechanical stresses, includes the bending stress, straightening stress, roller-misalignment stress, the stress of rollers acting on the slab, and the static pressure of molten steel, determine the degree of slab broadening.

The bulging stress of a slab is defined as (Sheng Y et al ,1993) :

$$\delta = \frac{pl^4}{32E_x S^3} + \frac{pl^4}{32E_x S^3}(1\mathrm{sqrt(t)}), \tag{3}$$

where E_x is the equivalent elastic modular ratio, expressed as:(Sheng Y et al ,1993)

$$E_x = \frac{T_{So} - T_m}{T_{so} - 100} \times 10^6,$$

(4)

The stress of bending and straightening is expressed as:(Lei I I et al ,2007)

$$\varepsilon_i = \left(\frac{D_s}{2} - S_i\right)\left(\frac{1}{Ri - \frac{D_s}{2}} - \frac{1}{R_{i+1} - \frac{D_s}{2}}\right) \times 100\%,$$

(5)

The stress of disalignment is expressed as: (Chen J,1990)

$$\varepsilon_q = \frac{300 s_i \delta}{l^2} \times 100\%,$$

(6)

The values of these stresses calculated for the Q235 slab are listed in Tables 2 and 3. The calculations were based on the parameters of the continuous casters of Maanshan Iron and Steel Co. Ltd. A casting speed of 0.0167 m·s-1 was used. The bending zone of the continuous caster is at the 10th~15th rollers of the 2nd segments, and the straightening zone is at the 60th~65th rollers of the 9th segments. A negative stress indicates that a pushing stress acts on the contact surface between the slab and rollers whereas a positive stress indicates a tensile stress acting on the contact surface.

Roller ID	Slab shell thickness (10^{-2} m)	Casting speed (m·s⁻¹)	Bulging (10^{-2} m)	Bulging strain(%)	Disalignment strain(%)	Bending strain(%)
10	4.10	0.0167	0.101	0.115	0.0128	-0.0011
11	4.28	0.0167	0.097	0.115	0.0134	-0.0011
12	4.45	0.0167	0.093	0.115	0.0139	-0.0012
13	4.62	0.0167	0.090	0.115	0.0144	-0.0013
14	4.78	0.0167	0.087	0.115	0.0149	-0.0011
15	4.93	0.0167	0.084	0.115	0.0154	-0.0010

Table 2. Comparison of slab stresses in the bending zone(FU JianXun et al.2011(a))

Roller ID	Slab shell thickness (10^{-2} m)	Casting speed (m·s⁻¹)	Bulging (10^{-2} m)	Bulging strain(%)	Disalignment strain(%)	Bending strain (%)
60	10.4	0.0167	0.14	0.196	0.0160	0.0160
61	10.5	0.0167	0.09	0.152	0.0197	0.0197
62	10.6	0.0167	0.08	0.149	0.0198	0.0198
63	10.7	0.0167	0.08	0.147	0.0200	0.0200

Roller ID	Slab shell thickness (10^{-2} m)	Casting speed (m·s^{-1})	Bulging (10^{-2} m)	Bulging strain(%)	Disalignment strain(%)	Bending strain (%)
64	10.8	0.0167	0.08	0.145	0.0202	0.0202
65	10.9	0.0167	0.08	0.142	0.0203	0.0203

Table 3. Comparison of slab stresses in the straightening zone(FU JianXun et al.2011(a))

Tables 2 and 3 reveal that the bending, straightening, and disalignment stresses are far lower than the stress of bulging. Therefore, the stresses of bending, straightening, and disalignment do not cause the broadening of a slab.

3. Model and parameters

3.1. Finite element model

Building a satisfactory three-dimensional (3D) finite element model for the numerical simulation of continuous casting in the secondary cooling zone is quite complex. Thus, to simplify the problem, the following assumptions are made, as in our previous work((FU JianXun et al. 2010(b-c); 2011(b)):

1. The bending and straightening effects of the slab are ignored, and the slab is considered to be a linear object.
2. In the simulations, time, space, the characteristics of steel, and the temperature field in the slab are continuous, and the effects of the initial mechanical conditions of the slab on the deformation are ignored. The continuous caster in the secondary cooling zone is divided into several stages.
3. Because of symmetry, 1/4 of the slab and rollers on one side is used for the calculation.
4. The slab is deformable, the rollers are stiff, and the gap between rollers is variable. The calculation boundaries are placed at the rollers.

Based on these assumptions, the thermal-mechanical coupled model of the whole secondary cooling zone is divided into 6 independent sub-models for calculation. The 15 segments of the secondary cooling zone are divided into 6 groups. The first 5 groups each contain 2 segments; the remaining 5 segments make up the last group as a completely solidified slab. A 2-m slab is used for the simulation. The slab goes through the roll arrangement at a given speed. The simulation is performed continuously from the first group to the last group, and the results of a group of rollers are taken as the initial inputs for the subsequent group.

Eight-node isoparametric elements are used for the geometric discretization of the computational domain in the model. The slab comprises 4500 elements and 5250 nodes. Figure 2 shows the finite element models of the rollers and the slab in the caster. Figure 3(a) shows the rollers and the slab in the 3rd independent sub-model. Figure 3(b) shows the 6th independent sub-model.

Due to the symmetry of the slab in the width direction, one half of the slab was simulated. The grid units at the start plane of the slab move forward at a given speed. The static pressure of molten steel is taken as a mechanical boundary condition. The boundary is applied to the solidifying front of the slab, which is defined as the position with zero-strength temperature (ZST). Considering the effects of solidification-induced segregation and solid fraction (fs), the temperature of the units is the ZST where fs is equal to 0.8, and the units are considered a solidified shell where $fs \geq 0.0$. T_{80} denotes the temperature at the boundary between the solid phase and the liquid phase (T_{80}=ZST). Static pressure acts on the units where the temperature is higher than T_{80}. The boundary conditions of heat transfer and contact are also applied to the model.

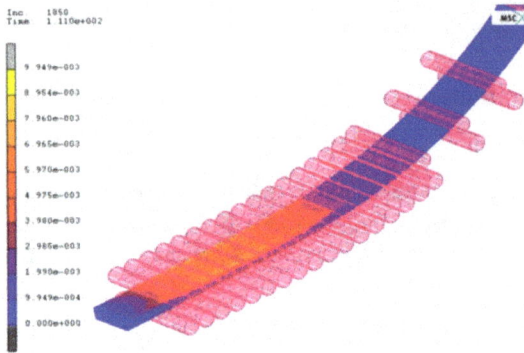

Figure 2. Finite element model of all the rollers and the slab (FU JianXun et al. 2011(b))

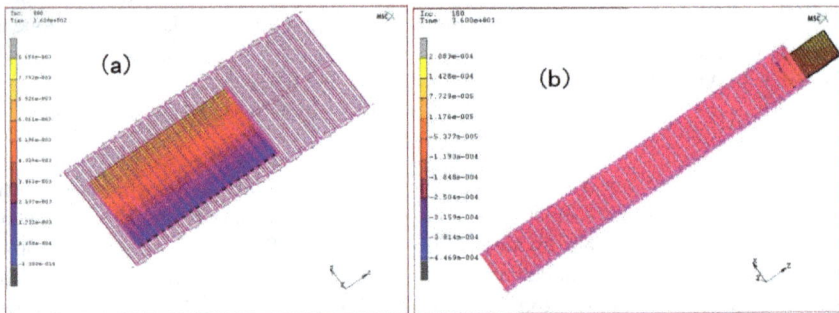

Figure 3. Finite element models of (a) the third group of rollers and the slab (b)the sixth group of rollers(FU JianXun et al. 2010(c))

3.2. Constitutive equations

The key factors that determine the accuracy of a model for analyzing the stress in a slab are included in the constitutive equation of the slab. These factors are heat transfer, mechanical load, stress relaxation, and plastic strain, all of which are time-dependent. The constitutive equation of steel at high temperature, which determines the accuracy of numeric simulations, is expressed as:

$$\varepsilon_{ij} = \varepsilon_{ij}^{e} + \varepsilon_{ij}^{ie} + \varepsilon_{ij}^{T}. \tag{7}$$

where:

ε_{ij} is the total strain;
ε_{ij}^{e} is the elastic strain;
ε_{ij}^{ie} is the non-elastic strain;
ε_{ij}^{T} is the thermal strain, and ij is the strain tensor.

The non-elastic strain ε_{ij}^{ie} is composed of time-independent inelastic strain and time-dependent creep deformation. A viscoelastic-plastic model is used to describe the solidifying behavior of the slab under the conditions of continuous casting, which is expressed as: (Chen J,1990)

$$\dot{\varepsilon}_{ij}^{ie} = f(\sigma_{ij}, T, \varepsilon_{ij}^{ie}). \tag{8}$$

where: $\dot{\varepsilon}_{ij}^{ie}$ is the non-elastic strain ratio; σ_{ij} is the stress; and T is the temperature.

This equation indicates that the non-elastic strain ratio is a function of the stress, temperature, and non-elastic strain.

The constitutive equation of time-dependent plastic deformation is used to describe the stress of carbon steel under various temperatures and strain ratios; it is expressed as: (S. Kobayashi et al.,1988)

$$\begin{cases} \dot{\bar{\varepsilon}}_P = A\exp(-Q/RT)[\sinh(\beta K)]^{1/m}, \\ \bar{\sigma} = K\cdot(\bar{\varepsilon}_p)^n, \end{cases} \tag{9}$$

where:

$\dot{\bar{\varepsilon}}_P$ is the equivalent plastic strain ratio;
R is the gas constant;
Q is the activation energy of deformation;
σ is the equivalent stress;
$\bar{\varepsilon}_P$ is the equivalent plastic strain;
K is the strength factor;
n is the factor of hardening; and A, β, m are constants.

When carbon steel becomes plastic, strain hardening is observed. The coefficient of strain hardening can be obtained from the following equation:

$$H = K\cdot n\cdot(\bar{\varepsilon}_p)^{n-1}. \tag{10}$$

In this work, the user program includes the strain hardening coefficient in the elastic-plastic model of the MSC. Marc solver to describe the viscoelastic-plastic behavior of the cast slab under high temperature. The work hardening of carbon steel is described by the equations given by Sorimachi and Brimacombe (K. Sorimachi et al.,1977);

$$K_1 / E = 0.13\exp(-0.023\theta), \tag{11}$$

$$K_2 / E = \begin{cases} 0.045 - 3.87 \times 10^{-5}\theta \, , \ \theta < 1050°C \\ 0.385 \cdot \exp(-0.00422\theta) \, , \ \theta \geq 1050°C \end{cases} \tag{12}$$

$$K_3 / E = \begin{cases} 0.0197 - 1.68 \times 10^{-5}\theta \, , \ \theta < 1050°C \\ 0.0226\exp(-0.00223\theta) \, , \ \theta \geq 1050°C \end{cases} \tag{13}$$

where: E is the elastic modulus; K_1~K_3 are factors of hardening; Θ is the temperature (°C).

Equations (11)~(13) are applied when the strain is smaller than 0.01~0.02, and greater than 0.02, respectively. The factors of hardening are incorporated into the elastic-plastic model in the software package Marc by the user programs.

3.3. Parameters

All simulation parameters are taken from the technological parameters of the #2 continuous caster (SMS-Demag) of Maanshan Iron and Steel Co. Ltd. The parameters of the continuous caster are listed in Table 4.

Segment No.	Shrinkage between rollers (mm)	Slab thickness (m)	Roller diameter(m)	Slit width between rollers (m)	Distance from meniscus (m)
1-2	0.20/0.46	0.2375	0.200	0.240	0-4.374
3-4	0.46/0.46	0.2370	0.245	0.284	4.374-8.388
5-6	0.46/0.44	0.2362	0.255	0.297	8.388-12.592
7-8	0.44/0.44	0.2354	0.265	0.310	12.592-16.992
9-10	0/0.30	0.2346	0.283	0.322	16.992-21.254
11-15	0.30/0.30	0.2343	0.300	0.335	21.254-33.249

Table 4. Parameters of the slab and caster at various segments

Casting temperature: T=1533 °C;
Liquidus temperature T_l=1513 °C;
Solidus temperature T_s =1446.0 °C;
T_{80} = 1459.6 °C;
Environmental temperature: 25 °C;
Roller temperature: 100 °C;
Coefficient of contact heat transfer: 25.0 W/(m·K) (Y. S. Xi and H. H. Chen,2001) ;
Coefficient of fraction: 0.3; Distance tolerance: 0.01 (Y. S. Xi and H. H. Chen,2001).

When the casting speed is in the range of 1.0~1.2 m/min, the SMS-Demag casting machine uses a fixed cooling water intensity in the secondary cooling zone. The parameters of the SPHC steel and Q235 steel are listed in Table 5.

Steel	C (%)	S i(%)	Mn(%)	P(%)	S(%)	Al(%)	Tₗ(°C)	Tₛ(°C)
SPHC	0.05	0.05	0.20	≦0.02	≦0.012	0.03	1528.9	1493.0
Q235	0.18	0.20	0.40	≦0.025	≦0.022		1517.0	1446.0

Table 5. Compositions of SPHC steel and Q235 steel

The coefficient of thermal expansion, Young's modulus of elasticity, and Poisson's ratio of the steel as functions of temperature are required for simulation. The elastic modulus of carbon steel for various temperatures during continuous casting is given in equations (14) and (15).(Ueshima Y et al; 1986, I.Ohnaka,1986)

$$\begin{cases} E = (347.6525\text{-}0.350305^*\text{T}) * 10^9 & , \ T < 900 , \\ E = (968 - 2.33 * T + (1.9 * 10^{-3})^*\text{T}^2 - (5.18 * 10^{-7}) * T^3) * 10^9 & , \ Ts > T \geq 900, \end{cases} \tag{14}$$

$$E = \frac{(f_S - f_{ZST}) \cdot E_S + (1 - f_S) \cdot E_{ZST}}{1 - f_{ZST}}, \ T_{80} > T > T_S (f_{ZST} \leq f_S \leq 1), \tag{15}$$

When $E=E(\text{Ts})$ and $Ezst=E(\text{Tzst})$, Ezst takes a small non-zero value in order to restrain the deviatoric stress in the liquid phase region to maintain hydrostatic pressure. When the temperature is lower than Ts, the elastic modulus is expressed by (14); when it is higher than T_s, it is expressed by (15).

When the temperature is lower than Ts, Poisson's ratio can be defined as equation (14); when the temperature is higher than Ts, with decreasing f_s, Poisson's ratio gradually increases from the value at Ts to a certain value which is close to 0.5; it remains at this value above ZST, as expressed by equations (16) and (17). (Uehara M et al.,1986)

$$v = \frac{(f_s - f_{ZST}) \cdot v_s + (1 - f_s) \cdot v_{ZST}}{1 - f_{ZST}}, \ T > T_L \left(f_{ZST} \leq f_S \leq 1 \right), \tag{16}$$

$$v = v_{ZST} , \quad T \geq T_{80} \quad , \quad (f_S \geq f_{ZST}). \tag{17}$$

Where f_{ZST} is the solid phase ratio at ZST, often taken as 0.80. v_{ZST} is the Poisson's ratio at ZST; it is very close to 0.5. The Poisson's ratio of the steel for various temperatures is shown in Figure 4(a). The coefficient of thermal expansion values of Q235 and SPHC are taken from the local measurement results shown in Figure 4(b).

4. Effects of casting speed on slab broadening (Jian-Xun Fu et al , 2011b)

4.1. Numerical simulation

By tracing one node of the slab at the side-face and recording its width, the width of the slab at various positions in the secondary cooling zone can be obtained. The RUB can then be derived from the simulated width of the slab. The simulated RUB values of Q235 and SPHC steels at three casting speeds are shown in Figure 5(a) and (b), respectively.

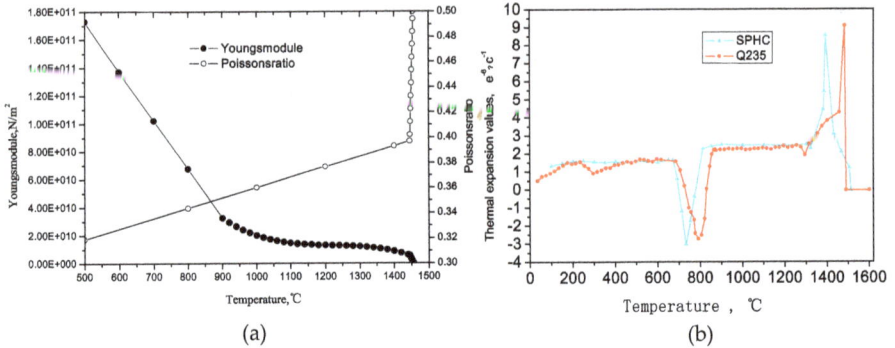

(a)   (b)

Figure 4. (a) Young's modulus of elasticity and Poisson's ratio of Q235 steel for various temperatures; (b)Coefficient of thermal expansion of Q235 and SPHC. (Jian-Xun Fu et al , 2011b)

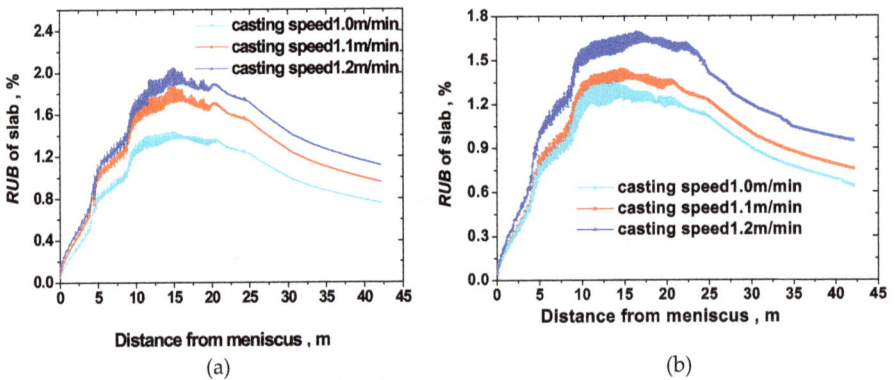

(a)   (b)

Figure 5. (a) RUB values versus distance from meniscus of a Q235 steel;.(b) RUB values versus distance from meniscus of a SPHC steel at three casting speeds. (FU JianXun et al. 2011(b))

Slab broadening for Q235 and SPHC steels at three casting speeds shows similar characteristics. The values of the RUB at the three casting speeds are all positive in the whole secondary cooling zone, which means that slab broadening existed for Q235 and SPHC steels at these speeds. The RUB changed from one segment to another for the first five segments. The RUB increased and then gradually decreased after reaching its maximum at the fifth and sixth segments. Near the tenth segment, the RUB decreased smoothly; the slab became completely solidified at this location.

The simulations of Q235 and SPHC steels produced similar results. The RUB increased with increasing casting speed. For Q235 steel, when the casting speeds were 1.0, 1.1, and 1.2 m/min, the maximum RUB values were 1.44%, 1.88%, and 2.04 %, respectively, and the RUB values at the exit of the caster were 0.76%, 0.96%, and 1.14%, respectively. For SPHC steel, when the casting speeds were 1.0, 1.1, and 1.2 m/min, the maximum RUB values were 1.34%, 1.44%, and 1.69%, respectively, and the RUB values at the exit of the caster were 0.64%, 0.76%, and 0.95%, respectively.

Slab broadening is closely correlated with casting speed, which may be due to the slab's temperature changing with casting speed. When the casting speed increased, the liquid core length and temperature of the slab both increased. With increasing temperature of the slab, the high-temperature mechanical properties of the slab changed; ductility increased and the strength and resistance to external pressure decreased, increasing the RUB.

With increasing casting speed, a given cross section of the slab takes up the same amount of space. At 25 m away from the meniscus, the surface temperature in the wide face at a casting speed of 1.2m/min is 11.6 °C and 21.7 °C higher on average than those at casting speeds of 1.1 and 1.0 m/min, respectively (see Figure 6). Under a given set of conditions, increasing the casting speed increases production. However, high casting speed can lead to slab broadening.

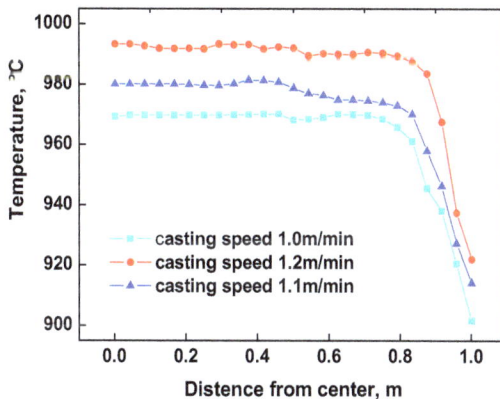

Figure 6. Surface temperature on the slab at various positions.(FU JIanXun et al. 2011(b))

4.2. Verification of simulation results

For continuous caster #2 at Maanshan Iron and Steel Co. Ltd., the slab widths of two steel grades were tracked online at the exit of the caster (the end of the 15th segment); the slab width was measured once per minute. For each grade of steel, measurements were taken for more than 70 minutes. The data are shown in Figures 7(a), and (b), respectively.

Figure 7(a) shows the slab width and the RUB of SPHC steel at various moments. Slab broadening can be clearly seen. The RUB of SPHC steel ranges from 1.4% to 2.4%, with an average of 1.96%. The average RUB is greater than the ratio of linear shrinkage, indicating that the width of the slab after cooling was greater than the top width of the mold. This result shows that slab broadening occurred in the secondary cooling zone.

The slab width changed smoothly except from 45 to 55 min, during which time a sharp trough appears on the RUB curve. In the initial 6 minutes of this period, the RUB decreased to 1.4% from 2.25%, and in the following 4 minutes, the RUB increased to 2.1% from 1.4%. This trough was caused by the changing of the tundish, during which the casting speed decreased sharply, and then quickly recovered to normal, i.e., the change in casting speed caused the change in slab broadening.

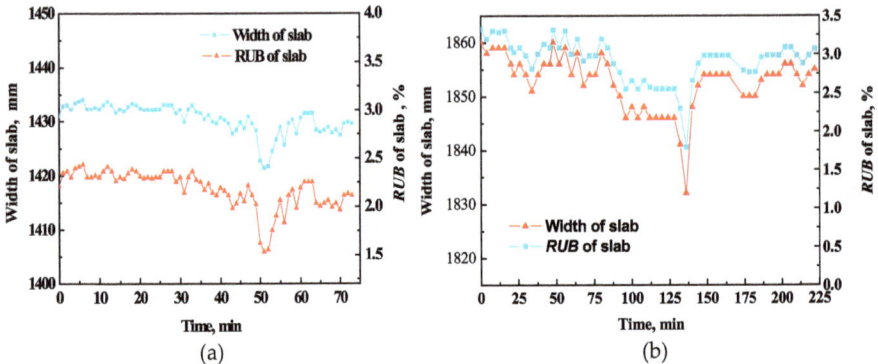

Figure 7. (a) Width of slab and RUB for SPHC steel at various moments; (b) Width and RUB for Q235 steel at various moments. (FU JianXun et al. 2011(b))

Figure 7(b) shows the slab width and the RUB for Q235 steel at various times. The RUB for Q235 steel ranges from 0.77% to 2.91%, with an average of 2.04%. There are five sharp corners on the RUB curve for Q235 steel. By comparing the curve with the production process of Q235, it was found that each sharp corner corresponds to an unsteady production stage. The biggest one corresponds to the changing of the tundish, the last one corresponds to the end of casting, and the remaining three correspond to the changing of ladles.

Figure 8 shows the relationship between the RUB and the casting speed for Q235 steel. The shapes of the RUB curve and the casting speed curve are very similar. When the tundish was changed, the casting speed decreased to 0.5 m/min over a 10-minute period and then recovered to normal in 5 minutes; this change formed a sharp trough in the casting speed curve. At nearly the same time, the RUB decreased to 1.91% from 0.77% in 10 minutes and then increased to 2.1% in 5 minutes, producing a sharp trough in the curve. When the ladle was changed, a similar change happened. When the casting speed was maintained at 1.0 m/min, the RUB remained stable at about 2.0%. The RUB is thus closely correlated with casting speed.

There is a small lag between the RUB curve and the casting speed curve in Figure 12. The change in casting speed curve occurred earlier than that in the RUB curve. For example, the casting speed curve exhibits a sharp trough at about 100 minutes; a sharp trough appears in the RUB curve at about 110 minutes. Comparing Figure 7 and Figure 8, it can be seen that the simulation results generally agree with the industrial measurement results.

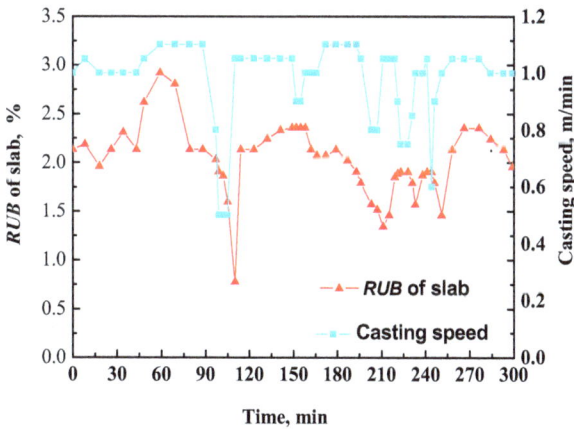

Figure 8. Relationship between the RUB and casting speed. (FU JianXun et al. 2011(b))

5. Effects of width and thickness on slab broadening (FU JianXun et al. 2010(b))

5.1. Numerical simulation

One node of the slab was traced and the width was recorded at various positions of the secondary cooling zone. The RUB was derived from the calculated width of the slab.

The calculated RUB of Q235 steel slab with a cross section 2000 mm × 230 mm at speed of 1.0m/min is shown in Figure 9(a). The RUB changes from one segment to another; its value is over 0 throughout the secondary cooling zone, indicating slab broadening. The RUB increases in the first five segments, and then drops down gradually after reaching its maximum in the sixth segment. In the sixth segment, the width of the slab reaches its maximum with a large fluctuation due to the bulging of the slab in the direction of thickness. Figure 9(b) shows the simulated deformation of the slab in this direction. The shell of the slab has low yield strength and high plasticity; thus, the slab at the points contacting the rollers is depressed and bulges at the slit between the two rollers. Similar to the periodicity of bulging, the width of the slab fluctuates periodically.

The simulated broadening and bulging of the slab in the sixth segment are shown in Figure 10. There is an obvious correlation between broadening in the width direction and bulging in the thickness direction. The position in the slab where the smallest bulging is observed has the greatest broadening. This is due to the depression of slab in the thickness direction contributing to slab broadening in the width direction.

5.2. Effects of slab width on broadening (FU JianXun et al. 2010(b))

230-mm-thick slabs of Q235 with various widths were simulated at a casting speed of 1.0 m/min. The RUB values for various segments are shown in Figure 11. It shows that the simulated RUB of slab slightly increases with the increase of width. The maximum values are 1.27 %, 1.36 %, and 1.44 %, respectively. The RUBs at the exit of caster are 0.63 %, 0.70 %, and 0.76 %, respectively. There is no obvious increase of RUB for slabs with increasing the width, but the increase of broadened size is noticeable. In conclusion, slabs with great width have great broadening.

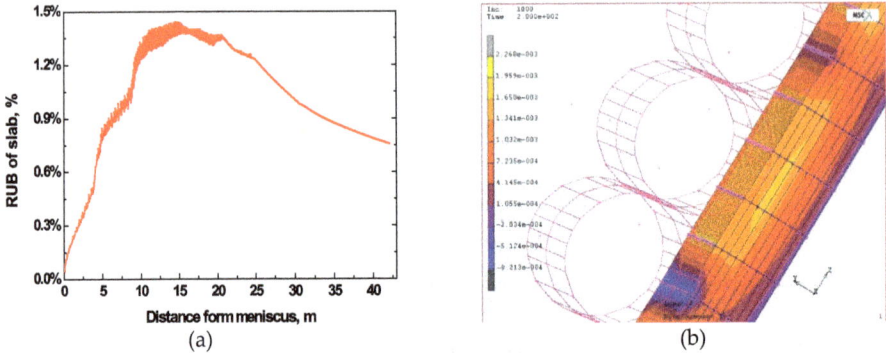

(a)   (b)

Figure 9. (a) Calculated RUB of Q235 steel in the secondary cooling zone; (b)Calculated deformation of slab between rollers. (FU JianXun et al. 2010(b))

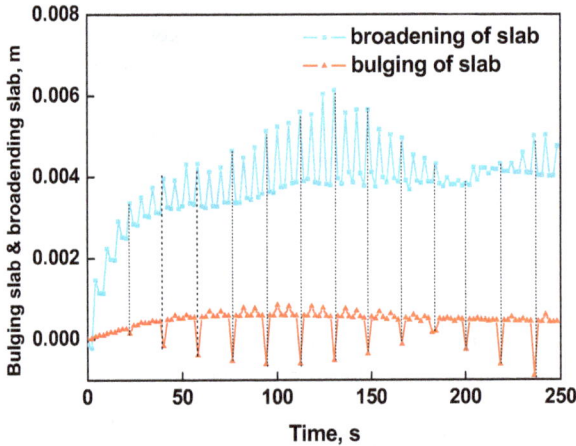

Figure 10. Broadening and bulging of slab in the sixth segment. (FU JianXun et al. 2010(b))

Under the same conditions, the wide slab has greater broadening than narrow slab because of compound effects of temperature and stress. Compared with wide slab, narrow slab has a larger range for heat flow distribution and hence the greater equivalent von Mises stress. But the wider slab has more enthalpy to be removed. So in the same position of caster, the narrow slab has higher yield strength and lower plasticity, and the solidified shell is able to resist great stress.

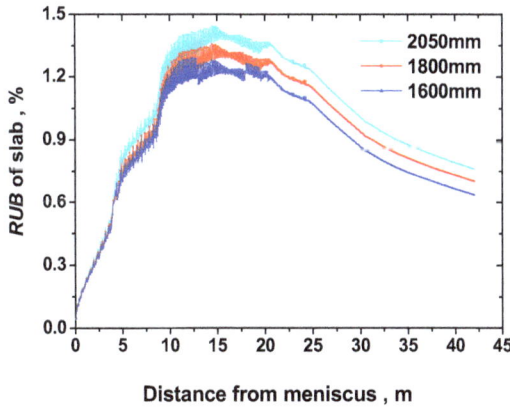

Figure 11. Calculated RUB values for slabs of various widths . (FU JianXun et al. 2010(b))

5.3. Effects of slab thickness on broadening

To study the effect of slab thickness on broadening, 2050 mm-thick Q235 slabs with thicknesses of 230 and 250 mm, respectively, were simulated at a casting speed of 1.0 m/min; the results are shown in Figure 12.

The calculated broadening values for the two slabs are slightly different. The maximum RUB values are 1.4% and 1.38% for 250- and 230-mm-thick slabs, respectively. The RUB values are 0.74% and 0.71% at the exit of the continuous caster, respectively. The difference of broadening is just 0.6 mm between the two slabs. This is because the bulging changes little with increasing thickness.

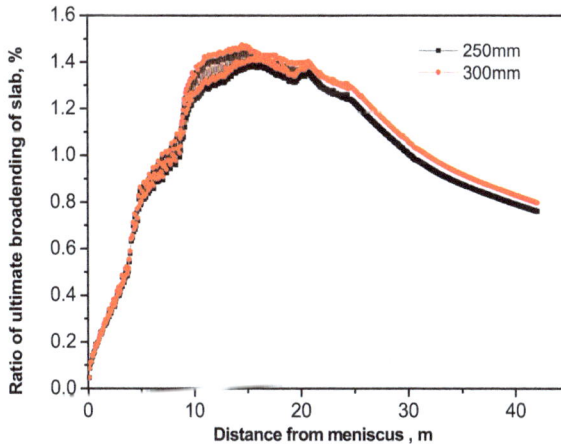

Figure 12. Calculated RUB with different thicknesses. (FU JianXun et al. 2010(b))

5.4. Verification of simulation results

To verify the obtained simulation results, the slab broadening cast on the #2 caster in Maanshan Iron and Steel Co. Ltd was measured. The online measuring system was designed to measure the width of the slab. The digital camera was fixed above the exit of the caster. Q235 steel was used for the experiments. The parameters of the continuous caster and measured results are listed in Table 6.

The online measured RUB values are greater than the simulation results for all the experimental slabs. This is because the preset width of a cold slab in the experiments was the upper width of the mold. The upper width is always greater than the defined width. For the slab with a preset width of 2050 mm, the upper width of the mold is 2081.3 mm (a broadening of 1.56%). With this difference taken into account, the experimental results well agree with those of the simulation.

Defined width (mm)	Upper width (mm)	Lower width of mold (mm)	Measured width (mm)	Measured RUB (%)	Measured broadening (mm)	Calculated broadening (mm)	Deviation rate (%)
1600	1623.8	1610.4	1630.4	1.90	6.6	7.04	6.7
1850	1877.7	1867.3	1885.9	1.94	8.2	9.25	12.8
2050	2081.3	2067.5	2091.4	2.02	10.3	11.69	13.5

Note: the measured broadening of the slab is the difference between the measured width of the slab and the upper width of the mold, and the calculated broadening of the slab is that between the calculated width and the defined width.

Table 6. Measured and calculated widths of slabs

6. Analysis of slab broadening (FU JianXun et al.2011(a))

6.1. Change of mold size

The slabs broaden in width, which varies with the operating parameters of steel produced. The statistical data of 76 taper samples of the mold revealed that the change is very small for the taper of the mold. The average change of a one-sided taper was 0.37 mm, and only a few samples had changes of 1~2 mm. The slight change of the taper is due to metering errors, wear, and deformation. Slab broadening is thus independent of the mold size.

6.2. Exception of equipment or operating parameters

The secondary cooling process is the most important procedure in continuous casting. The temperature field of the slab was checked with the data provided by the producer of the caster. A good agreement was found, indicating that the caster worked well in the secondary cooling process. The monitoring records obtained in a controlled room also reveal that the caster worked well. However, the width of the produced slabs exhibited obvious broadening during the process. It is thus concluded that slab broadening is independent of exceptions of equipment or operating parameters.

6.3. Soft reduction

Soft reduction may strengthen slab broadening and even cause side bulging. For SPHC and Q235 steels, the slabs were broadened in the process of continuous casting with soft reduction set to 0.5~2.5 mm. The broadening width ranges are 2~19 mm and 2~8 mm for SPHC and Q235 steels, respectively. The ratios of broadening are 0.1%~1.46% and 0.15%~0.62% for SPHC and Q235 steels, respectively. Therefore, soft reduction contributes to slab broadening, but is not the main cause.

6.4. Contraction of roll gap

For the continuous caster #2 in Maanshan Iron and Steel Co. Ltd, the ultimate thickness of a produced slab is 230 mm, and the bottom thickness of the mold is 237.5 mm. With a casting speed of 1.1 m/min, the molten steel completely solidifies at the start of the 11th sector where the thickness of the slab is 234.3 mm and the roll gap contraction is 3.2 mm. In this zone, the linear shrinkage ratio is 0.5%~0.7% (1.2~1.7 mm) due to the drop of temperature. Without the contribution of the temperature drop, the roll gap contraction is 1.5~2.0 mm. This amount of shrinkage equals soft reduction of medium or light scale. The roll gap contraction is uniformly distributed. The roll gap contraction acts on the slab and affects the fluctuation of the liquid level of molten steel. However, the slab broadening is far less than that induced by soft reduction. So roll gap contraction is not the main cause of broadening.

6.5. Summary

The static pressure of the molten steel core and the force of the driving rollers may be the main cause of slab broadening.

When there is no support on the narrow face of a slab, the slab deforms in the width direction under the static pressure of molten steel. The high-temperature mechanical properties of the slab are worse than those under normal temperature(Lei H et al,2007; Chen J,1990; S. Kobayashi et al,1988). The slab has good ductility under high temperature and is unable to resist the static pressure of molten steel in the width direction. Therefore, the slab greatly deforms at the edges, and thus the width is broadened. Previous studies found that the hardness of the solidified shell and the ability to resist the static pressure of molten steel are determined by the thickness of the shell and the formation of ferrite-austenite with a dual phase.(Mizukami H et al,1977; Uehara M et al, 1986; Ramacciotti A,1988)

The shell of the slab is clamped under the pressure of the driving roll cylinders so that it moves forward with the rotation of the driving rollers. The solidifying and soft slab is extended and broadens under the pressure of the driving rollers when passing through the cast-rolling segment. The degree of extension and broadening increases with casting speed.

6.6. Creep deformation

The forces acting on the slab shell in the secondary cooling zone can be modeled as the bending of a rectangular thin plate under loading (i.e., static pressure of molten steel). One segment of the slab along the strand direction is taken to build the model. In the model, the slab is a rectangular thin plate fixedly supported along two sides and simply supported along the other two sides. In addition, the thin plate is subjected to lateral loads, and the temperature field linearly changes in the thickness direction of the slab. Because the width of the slab is much greater than the gap between the rollers, the effects of the slab boundary on the internal side of the slab can be ignored according to the Saint-Venant principle. According to plate theory, the slab shell is viscoelastic at high temperature, and the stress and deformation satisfy the Maxwell creep law. As shown in Figure 13.

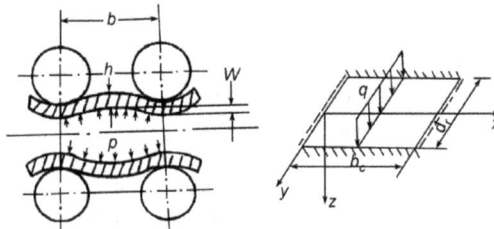

Figure 13. Model of slab shell and the force model of slab creep. (Sun J et al,1996)

In the secondary cooling zone, the total stress equals the sum of elastic strain and creep strain, when the slab shell creepily bends under the static pressure of molten steel. The elastic strain changes little with time. The elastic deflection is expressed as: (Sun J et al,1996):

$$W_e(x,y) = W_{qe} + W_{Te} = \frac{4qa^4}{\pi^5 D} \sum_{i=1,3...}^{\infty} \frac{1}{m^5} \times$$

$$\left(1 - \frac{\alpha_m ch\alpha_m + sh\alpha_m}{\alpha_m + sh\alpha_m ch\alpha_m} ch\frac{2\alpha_m}{b}y + \frac{\sigma_m sh\alpha_m}{\alpha_m + sh\alpha_m ch\alpha_m} \cdot \frac{2y}{b} sh\frac{2\alpha}{b}y\right) \tag{18}$$

$$\times\sin\frac{m\pi}{\alpha}x + \frac{48\alpha aT_0}{\pi^3 h^3(1-2\gamma)(1-\gamma)} \times \sum_{i=1,3...}^{\infty} (1 - \frac{ch\frac{2\alpha_m}{b}y}{ch\alpha_m})\sin\frac{m\pi}{a}x.$$

where: $T_0 = \int_{-h/2}^{h/2} T(z)\cdot z \cdot dz$.

The expression of elastic deflection has the series of hyperbolic function, and converges rapidly, thus setting m=1 is sufficiently accurate for calculation. Using equation (18) and the Cauchy equation, the creep deformation of slab shell on the narrow side can be derived as:

$$\begin{cases} \varepsilon_{xe} = -\dfrac{\partial^2 W_e}{\partial x^2} z, \\[2mm] \varepsilon_{ye} = -\dfrac{\partial^2 W_e}{\partial y^2} z, \\[2mm] \gamma_{xye} = -\dfrac{\partial^2 W_e}{\partial x \partial y} z. \end{cases} \qquad (19)$$

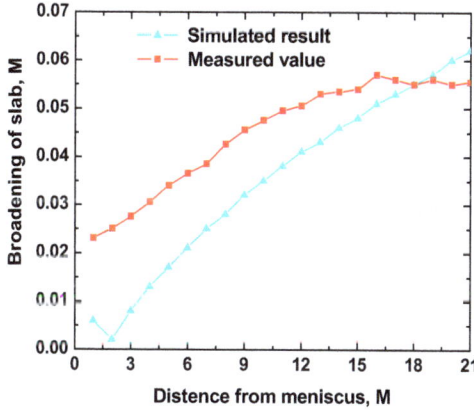

Figure 14. Comparion of calculated and measured side creep results. (FU JianXun et al.2011(a))

The amount of creep deformation for the narrow side of the slab was calculated using Matlab software; the results are shown in Figure 14. The figure also shows the measured results from the experiments of a stagnant slab. The agreement between the calculated results from the Maxwell model and the measured results illustrates that the Maxwell model is able to reveal deformation behavior at high temperature.

6.7. Industrial experiments of stagnant slab (FU JianXun et al.2011(a))

If the static pressure of molten steel is the main reason for the broadening of a slab, the broadening must happen at the forepart of the continuous caster where the slab has a high temperature and a thin shell. If the stress of the rollers is the main reason for broadening, the broadening must happen at the middle part of the continuous caster, specifically the position near the completely solidified zone. Because the molten steel is fluidic before this part(Lin Q Y et al ,2004), decreasing the roller gap does not broaden the slab.

Since the continuous caster is a vertical bow type, it is very dangerous to keep close to it, and thus it is impossible to measure the width of the slab directly. Therefore, when the caster stopped to for the tundish replacement, the width of the stagnant slab was measured to determine where slab broadening happens. When the tundish is to be replaced, the casting speed gradually slows down to zero. This process takes about 4-5 min to form a stagnant slab, which is cooled down continuously by secondary cooling water. The slab in the continuous caster is composed of three parts:

1. For the fully solidified part during casting, the slab broadening is the sum of broadening of the molten steel core.
2. For the part solidified during the stopping period, because it reveals the slab broadening at specific position, and corresponds to the real broadening amount of slab, this part is focused on in our experiments.
3. For the unsolidified part until restarting the casting, the slab broadening continues during subsequent casting, as the molten steel core still exists in the slab shell. However, because the slab shell is very thick, little broadening happens.

Using the square-root law of solidification, the fully solidified normal position and stagnant position can be derived, and thus the above three parts of the slab could be determined.

The slab was Q235 steel, the casting speed was 0.0167 m·s^{-1}, the cross section of the slab was 2.050 m × 0.230 m, the upper width of the mold was 2.0813 m, the lower width of the mold was 2.0675 m, the casting temperature was 1533°C, T_l was 1513°C, and T_s was 1546°C.

The width of the front slab was also traced. It remained at 2.040 m, indicating that nearly no broadening of the slab happened at this position. It may be because it was cooled so rapidly that there was no time for broadening. Therefore, the width of the front slab was used as the standard width for assessing slab broadening.

The absolute broadening of the slab was derived from the slab width, which was measured while the slab was pushed through the exit of the continuous caster, subtracting the width of the front slab. The broadening values of the slab are shown in Figure 15(a) and (b).

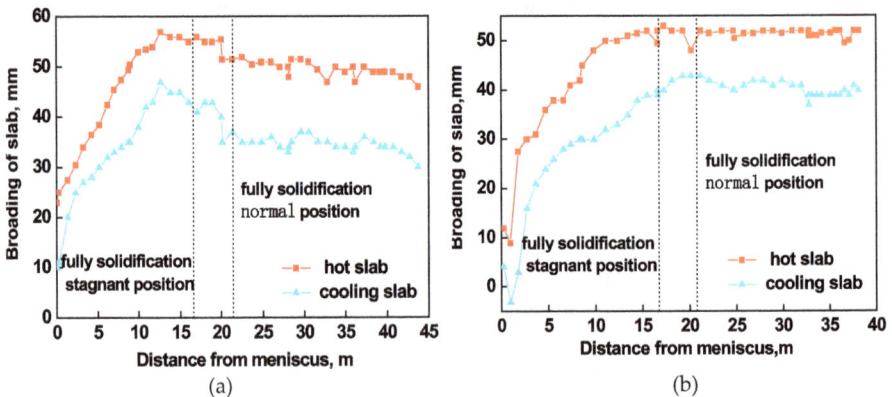

(a)   (b)

Figure 15. (a) Absolute broadening of slab in the first strand of stagnant slab; (b) Absolute broadening of slab in the second strand of stagnant slab.(FU JianXun et al.2011(a))

Slab broadening mainly happened in the front 6 segments (before 12.6 m). In these sectors, the broadening increases linearly with the distance from the meniscus. At the position of

12.6 m, the slab broadening was at its maximum, and then decreased slowly with distance from the meniscus. These results confirm that the stress of the roller is not the main reason for broadening. Otherwise, the slab broadening will happen before the slab is fully solidified near the 9th and 10th segments. The trend of slab broadening is consistent with that of the static pressure of molten steel, which confirms that the slab broadening is dependent on the static pressure of molten steel.

7. Mechanism of slab broadening

The static pressure of molten steel deforms the slab shell. The coupled thermo-mechanical viscoelastic-plastic 3D finite element model was built with the secondary development of the commercial software MSC.Marc. The calculated and measured results of slab width are shown in Figure 16.

The figure reveals that the calculated deformation agrees very well with the measured deformation. Slab broadening is the result of slab deformation under the pressure of static melting at high temperature. The deformation of the slab in the direction of thickness is shown in Figure 17(a). The temperature field of the slab is shown in Figure 17(b).

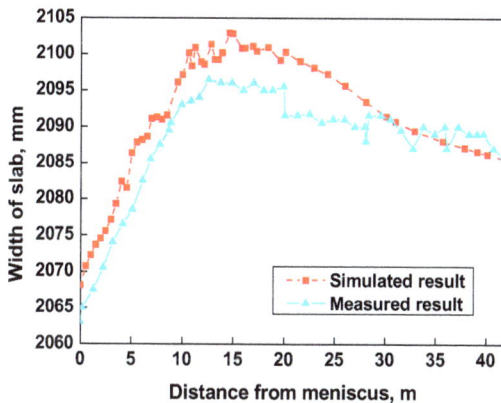

Figure 16. Simulated and measured widths of slab;(FU JianXun et al.2011(a))

The on-site investigation, force analysis, calculation from Maxwell creep model, and numerical simulation from the coupled thermo-mechanical viscoelastic-plastic 3D finite element model reveal that the slab broadening is due to slab deformation under the static pressure of molten steel. The slab shell deforms without constraints on the narrow side.

Creep deformation appears when the material plastic gradually deforms with time under certain conditions. Plastic deformation only happens when the stress exceeds the elastic limit. However, creep deformation happens when the acting time of stress is sufficiently long, even if the stress is very small. The creep deformation of metal is obvious only if the temperature is over the creep temperature (about $0.3\ T_m$). The slab deforms for a long time under the pressure of static molten steel at high temperature. The creep rate depends on the composition of the compound metal, and the processes of refining and thermal treatment. Creep deformation causes slab broadening because it makes the material keep stress relaxed, reduces hardness, and enhances plasticity.

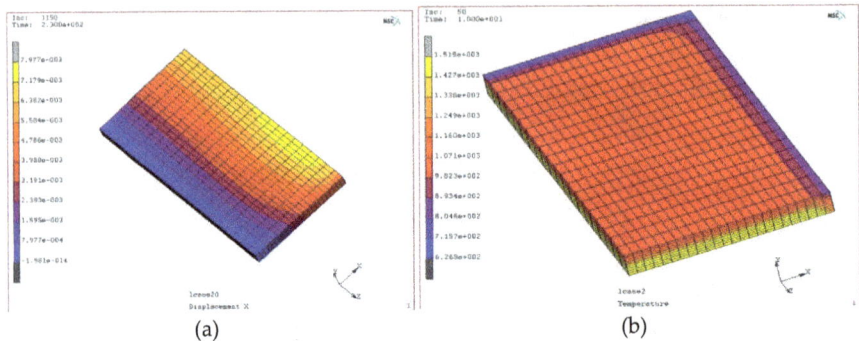

(a)   (b)

Figure 17. (a) Deformation of slab in the direction of thickness at 230 mm; (b) Temperature field of slab at 1150 s. (FU JianXun et al. 2010(c))

The amount of broadening depends on the forces acting on the slab and the properties of the slab material, especially those at high temperature. Specifically, it depends on the static pressure of molten steel, the high-temperature mechanical properties of steel, the composition of the slab material, the thickness of the slab shell, secondary cooling intensity, casting speed, and the constitution of the caster.

The static pressure of molten steel is the driving force for the deformation of the slab shell. It is related to the type and constitution of caster. At present, vertical-bending casters are most common. For these casters, the static pressure is related to the density of molten steel and the height of the caster.

Under the conditions of high casting speed and constant cooling water, the fully solidified zone extend, the length of molten core increases, and the shell becomes thinner. Because of the higher temperature, the slab shell also has lower yield strength and better malleability. Consequently, the slab broadening increases. However, if the cooling water supply is changed when the casting speed is increased, the problem will become sophisticated.

The effects of steel grade on the broadening result from differences in material properties at high temperature, and hence differences in resistance to plastic deformation and creep deformation. With an increase in the carbon percentage, the ratio of ferrolite and austenite in the two phase regions changes. The increase in austenite is helpful to the reduction of slab broadening.

Intracell dislocation climb and intercell slide are two forms of creep deformation. Solution strengthening, precipitation strengthening, and dispersion strengthening insert a lot of defects into the crystal structure of steel, which hinder dislocation movement and thus reinforce steel. Thus, micro-alloying of steel could enhance the hardness of the slab and reduce slab broadening.

In summary, higher casting speed, lower intensity of secondary cooling, thinner slab shell, larger static pressure of molten steel, and lower hardness of steel at high temperature increase slab broadening.

8. Conclusion

1. The mechanism of slab broadening is that the slab with high temperature exposes to no constraint at the direction of narrow face, and because of the static pressure of molten steel, the slab deforms in this direction.
2. Slab broadening is a common problem in continuous casting. The average RUB for the three grades of steel studied was in the range of 1.27%~3.00%, with a maximum of 4.4%.
3. Stagnant slab measurement experiments reveal that slab broadening happens in the 6 front segments, and that roller compaction is not responsible for slab broadening.
4. The agreement between the calculated results from the Maxwell model and the measured results illustrates that the Maxwell model is able to reveal the deformation behavior of a slab at high temperature.
5. Higher casting speed, lower intensity of secondary cooling, thinner slab shell, larger static pressure of molten steel, and lower hardness of steel at high temperature increase slab broadening. The micro-alloying of steel improves the hardness of the slab and reduces slab broadening.

Author details

Jian-Xun Fu and Weng-Sing Hwang

Research Center for Energy Technology and Strategy & Department of Materials Science and Engineering, National Cheng Kung University, Tainan 701, Taiwan

Acknowledgement

The authors gratefully acknowledge the guidance of Prof. Jingshe Li, Prof. Hui Zhang, Prof. Xingzhong Zhang, et al. The authors would like to thank the National Science Council of Taiwan (NSC100-2221-E-006-091-MY3) for funding this work.

9. References

Chen J. Hand Book of Continuous Casting (in Chinese). Beijing: Metallurgical Industry Press, 1990,

Fu JianXun, Li Jingshe, Zhang Hui, Zhang Xing-zhong. Industrial Research on broadening of slab in continuous casting. JOURNAL OF RON AND STEEL RESEARC INTERNATIONAL. 2010. 17(8):20-24,

Fu JianXun, Li Jingshe, Zhang Hui, Zhang Xingzhng, Mechanism of broadening of slab in the secondary cooling zone of continuous casting. SCIENCE CHINA- Technological Sciences. 54 (2011), No.5: 1228–1233,

Fu JianXun, LI Jingshe, ZHANG Hui,. Effects of Width and Thickness of Sab on Broadening in Continuous Casting. International Journal of Minerals, Metallurgy and Materials. 17 (2010), No. 6, 723,

Fu JianXun, Li Jingshe, Zhang Hui, Viscoelastic–Plastic Analysis of Broadening of Slab in the Secondary Cooling Zone of Continuous Casting. Acta Metallurgica Sinica, 46 (2010), No. 1, 91-96 (in chinese),

Jian-Xun Fu, Weng-Sing Hwang, Jing-she Li, and Zhang Hui. Effect of Casting Speed on Slab Broadening in Continuous Casting. steel research int. 2011 (82) No. 11. 1266- 1272,

I.Ohnaka. Mathematical Analysis of solute redistribution during solidification with diffusion in solid phase. Transactions of ISIJ, 1986, 26: 1045-1051,

K. Sorimachi, J. K. Brimacombe: Ironmak Steelmak, 1977; 4: 240,

Lei H, Yang L D, Zeng J, et al. Discussion of computational methods for the forces incurred in the bending phase in slab casting (in Chinese).Heavy Mach, 2007, (3): 41–45,

Lin Q Y, Jiang H J, Zhu M Y. Analysis of reduction parameters of dynamic soft reduction in continuous casting. J Mater Metall, 2004, 3(4): 261–265,

Miyazawa K, Schwerdtfeger K. Computation of Bulging of Continuously Cast Slab With Simple Bending Theory. Ironmaking and Steelmaking, 1979(2) : 68,

Mizukami H , Murakami K, Miyashita Y. Mechanical Proper ties of Continuously Cast Steel at High Temperatures [J]. Tetsu-to-Hagane. 1977. 63: 146 (in Japanese),

Ramacciotti A. Thermo-Mechanical Behavior of the Solidified Shell in a Funnel-Shaped Mold for Continuous Casting of Thin Slabs rJ1. Steel Research, 1988. 5gr 438,

Sheng Y, Sun J, Zhang M. Calculation for bulging deformation of continuously casting slab (in Chinese). Iron Steel, 1993, 28(3): 20–25,

S. Kobayashi, T. Nagamichi and K. Gunji: Transactions of ISIJ, 28(1988), 543,

Sun J, Sheng Y, Zhang X. Analysis of bulging deformation and stress in continuous cast slabs (in chinese). J Iron Steel Res, 1996, 8 (1): 11–15,

Uehara M. Samarasekera I V, Brimacornbe J K. Mathematical Modeling of Unbending of Continuously Cast Steel Slabs. Ironmaking and Steelmaking, 1986(12) : 138,

Ueshima Y, Mizoguchi S, Matsumiya T, et al. Analysis of solute distribution in dendrites of carbon steel with δ/γ transformation during solidification. Metallurgical and Materials Transactions B, 1986, 17B(4): 845-859,

Xiong Yi-gang. Continuous Casting of Slab [M]. Beijing: Metallurgical Industry Press, 1994 (in Chinese),

Y. S. Xi and H. H. Chen: MSC.Marc Manual for Analysis Using Temperature Field and the Coupled Field, MSC. Software user manual, 2001 (Iin Chinese),

Multidimensional Numerical Simulation of Ignition and Propagation of TiC Combustion Synthesis

A. Aoufi and G. Damamme

Additional information is available at the end of the chapter

1. Introduction

The purpose of this book chapter is to analyze from a numerical point of view a reaction diffusion mathematical modelling of titanium carbide combustion synthesis from a mixture of titanium and carbide reactive powders thanks to self-propagating high temperature synthesis process. This modelling results in the coupling between a nonlinear parabolic equation expressing the enthalpy balance of the system with radiative boundary conditions and a nonlinear differential equation describing the exothermic chemical reaction in the system. An another multiphysics coupling was analyzed in [3]. This Self-propagating High temperature Synthesis (SHS) process was discovered in 1965 by Merzhanov [7], [8] and uses the energy released by the exothermic reaction itself to ensure its self-propagation inside the material after a localized heat supply has been performed for several seconds on the surface of the solid mixture. The stoichiometric solid mixture is made of several kinds of reactive powders. We analyze in this book chapter, the influence of radiative boundary conditions *related to the heat supply* over the ignition and eventual propagation of a combustion front inside the material.

Four sections are used to present our numerical simulation work. Section two presents the governing equations of the modelling. Section three outline the main aspects of the numerical scheme. Section four analyzes and discusses the main numerical simulation results of the combustion synthesis process. A conclusion summarizes the results that were obtained.

2. Mathematical modelling

This section describes the main features of the modelling which expresses the coupling between a reaction-diffusion written for the enthalpy balance and a differential equation written for the exothermic chemical kinetics. SHS (Self-propagation High-temperature Synthesis) process is a condensed phase process which converts a mixture of powders into an end product. In this paper we consider the synthesis of titanium carbide in solid phase

((s) subscript) thanks to the exothermic reaction $\text{Ti}_{(s)} + C_{(s)} \rightarrow \text{TiC}_{(s)}$. We assume that we have a stoichiometric and isotropic mixture of compacted powders of titanium and carbide. Let us denote by $T = T(M, t)$ (*resp.* $\xi = \xi(M, t)$) the temperature (*resp. conversion rate*) at point $M \in \Omega$ at time t. The system is composed of the fraction ξ of titanium carbide **TiC** end product and the fraction $(1 - \xi)$ of remaining powders of titanium and carbide.

2.1. Exothermic kinetics modelling

A first order exothermic kinetics is used to describe the synthesis of titanium carbide, hence a single variable, the conversion rate $\xi \in [0, 1]$ of the reaction is defined. When $\xi = 0$ (*resp.* $\xi = 1$) the reaction has not started (*resp. is ended*). The velocity of the reaction may have or not a cutoff temperature $T_s = 1166K$, which corresponds to the first allotropic phase transition of titanium α, Ti_α to titanium β, Ti_β. An Arrhenius type equation is used

$$\frac{d\xi}{dt} = k_d(T)(1 - \xi),$$ (1)
$$\xi(., 0) = 0.$$

The velocity constant $k_d(T)$ follows an Arrhenius law in temperature such that

$$k_d(T) = k_d^* T^d \exp\left(-\frac{E^*}{R.T}\right).$$ (2)

In both cases, (with or without cut-off temperature), k_d^* is a frequency factor in s^{-1}, E^* is the activation energy of the reaction in J/mol, $R = 8.314J/(mol.K)$ is the perfect gas constant and $d \in [0, 2]$ is the degree, which is not necessarily an integer. Such an expression is commonly used in literature. It is worth mentioning that the activation temperature $T^* = E^*/R = 4000\ K$ is high and gives an idea of the stiffness of the reaction-diffusion/kinetics coupling.

2.2. Heat transfer modelling

This subsection presents successively the expressions used for the heat capacity and the thermal conductivity. It then analyzes the enthalpy balance that expresses the coupling between the heat transfer by thermal diffusion with the exothermic kinetics and provides the boundary conditions to close the mathematical modelling of the system. We also define and compute the adiabatic temperature of the exothermic kinetics. Due to the high temperatures involved in the process , up to 3000K, two phase changes occur

- at $T = T_{\alpha,\beta} = 1166K$, the allotropic phase transition of Ti_α to Ti_β, characterized by its mass latent heat $L_{\alpha,\beta}$,[17], and $f_{\alpha\beta} \in [0, 1]$ is the fraction of β phase,
- at $T = T_{sl} = 1943K$, titanium melting is characterized by its mass latent heat L_{sl},[17], and $f_{sl} \in [0, 1]$ is the fraction of liquid phase.

2.2.1. Equation used for the heat capacity

The mass heat capacity at constant pressure is a function of temperature, conversion rate and porosity. Assuming that the mass of the system remains constant during the process, we can therefore apply a linear mixing law between the heat capacity of reactants $C_{p_{\text{Ti}}}(T), C_{p_C}(T)$

and the heat capacity of the product $C_{p_{TiC}}(T)$ weighted by the conversion rate of the reaction ξ to obtain

$$C_p(T,\xi) = \xi . C_{p_{TiC}}(T) + (1-\xi) . \frac{M_{Ti} . C_{p_{Ti}}(T) + M_C . C_{pc}(T)}{M_{TiC}}, \tag{3}$$

where M_{Ti}, M_C et M_{TiC} denote respectively the molar mass of titanium, carbide and titanium-carbide. The expression given by Eq.(3) is also used in [5],[6].

2.2.2. Equation used for the thermal conductivity $\lambda (T, \xi, f_{sl})$

Thermal conductivity $\lambda (T, \xi, f_{sl})$ in $J.m^{-1}.K^{-1}.s^{-1}$ is a key parameter governing the propagation of the combustion front. Following [14] we define the effective thermal conductivity $\lambda (T, \xi, f_{sl})$ by

$$\lambda (T, \xi, f_{sl}) = \lambda_{cond}(T, \xi, f_{sl}) + \lambda_{fus}(f_{sl}) + \lambda_{rad}(T) + \lambda_{conv}(T), \tag{4}$$

where

- $\lambda_{cond}(T, \xi)$ is the thermal conductivity of the system which is a non-linear weighting between the thermal conductivity of the reactants λ_{Ti+C} and the product $\lambda_{TiC}(T_a)$ according to [17]. It is given by

$$\lambda_{cond}(T, \xi) = \left\{ \lambda_{Ti+C}(T) + (\lambda_{TiC}(T) - \lambda_{Ti+C}(T)) . \sqrt{\xi} \right\} . f(T), \tag{5}$$

  while the expression of function $f(T)$ is given in [17]. Expression used for $\lambda_{TiC}(T), \lambda_{Ti+C}(T)$ are given in [13].

- $\lambda_{fus}(f_{sl}) = f_{sl}(1-\xi)$ accounts for the appearance of a liquid phase when temperature $T = T_{sl} = 1943K$, i.e. the temperature for the fusion of titanium. In this case, the fraction f_{sl} will increase from 0 to 1. Moreover this temperature is reached thanks to the exothermic kinetics while a fraction $(1-\xi)$ of titanium has already been consumed during the kinetics. This expression represents the fraction of fused titanium that has not yet reacted.

- $\lambda_{rad}(T)$ expresses the contribution of the radiative heat transfer between the grains and depends on the diameter d_p of the titanium/carbide particules, their emissivity ϵ, their porosity p. The expression used is taken from [18]. It was reported in [15] that this term contributes significantly to the velocity of the combustion wave.

$$\lambda_{rad}(T) = 4 \, \epsilon \, \sigma \, T^3 \, d_p \, p^{-3}. \tag{6}$$

- $\lambda_{conv}(T)$ expresses the contribution of gaz that are present in the porous system. $\lambda_{air}(T)$ is taken from [13].

$$\lambda_{conv}(T) = \lambda_{air}(T) \, p^{-3}. \tag{7}$$

2.2.3. Enthalpy balance

The enthalpy balance expressing the local conservation of internal energy can be written as

$$\rho \frac{dh}{dt} = \nabla . (\lambda \nabla T), \tag{8}$$

where ρ in $kg.m^{-3}$ is the density of the system, h the mass enthalpy and λ the thermal conductivity. Temperature T and conversion rate ξ are two thermodynamical variables of the problem. Moreover two scalar variables $f_{\alpha\beta}$ and f_{sl}, representing the fraction of solid and liquid phases are also used for the description of the mass enthalpy of the system. We have therefore a set of four independent variables $T, \xi, f_{\alpha\beta}, f_{sl}$ that describe the evolution of the system. The mass enthalpy is defined by

$$h\left(T, \xi, f_{\alpha\beta}, f_{sl}\right) = \xi\,(\Delta_r H)_{T_a} + \int_{T_a}^{T} C_p\,(\theta, \xi)\,\,d\theta + f_{\alpha\beta}\,L_{\alpha\beta} + f_{sl}\,L_{sl}. \tag{9}$$

The computation of the partial derivatives $\dfrac{\partial h}{\partial T}$, $\dfrac{\partial h}{\partial \xi}$, $\dfrac{\partial h}{\partial f_{\alpha\beta}}$ and $\dfrac{\partial h}{\partial f_{sl}}$ leads to

$$\frac{\partial h}{\partial T} = C_p\,(T, \xi), \qquad \frac{\partial h}{\partial f_{\alpha\beta}} = L_{\alpha\beta}, \qquad \frac{\partial h}{\partial f_{sl}} = L_{sl},$$

$$\frac{\partial h}{\partial \xi} = (\Delta_r H)_{T_a} + \int_{T_a}^{T} \frac{\partial}{\partial \xi}\left[C_p\,(\theta, \xi)\,\right]\,d\theta. \tag{10}$$

Thanks to the definition of $C_p\,(T, \xi)$ given by Eq.(3), we obtain

$$\frac{\partial h}{\partial \xi} = (\Delta_r H)_{T_a} + \int_{T_a}^{T}\left[C_{p_{TiC}}\,(\theta) - \frac{M_{Ti}.C_{p_{Ti}}\,(\theta) + M_C.C_{pc}\,(\theta)}{M_{TiC}}\right]\,d\theta, \tag{11}$$

where the integral $I(T)$ is computed with the trapezoidal rule.

$$I\,(T) = \int_{T_a}^{T}\left[C_{p_{TiC}}\,(\theta) - \frac{M_{TiC}\,C_{p_{Ti}}\,(\theta) + M_C\,C_{pc}\,(\theta)}{M_{TiC}}\right]d\theta. \tag{12}$$

Fig.1 represents the evolution of $\dfrac{I(T)}{\Delta_r H}$ for $T \in [300, 3500]$. It is observed that

$\displaystyle\max_{300 \leq T \leq 3500}\left|\frac{I\,(T)}{(\Delta_r H)_{T_a}}\right| \leq 2.5\,10^{-5}$. The contribution of $\displaystyle\int_{T_a}^{T} \frac{\partial}{\partial \xi}\left[C_p\,(\theta, \xi)\right]d\theta$ to $\dfrac{\partial h}{\partial \xi}$ is therefore neglected.

Finally the enthalpy balance for the system is written by

$$\rho.\frac{\partial h_s\,(T, \xi)}{\partial t} = \nabla.\,(\lambda\,\nabla T) + \rho.\,(-\Delta_r H)_{T_a}.\frac{\partial \xi}{\partial t} - \rho.L_{\alpha\beta}.\frac{\partial f_{\alpha\beta}}{\partial t} - \rho.L_{sl}.\frac{\partial f_{sl}}{\partial t}, \tag{13}$$

where

$$h_s\,(T, \xi) = \int_{T_a}^{T} C_p\,(\theta, \xi)\,\,d\theta. \tag{14}$$

Radiative boundary conditions are defined over $\partial\Omega$ for the system and take the form

$$-\lambda\,\frac{\partial T}{\partial n} = \varepsilon.\sigma.\left(T^4 - T^4_{\infty\partial\Omega}\right), \tag{15}$$

where $\varepsilon = 0.7$ is the emissivity of the material while σ is the Stefan-Boltzman constant. The value of $T_{\infty\partial\Omega}$ depends on the boundary $\partial\Omega$.

The initial condition given on Ω states that the sample is at room temperature.

$$T\,(M, 0) = T_a. \tag{16}$$

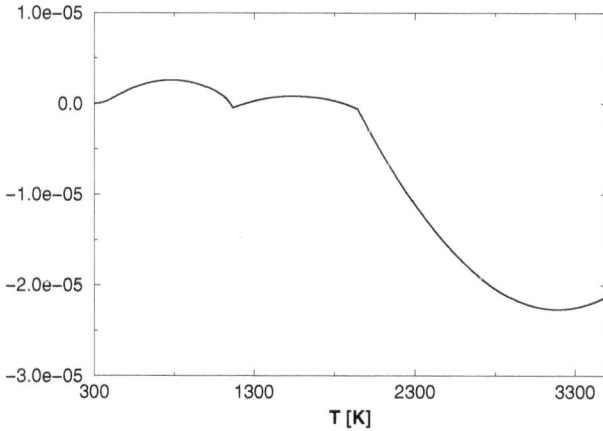

Figure 1. Evolution of $\dfrac{I(T)}{\Delta_r H}$ for $T \in [300, 3500]$ K.

2.2.4. Adiabatic temperature of the reaction T_{ad}

The adiabatic temperature T_{ad} of the reaction, or flame temperature, is the maximum temperature when the system is adiabatic. Titanium-carbide is obtained from titanium and carbide elements at temperature T_a. The enthalpy of the reaction corresponds to the heat of formation of TiC at temperature T_a, hence $(\Delta_r H)_{T_a} = \Delta_f H^0 (\text{TiC})_{T_a}$. Considering a thermodynamical path composed of the two steps:

(i) Titanium and carbide reactants are transformed into titanium-carbide, -TiC- at T_a,

(ii) End product TiC is heated up from T_a to T_{ad}.

One obtains that

$$\Delta_f H^0 (TiC)_{T_a} + \int_{T_a}^{T_{ad}} C_p (\text{TiC}) \ dT = 0. \tag{17}$$

We use Janaf tables [11] for the expression of the heat capacity as a function of temperature. When $T_a = 298K$ (room temperature value), we solve numerically the integral equation Eq.(17), and obtain that the value of the adiabatic temperature is $T_{ad} = 2900K$. According to [17], we express $-\Delta_r H = \dfrac{\Delta H_f}{M_{TiC}}$.

2.3. Mathematical modelling of SHS process

The purpose of the numerical simulation is to determine at each point $M \in \Omega$, and at each time t, temperature field $T(M, t)$, conversion rate $\zeta(M, t)$, solid fraction $f_{\alpha\beta}(M, t)$ and liquid fraction $f_{sl}(M, t)$ that will permit to determine spatial and temporal location of the reaction ignition. The mathematical modelling given by Eq.(1),Eq.(2) and Eq.(13),Eq.(14),Eq.(15) is well posed and will be solved numerically by the methods described in the next section.

3. Numerical discretization scheme

In this section we present the main features of the fully implicit finite-volume discretization scheme such as discrete maximum principle for both the reaction-diffusion and the differential equation discretization [1]. Error estimates are given in [2]. The iterative solution of penta-diagonal sparse matrix for 2D computations and hepta-diagonal sparse matrix for 3D computations is done thanks to SSOR and SIP methods. A fixed point technique is used to solve the coupled nonlinear system. A first order linearization of the exponential Arrhenius term is used which enhances the diagonal dominance of the matrix and accelerates the iterative SSOR/SIP solvers. We have used the enthalpy method to compute the phase change for which a detailed description can be found in [10]. This will be omitted from now, since we will mainly focuss our attention on the construction of the numerical scheme for the reaction-diffusion equation on a non uniform structured mesh and discard formally the specifics of the enthalpy method.

3.1. One-dimensional finite-volume discretization

We present the set of discrete equations that arise from the implicit finite-volume discretization of the enthalpy balance (reaction-diffusion equation) and the exothermic chemical reaction (differential equation).

3.1.1. Numerical discretization of the enthalpy balance

We integrate the enthalpy balance over a space-time finite volume $\Omega_i \times [t_n, t_{n+1}]$, where $\Omega_i = [r_i, r_{i+1}]$ is a control volume. A cell-centered approximation is used. The unknown T_i^n denotes the mean value of $T(x, t)$ at time t_n over Ω_i. $m_i = \int_{r_i}^{r_{i+1}} r^g \, dr$ is the discrete "mass" of control volume Ω_i.

$$\int_{t_n}^{t_{n+1}} \int_{\Omega_i} \rho \cdot \frac{\partial h_s(T, \xi)}{\partial t} \, d\Omega \, dt = \int_{t_n}^{t_{n+1}} \int_{\Omega_i} \left(\nabla \left(\lambda \left(T, \xi, f_{sl} \right) \nabla T \right) + \frac{\rho . \Delta H_f}{M_{TiC}} k\left(T\right) \left(1 - \xi\right) \right) \, d\Omega \, dt. \tag{18}$$

Backward Euler implicit scheme is used for the temporal discretization since it is unconditionally stable, robust and well adapted for the discretization of such problems. Δt is the time-step.

We give only for internal control volumes, the finite-volume discretization of the enthalpy balance in a one-dimensional cartesian ($g = 0$), cylindrical ($g = 1$) and spherical ($g = 2$) coordinate system

$$m_i \rho \frac{h_s\left(T_i^{n+1}, \xi_i^{n+1}\right) - h_s\left(T_i^n, \xi_i^n\right)}{\Delta t} = \lambda_{i+\frac{1}{2}}^{n+1} r_{i+1}^g \frac{T_{i+1}^{n+1} - T_i^{n+1}}{d_{i+\frac{1}{2}}} - \lambda_{i-\frac{1}{2}}^{n+1} r_i^g \frac{T_i^{n+1} - T_{i-1}^{n+1}}{d_{i-\frac{1}{2}}}$$

$$+ m_i \frac{\rho \Delta H_f}{M_{TiC}} k_d(T^{n+1}) \left(1 - \xi_i^{n+1}\right). \tag{19}$$

Here $d_{i+\frac{1}{2}}$ is the distance between the center of two adjacent control volumes, and denoting by $h_i = r_{i+1} - r_i$ the length of control volume Ω_i, then $d_{i+\frac{1}{2}} = (h_{i+1} + h_i)/2$.

As pointed out by [9], in order to ensure the conservativity of the heat flux at the interface between two adjacents control volumes, $\lambda^{n+1}_{i+\frac{1}{2}}$ is computed thanks to the following harmonic mean formula:

$$\lambda^{n+1}_{i+\frac{1}{2}} = \frac{h_i + h_{i+1}}{\frac{h_{i+1}}{\lambda^{n+1}_{i+1}} + \frac{h_i}{\lambda^{n+1}_i}}, \tag{20}$$

where $\lambda^{n+1}_i = \lambda\left(T^{n+1}_i, \xi^{n+1}_i, f_{sl\,i}^{\ n+1}\right)$. A decoupled iterative solution of the nonlinear system is achieved thanks to the fixed point method. A first-order linearization of both the stiff Arrhenius term, and the enthalpy term is done. This reinforces the strictly dominant three-diagonal matrix that is inverted at each step of the non-linear solver thanks to the direct (α, β) Gauss algorithm, i.e. the TDMA algorithm [9]. The numerical solution cost of the three-diagonal linear system by this method increases linearly with the number of unknowns.

3.1.2. Numerical discretization of the exothermic kinetics

We integrate the differential equation over a space-time finite volume $\Omega_i \times [t_n, t_{n+1}]$. A cell-centered approximation is used.

$$\int_{t_n}^{t_{n+1}} \int_{\Omega_i} \frac{d\xi}{dt} - \int_{t_n}^{t_{n+1}} \int_{\Omega_i} k_d\,(T)\,(1 - \xi). \tag{21}$$

The backward Euler scheme, first order accurate fully implicit scheme is used. Denoting by ξ^n_i the mean value at time t_n over Ω_i of $\xi(x,t)$, we obtain the following discrete equation

$$\xi^{n+1}_i = \frac{\xi^n_i + \Delta t\,k_d(T^{n+1}_i)}{1 + \Delta t\,k_d(T^{n+1}_i)}. \tag{22}$$

It was proved in [1] that the numerical scheme is L^∞ stable, moreover for each time index n and for all control volume index $i = 1, \ldots, I : \xi^n_i \in [0,1]$. For a given control volume index i, the time sequence $(\xi^n_i)_n$ is increasing, hence the discrete finite-volume approximation mimics the behavior of the continuous solution [2].

3.2. Two-dimensional finite-volume discretization

We will not detail the set of discrete equations for the chemical kinetics, since it is a straightforward extension from the 1D case. We now give the discrete equations related to the finite-volume approximation of the enthalpy balance over a space-time finite volume $\Omega_{i,j} \times [t_n, t_{n+1}]$. Each rectangular structured non uniform control volume $\Omega_{i,j} = [r_i, r_{i+1}] \times [z_j, z_{j+1}]$, has surface $m_{i,j}$ defined by :

$$m_{i,j} = \int_{\Omega_{i,j}} d\Omega = \int_{z_j}^{z_{j+1}} \left(\int_{r_i}^{r_{i+1}} r^g\,dr \right)\,dz = \frac{(r_{i+1})^{g+1} - (r_i)^{g+1}}{g+1} \cdot \left(z_{j+1} - z_j \right). \tag{23}$$

The enthalpy balance is written for 2D cartesian geometry ($g = 0$) or 2D cylindrical geometry ($g = 1$).

$$\frac{m_{i,j}}{\Delta t} \cdot \rho \cdot \left(h_s \left(T_{i,j}^{n+1}, \varsigma_{i,j}^{n+1} \right) - h_s \left(T_{i,j}^{n}, \varsigma_{i,j}^{n} \right) \right) = \lambda_{i+\frac{1}{2},j}^{n+1} r_{i+1}^{g} \frac{T_{i+1,j}^{n+1} - T_{i,j}^{n+1}}{d_{i+\frac{1}{2}}^{r}} - \lambda_{i-\frac{1}{2},j}^{n+1} r_{i}^{g} \frac{T_{i,j}^{n+1} - T_{i-1,j}^{n+1}}{d_{i-\frac{1}{2}}^{r}}$$

$$+ \lambda_{i,j+\frac{1}{2}}^{n+1} \frac{T_{i,j+1}^{n+1} - T_{i,j}^{n+1}}{d_{j+\frac{1}{2}}^{z}} - \lambda_{i,j-\frac{1}{2}}^{n+1} \frac{T_{i,j}^{n+1} - T_{i,j-1}^{n+1}}{d_{j-\frac{1}{2}}^{z}} + m_{i,j} \cdot \frac{\rho \cdot \Delta H_f}{M_{TiC}} k \left(T_{i,j}^{n+1} \right) \left(1 - \varsigma_{i,j}^{n+1} \right).$$

The expression used for $\lambda_{i+\frac{1}{2},j}^{n+1}$, $\lambda_{i-\frac{1}{2},j}^{n+1}$, $\lambda_{i,j+\frac{1}{2}}^{n+1}$, $\lambda_{i,j-\frac{1}{2}}^{n+1}$ is a straightforward adaptation of Eq.(20). We use the same decoupled iterative solution of the nonlinear system as in the one-dimensional case. A first-order linearization of both the stiff Arrhenius term, and the enthalpy term is done. This reinforces the strictly dominant sparse penta-diagonal matrix that is inverted at each step of the non-linear solver. It is known that the more strictly dominant a matrix is, the faster iterative solver such as the SSOR, (successive over relaxation method) converges.

3.3. Three-dimensional finite-volume discretization

We now give the discrete equations related to the finite-volume approximation of the enthalpy balance over a space-time finite volume $\Omega_{i,j,k} \times [t_n, t_{n+1}]$. Each parallelepipedic structured non uniform control volume $\Omega_{i,j,k} = [x_i, x_{i+1}] \times [x_j, y_{j+1}] \times [z_k, z_{k+1}]$, has volume $m_{i,j,k} = \int_{\Omega_{i,j,k}} d\Omega$.

$$\frac{m_{i,j,k}}{\Delta t} \cdot \rho \cdot \left(h_s \left(T_{i,j,k}^{n+1}, \varsigma_{i,j,k}^{n+1} \right) - h_s \left(T_{i,j,k}^{n}, \varsigma_{i,j,k}^{n} \right) \right) \qquad (24)$$

$$= \lambda_{i+\frac{1}{2},j,k}^{n+1} \frac{T_{i+1,j,k}^{n+1} - T_{i,j,k}^{n+1}}{d_{i+\frac{1}{2}}^{x}} - \lambda_{i-\frac{1}{2},j,k}^{n+1} \frac{T_{i,j,k}^{n+1} - T_{i-1,j,k}^{n+1}}{d_{i-\frac{1}{2}}^{x}}$$

$$+ \lambda_{i,j+\frac{1}{2},k}^{n+1} \frac{T_{i,j+1,k}^{n+1} - T_{i,j,k}^{n+1}}{d_{j+\frac{1}{2}}^{y}} - \lambda_{i,j-\frac{1}{2},k}^{n+1} \frac{T_{i,j,k}^{n+1} - T_{i,j-1,k}^{n+1}}{d_{j-\frac{1}{2}}^{y}}$$

$$+ \lambda_{i,j,k+\frac{1}{2}}^{n+1} \frac{T_{i,j,k+1}^{n+1} - T_{i,j,k}^{n+1}}{d_{k+\frac{1}{2}}^{z}} - \lambda_{i,j,k-\frac{1}{2}}^{n+1} \frac{T_{i,j,k}^{n+1} - T_{i,j,k-1}^{n+1}}{d_{k-\frac{1}{2}}^{z}} + m_{i,j,k} \cdot \frac{\rho \cdot \Delta H_f}{M_{TiC}} k \left(T_{i,j,k}^{n+1} \right) \left(1 - \varsigma_{i,j,k}^{n+1} \right).$$

The expression used for $\lambda_{i+\frac{1}{2},j,k}^{n+1}$, $\lambda_{i-\frac{1}{2},j,k}^{n+1}$, $\lambda_{i,j+\frac{1}{2},j,k}^{n+1}$, $\lambda_{i,j-\frac{1}{2},k}^{n+1}$, $\lambda_{i,j,k+\frac{1}{2}}^{n+1}$, $\lambda_{i,j,k-\frac{1}{2}}^{n+1}$ is a straightforward adaptation of Eq.(20). We use the same numerical procedure as the one used for the two-dimensional case. It is worth mentioning that a sparse strictly dominant hepta-diagonal matrix is inverted at each step of the non-linear solver.

3.4. Numerical implementation

Due to space constraints, we only describe key aspects of our numerical software entitled Hephaïstos; a toolbox library for multidimensional numerical computation of combustion synthesis problems written in C/C++. Detailed documented results related to the implementation on various single core, multi-core and many-core architectures are given in [4]. Auto-parallelization and openmp based speedup results on core i7 quad-core using gnu gcc, sun studio and open64 compilers are reported. A similar study is done on cuda-core using nvidia nvcc compiler [4]. Efforts were done to achieve the highest possible performance on processors that use memory cache hierarchy such as MIPS R1X00 processors and INTEL Core i7 processors. As an example, all loops invariant quantities are pre-computed and stored into one-dimensional arrays. Dynamically allocated multidimensional arrays are not used, because of pointer aliasing problems. We prefer to convert such arrays into one-dimensional arrays. A storage similar to CRS(Compact Row Storage) is used that takes into account the penta (hepta)-diagonal sparsity of the matrices that are iteratively inverted by SSOR and SIP methods. The postprocessing of the software's results saved to vtk format is done thanks to mayavi software for temporal snapshot and Paraview for interactive analysis and movie production.

4. 1D/2D/3D numerical study

In this section we analyze the ignition and propagation of the combustion front during titanium carbide-TiC- combustion synthesis. It is worth mentioning that the propagation of the combustion wave is either stable i.e. at constant velocity or oscillatory and can be determined according to [16] by the computation of the Zeldovich number Ze defined by $Ze = \dfrac{(T_{ad} - T_a) \, E^*}{R \, T_{ad}^2}$. $T_{ad}[K]$ is the adiabatic temperature of the reaction, $T_a[K]$ is the ambient temperature and $(k_d^*[s^{-1}], E^*[kJ/mol])$ characterizes the first order exothermic kinetics. R is the gaz constant. Experimental observations were reported in [8] to sustain this theoretical analysis. Moreover in the constant velocity case resp.(*oscillatory case i.e. periodic oscillation of the velocity of the solid flame propagation around a mean velocity*), the synthesised titanium-carbide product has uniform resp.(*non-uniform*) physical properties. For $(k_0, E^*) = (3800 \, s^{-1}, 30.0 \, kcal.mol^{-1})$, the Zeldovich number $Ze = 4.52 < Ze_c = 2 \left(2 + \sqrt{5} \right)$, therefore the solid flame propagation is stable.

4.1. 1D numerical study

This subsection presents unpublished results related to the induction time τ_{ind} in planar/cylindrical/spherical geometry such as one depicted in Fig.(21).

The boundary conditions used to analyze the combustion front propagation in the reactive mixture of length R_e are

$$-\lambda \frac{\partial T}{\partial n} (0, t) = \epsilon \, \sigma \left(T(0,t)^4 - \left(T_{f_{x=0}} \right)^4 \right), \tag{25}$$
$$-\lambda \frac{\partial T}{\partial n} (R_e, t) = \epsilon \, \sigma \left(T(R_e,t)^4 - \left(T_{f_{x=R_e}} \right)^4 \right).$$

4.1.1. Tools used to analyse the numerical simulation results

We define the induction time, starting point and ending time, three notions that will be heavily used in this section.

- The induction time τ_{ind} is the required time for the build-up of a steady-state propagation regime of the chemical reaction inside the material. It is defined as the first instant for which there exists $x \in \Omega$ such that $\xi(x, \tau_{ind}) > 0$. It depends on various parameters such as ignition temperature, heat capacity, thermal conductivity, pre-heating time, mass density.

- Starting of the reaction is $\tau_{ind} > 0$ such that $\xi(x_{ind}, \tau_{ind}) = 0.5$,

- Ending time of the reaction is time $\tau_{end} > 0$ such that $\forall x \in \Omega, \xi(x, \tau_{end}) = 1$,

- Thickness of reaction zone. Assuming that the reaction starts at τ_{ind}, then we can follow the evolution of the combustion front propagation in the material thanks to the computation of spatial and temporal evolution of points $x_M^{0.01}(t)$, $x_M^{0.50}(t)$ et $x_M^{0.99}(t)$ such that $\xi(x_M^{0.01}(t), t) = 0.01$, $\xi(x_M^{0.50}(t), t) = 0.50$ et $\xi(x_M^{0.99}(t), t) = 0.99$,

- A each instant t, $x_{max}(t)$ denotes the location which has the maximum temperature in domain Ω,

- The synthesis temperature characterizes the thermal history of the material from the point of view of the kinetics and is defined for $x \in \Omega$ by

$$T_{syn}(x) = \int_0^{+\infty} T(x, t) \frac{d\xi}{dt}(x, t) \, dx. \tag{26}$$

It was defined and used in [2].

A meaningful numerical simulation is presented in Fig.(2) and in Fig.(3). We observe that the "thickness" of the reaction-zone defined by $x_M^{0.99}(t) - x_M^{0.01}(t)$ is nearly constant. Moreover, the slope of the three curves is nearly the same in the steady state regime. This confirms that a stable propagation anticipated thanks to the computation of the Zeldovich number is effectively obtained. Nevertheless, the temporal evolution of $x_{max}(t)$ gives an information about the thermal history. Fig.(2) and Fig.(3) represent the temporal evolution of $x_{max}(t)$, $x(\xi = 0.01, t)$ $x(\xi = 0.50, t)$ and $x(\xi = 0.99, t)$ for wall temperature $T_{f_{r=0}} = 1600$ K. Phase changes are taken (*resp. not taken*) into account in Fig.(3) (*resp in Fig.(2)*). One observes on both Fig.(2) and Fig.(3) that $x_{max}(t)$ is equal to zero during preheating time, moreover taking into account phase changes as seen on Fig.(3) modifies the velocity of $x_{max}(t)$. When the reaction starts, the front propagates, and $x_{max}(t)$ evolves differently than $x_M^{0.01}(t), x_M^{0.99}(t)$. Moreover when the front reaches the extremity of the sample, then $x_{max}(t)$ increases rapidly and soon after decreases because the synthesis is finished and the radiative heat losses contribute significantly to the decreasing of the temperature field inside the material.

4.1.2. Combustion front propagation

Fig.(4) shows that the propagation of the stiff combustion front is stable, since each profile in the steady state propagation regime are equally distant from each other.

Fig.(5) shows that the energy released by the exothermic kinetics is nearly constant during the propagation of the combustion wave inside the material.

4.1.3. Contribution of furnace temperature

A sample of length $R_e = 1$cm is placed inside a furnace maintained at temperature $T_f \in [800, 2400]$K. Fig.(6) shows that both induction time τ_{ind} and ending time τ_{end} increase exponentially when T_f decreases, whenever phase change is taken into account or not.

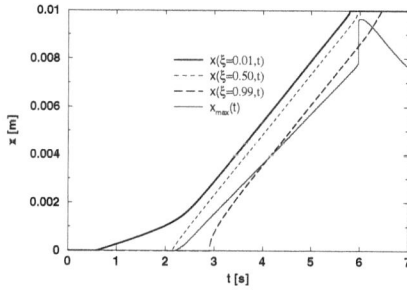

Figure 2. Temporal evolution of $x_{max}(t)$ and $x(\xi = 0.01, t)$ $x(\xi = 0.50, t)$ $x(\xi = 0.99, t)$ for wall temperature $T_{f_{r=0}} = 1600$ K. Phase changes are not taken into account.

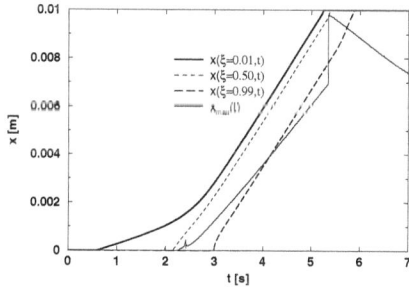

Figure 3. Temporal evolution of $x_{max}(t)$ and $x(\xi = 0.01, t)$ $x(\xi = 0.50, t)$ $x(\xi = 0.99, t)$ for wall temperature $T_{f_{r=0}} = 1600$ K. Phase changes are taken into account.

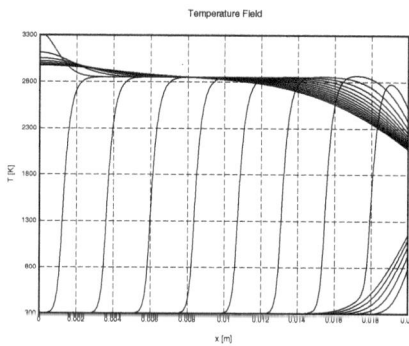

Figure 4. Temporal evolution of $T(x, t)$ at equally spaced instants. Phase change is not taken into account.

Energy released by the Kinetics

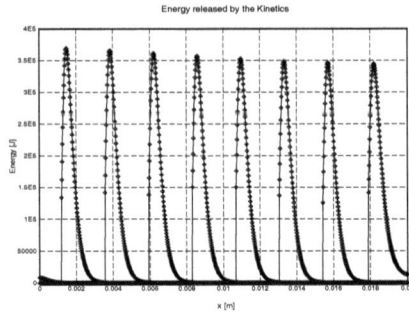

Figure 5. Temporal evolution of $\frac{\rho \Delta H_f}{M_{TiC}} k_d(T) (1 - \xi_i)$ at equally spaced instants. Phase change is not taken into account.

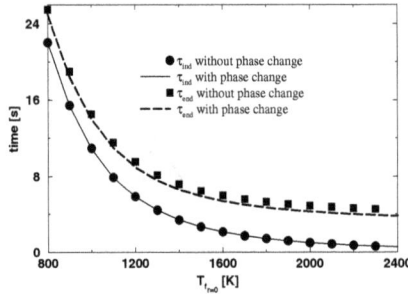

Figure 6. Evolution of τ_{ind} et τ_{end} as a function of temperature furnace $T_{f_{r=0}}$.

Fig.(7) shows that the evolution of $x_{ind}(t)$ is nearly similar whatever the furnace temperature T_f is. In fact, each curve can be translated from each other, so the behaviour is nearly independent from T_f, except from small values for which a slight curvature effect is observed. A similar conclusion can be drawn upon analyzing Fig.(8) which represents the time evolution of $T_{max}(t)$. It is also worth mentioning that a sharp peak is observed on each curve. An explanation comes from the fact that when the combustion front, at high temperature ends its propagation, it reaches the cold extremity, so an intense heat transfer occurs.

Fig.(9) shows the same behaviour for $x_{max(t)}$ whatever T_f is. The following analysis is proposed.

- A pre-heating stage, for which thanks to the heat supply at $x = 0$ (radiative boundary condition), there is a gradual increase in temperature. But $x_{max}(t) = 0$, so the temperature is below the ignition temperature.

- A combustion front propagation step $x_{max}(t)$ increases linearly with respect to time, so a constant velocity propagation of TiC synthesis is observed.

- An acceleration of the propagation which comes from the influence of the radaitive boundary condition at $x = R_e$. During a small time interval, the hot spot reaches the ending extremity of the sample.

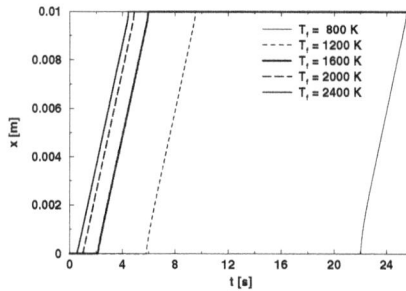

Figure 7. Temporal evolution of $x_{ind}(t)$ position , for various values of the furnace temperature $T_{f_{r=0}} \in [800, 1200, 1600, 2000, 2400]\ K$.

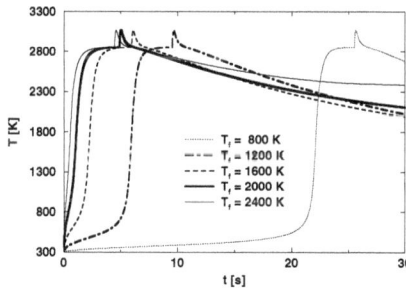

Figure 8. Time evolution of $T_{max}(t)$ for several values of temperature furnace $T_{f_{r=0}} \in [800, 1200, 1600, 2000, 2400]$.

- A cooling stage by thermal conduction, since the combustion synthesis is finished. The hot spot moves rapidly towards the center of the material thanks to the cooling. It is also observed that the velocity of the hot spot is a function of T_f.

Finally in the steady state propagation regime, each curve $x_{max}(t)$ can be translated from each other. The local heat supply induced by the boundary condition at $x = 0$ doesn't modify the characteristics of the combustion front propagation such as it's velocity. The time evolution of $T_{max(t)}$ depicted in Fig.(8) can be analyzed similarly.

Spatial distribution for different values of $T_{f_{r=0}}$ for temperature (*resp. conversion rate*) at $t = 1$s, is represented on Fig.(10). Using a suitable scaling, it is observed that their shape and stiffness is similar.

4.1.4. Energy released/absorbed by the boundary conditions

We analyze the time evolution of flux $\Phi_0(t)$, exchanged at $r = 0$ between the exterior and the material, defined by $\Phi_0(t) = -\left(\lambda \frac{\partial T}{\partial n}\right)(0, t) = \varepsilon \sigma \left(T(0, t)^4 - T_f^4\right)$. Three consecutive steps can be observed thanks to the analysis of $\Phi_0(t)$ represented by Fig.(12) and correlated to the time-evolution of $T(0, t)$

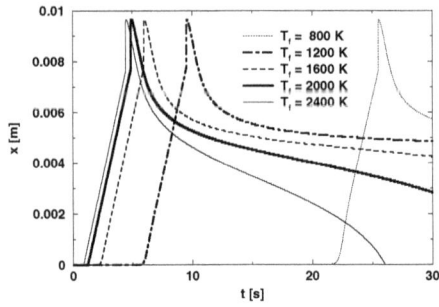

Figure 9. Time evolution of $x_{max}(t)$ for different values of furnace temperature $T_{f_{r=0}}$.

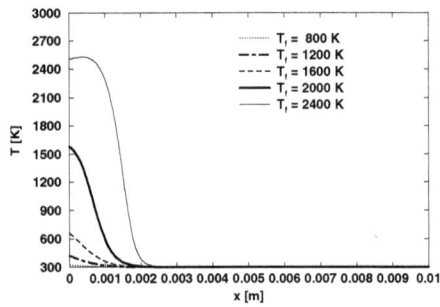

Figure 10. Spatial distribution of temperature profile at $t = 1$s, for each wall temperature $T_f \in \{800, 1200, 1600, 2000, 2400\}$ K.

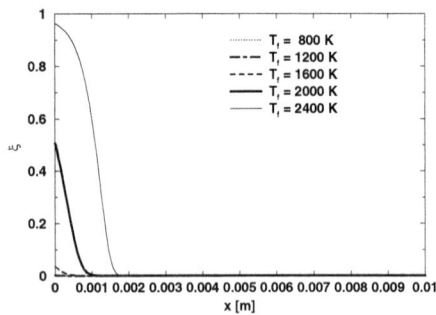

Figure 11. Spatial distribution of conversion rate profile at $t = 1$s, for each wall temperature $T_f \in \{800, 1200, 1600, 2000, 2400\}$ K.

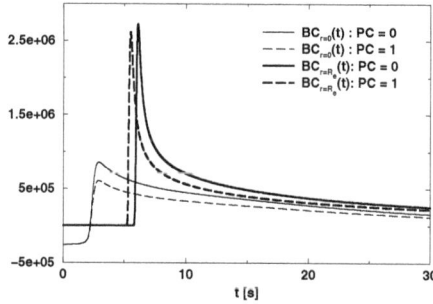

Figure 12. Time evolution of the energy released/removed by the radiative boundary conditions when phase change is taken into account or not.

(i) $T(0, t)$ increases from T_a to T_f thanks to the heat supplied by radiative heat transfer. $\Phi_0 (t)$ is negative and its absolute value decreases down to zero,

(ii) $T(0, t)$ is above T_f and reaches T_{ad} ($T_f < T_{ad}$), thanks to the exothermic reaction of titanium-carbide synthesis. The sign of $\Phi_0 (t)$ changes and its value increases to reach a maximum,

(iii)$T(0, t)$ decreases to T_f when the kinetics is ended. $\Phi_0 (t)$ decreases asymptotically to zero.

A similar analysis can be done for $\Phi_{R_e} (t) = - \left(\lambda \frac{\partial T}{\partial n} \right) (R_e, t) = \varepsilon \, \sigma \, \left(T(R_e, t)^4 - T_f^4 \right)$. It is worth mentioning that due to the thermal shock observed when the "hot" solid combustion front reaches the "cold" boundary, the amplitude $\Phi_{R_e} (t)$ is an order of magnitude higher than $\Phi_0 (t)$. Taking into account phase change doesn't modify the observation done thanks to Fig.(12).

4.1.5. Contribution of the cut-off temperature to the ignition and propagation

We assume that the chemical kinetics has a cut-off temperature T_s equal to the first phase transition $Ti_\alpha \to Ti_\beta$ temperature, $T_{\alpha\beta} = 1166K$. We observe on Fig.(13), that the discrepancy $\tau_{ign}^{T_s=0} - \tau_{ign}^{T_s=T_{\alpha\beta}}$ increases significantly when T_f decreases whenever the phase change is taken into account or not. In practice, $\left(\tau_{ign}^{T_s=0} \right)_{T_f=900\,K} = \left(\tau_{ign}^{T_s=T_{\alpha\beta}} \right)_{T_f=1600\,K}$. A similar conclusion is drawn on Fig.(14) for the difference $\tau_{end}^{T_s=0} - \tau_{end}^{T_s=T_{\alpha\beta}}$.

Below the cut-off temperature, the kinetics is not active, therefore the modelling reduces to a non-linear diffusion equation. The main phenomena is the pre-heating of the material. When the temperature reaches locally the ignition temperature for a certain amount of time, the chemical reaction starts.

We analyze the contribution of the cut-off temperature to the spatial stiffness of the combustion front through the temperature profile on Fig.(15) and conversion rate profile on Fig.(16). In both cases, the spatial resolution of the numerical discretisation scheme is satisfactory, and we observe that temperature and conversion rate profiles are stiffer when a cut-off temperature is taken into account.

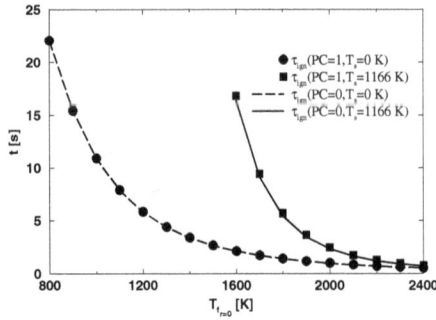

Figure 13. Evolution of τ_{ind} as a function of $t_{f_{r=0}}$ for a kinetics with/without a cut-off temperature. Phase change taken into account ($PC = 1$), or not ($PC = 0$).

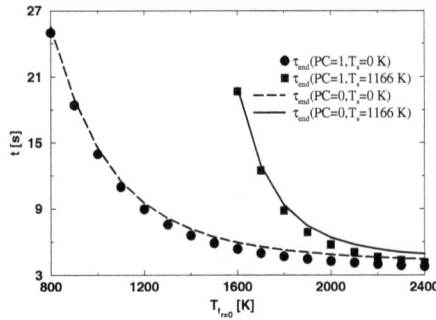

Figure 14. Evolution of τ_{ind} as a function of $t_{f_{r=0}}$ for a kinetics with/without a cut-off temperature. Phase change taken into account ($PC = 1$), or not ($PC = 0$).

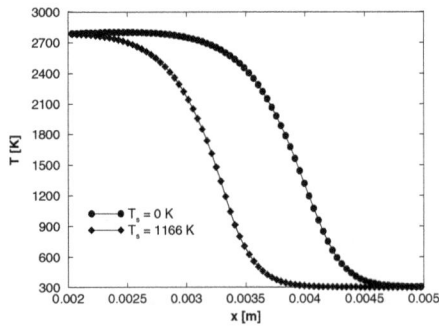

Figure 15. Spatial temperature profile at $t = 2$s, for chemical kinetic without cutoff temperature $T_s = 0$, and with cutoff temperature $T_s = T_{\alpha\beta}$.

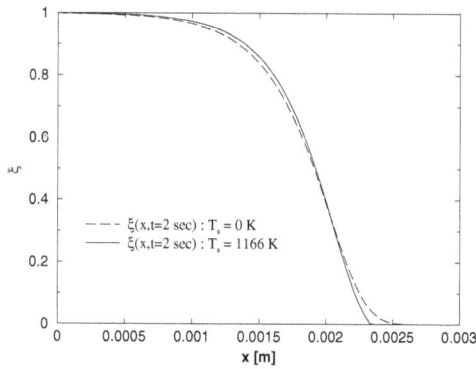

Figure 16. Spatial conversion rate profile at $t = 2$s, for chemical kinetic without cutoff temperature $T_s = 0$, and with cutoff temperature $T_s = T_{\alpha\beta}$.

4.1.6. Analysis of a double front propagation

We consider the situation when we heat identically both extremities of the cylinder with $T_{f_{r=0}} = T_{f_{r=R_e}} = T_f = 2400$ K. Obviously spatial profiles are similar and symmetrical with respect to $x = R_e/2$ as can be seen on Fig.(17)-(18).

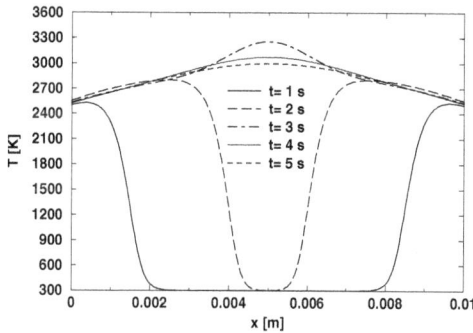

Figure 17. Spatial temperature profiles at several consecutive instants.

Moreover, when the two fronts meet each other at $x = R_e/2$, a liquid phase may appear. Temperature at the center of the material is significantly greater than the adiabatic temperature of the reaction represented in Fig.(19). The stiffness of the conversion rate in the double front case versus the single front case is also noticed in Fig.(20).

4.1.7. Contribution of the geometry to the induction time

Fig.(21) shows the evolution of $\log(\tau_{ind})(T_f)$ for slab, cylindrical and spherical geometry. The three profiles are similar, nevertheless it is observed that $\tau_{ind}^{slab} > \tau_{ind}^{cyl} > \tau_{ind}^{sph}$, for each value of T_f. Curvature effects may explain this result.

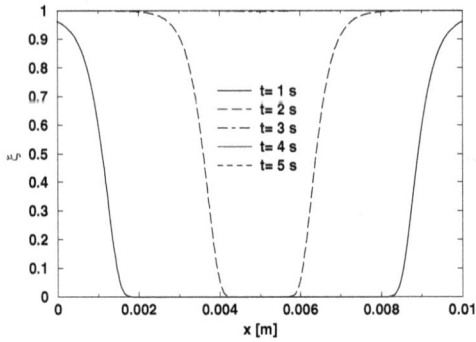

Figure 18. Spatial conversion rate profiles at several consecutive instants.

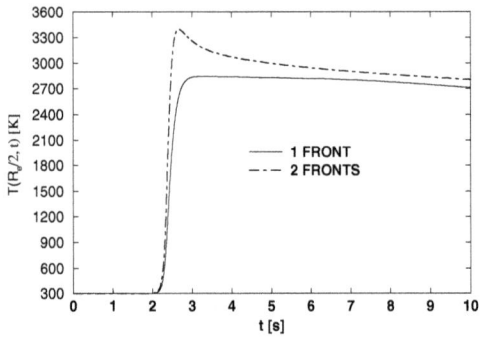

Figure 19. Temporal evolution of $T(R_e, t)$ for single and double fronts propagation.

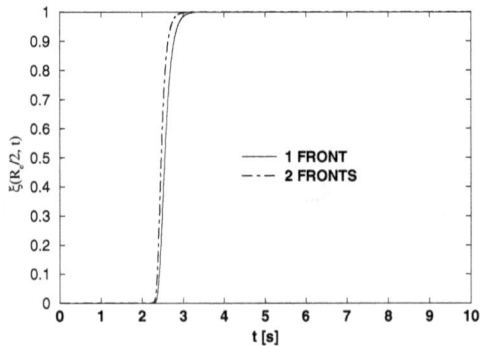

Figure 20. Temporal evolution of $\xi(R_e, t)$ for single and double fronts propagation.

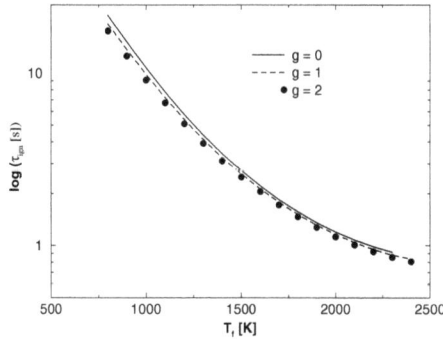

Figure 21. Evolution of the $\log(\tau_{ind})(T_f)$ for slab geometry $g = 0$, cylindrical geometry $g = 1$ and spherical geometry $g = 2$.

4.1.8. Contribution of the kinetics to the induction time

We analyze the contribution of the exponent d defined in Eq.(2) to the induction time τ_{ind} and the ending time τ_{end}. In order to obtain a better precision for $\tau_{ind}(d)$, we use fractional exponents $d \in [0, 2]$. Moreover to be able to compare the results obtained for various values of $k_d(T)$, we perform a normalization of the pre-exponential factor such that its value remain the same when $T = T_{ad}$ (the adiabatic temperature). We write therefore the equality between factors $k_0(T) = k_0^* exp\left(-\frac{E^*}{RT}\right)$ and $k_d(T) = k_d^* T^d exp\left(-\frac{E^*}{RT}\right)$ at $T = T_{ad}$

$$k_0^* = k_d^* . T_{ad}^d. \tag{27}$$

Fig.(22) represents the temperature dependance of $k_d(T)$. Ten numerical simulations are done for five equally-spaced values of $d \in [0, 2]$. For each value of d we take into account or not the phase change and use $T_{f_{r=0}} = 1600$K.

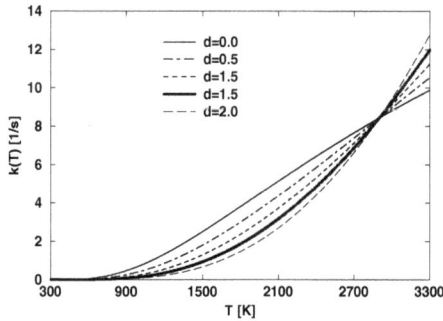

Figure 22. Temperature dependance of $k_d(T)$ for various values of exponent $d \in [0, 2]$.

We point out that

- Each curve $k_d(T)$ is an increasing function of temperature T and has the same concavity,
- When $T \leq T_{ad}$, then $k_d(T)_{d>0}$ is below $k_0(T)$,

- When $T > T_{ad}$, then $k_d(T) >> k_0(T)$. This imply that the heat released by the exothermic kinetics is significantly increasing when d increases. Temperature "overshoot" of $T_{max}(t)$ can be observed on Fig.(30).
- We conclude that the velocity of the combustion flame is higher when $d = 0$ than when $d > 0$, and the temperature obtained is super-adiabatic as seen on Fig.(30).

Fig.(23) shows that the increasing rate of the induction time τ_{ind} is super-linear, while it is linear for the ending time τ_{end} when degree d increases. It appears independent from the eventual phase change contribution. It is worth mentioning that all spatial temperature

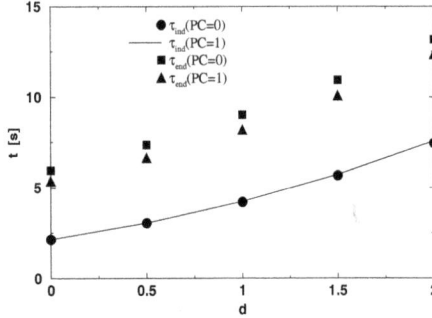

Figure 23. Comparison of induction time τ_{ind} and ending time τ_{end}, when phase change is taken into account (PC=1) or not (PC=0), for several values of exponent $d \in [0,2]$, with $T_{f_{r=0}} = 1600$ K and $T_{f_{r=R_e}} = 300$ K.

profiles represented on Fig.(24) have a discontinuity of the time-derivative when $T = T_{sl}$. This is explained by the contribution $\lambda_{fus}(f_{sl})$ of the titanium melting to the thermal conductivity given by Eq.(4).

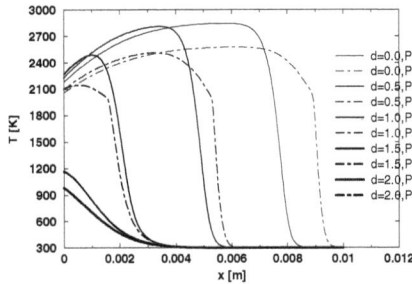

Figure 24. Spatial distribution of temperature field $T(x, t = 5s)$ for different values of exponent d, when phase change is taken into account (PC=1) or not (PC=0) with $T_{f_{r=0}} = 1600$ K and $T_{f_{r=R_e}} = 300$ K.

Conversion rate profiles represented on Fig.(25) are sharper when phase change is taken into account.

The spatial distribution of the synthesis temperature $T_{syn}(x, t = 30s)$ always increases when d increases, whether phase change is taken into account or not as seen on Fig.(26).

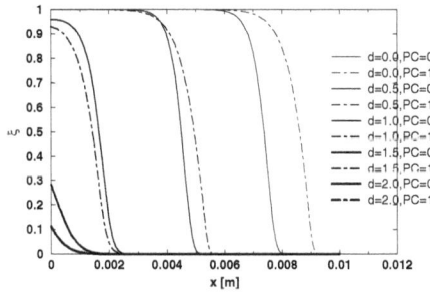

Figure 25. Spatial distribution of conversion rate field $\xi(x, t = 5\,s)$ for different values of exponent d, when phase change is taken into account (PC=1) or not (PC=0) with $T_{f_{r=0}} = 1600$ K and $T_{f_{r=R_e}} = 300$ K.

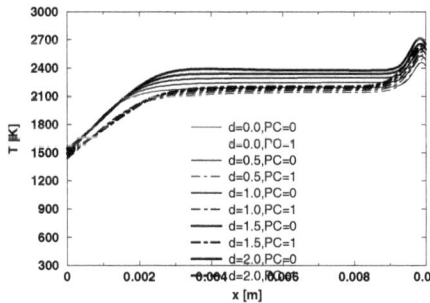

Figure 26. Spatial distribution of synthesis temperature $T_{syn}(x, t = 30\,s)$ for different values of exponent d, when phase change is taken into account (PC=1) or not (PC=0) with $T_{f_{r=0}} = 1600$ K and $T_{f_{r=R_e}} = 300$ K.

Fig.(27) shows that the time evolution of $x_{max}(t)$ has a similar shape for each value of d phase change is taken into account or not. Considering high values of d imply a significant delay for τ_{ind}. Moreover the spatial distribution of the heat released by the kinetics, as seen on Fig.(28) for $t = 5$s presents a maximum for $d = 0$, and a minimum for $d = 2$. This is correlated to the high values of induction time τ_{ind} when high values of d are used. Phase change contributes significantly to the time evolution of the front's position $x_{\xi=1/2}(t)$ as observed on Fig.(29). Without phase change, the evolution is nearly independent from the value of d since the slope of $x_{\xi=1/2}(t)$ is nearly the same. Time evolution of the front's maximum temperature $T_{max}(t)$ for different values of exponent d as seen on Fig.(30) is in fact nearly independent from d when a suitable time-translation is performed whenever phase change is taken into account or not.

4.2. 2D numerical study

This subsection computes the propagation of a dual radial/longitudinal front and a single longitudinal front for various values of the temperature furnace in which the cylindrical sample is placed. The main interest of this modelling with respect to the 1D case is to analyze the contribution of the lateral heat losses over the ignition and propagation of the

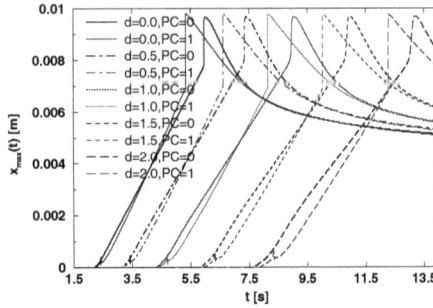

Figure 27. Time evolution of $x_{max}(t)$ for different values of exponent d when phase change is taken into account (PC=1) or not (PC=0) with $T_{f_{r=0}} = 1600$ K and $T_{f_{r=R_e}} = 300$ K.

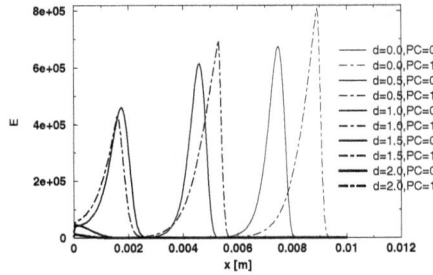

Figure 28. Spatial distribution of energy $E(x,t)$ released by the exothermic kinetics for different values of exponent d when phase change is taken into account (PC=1) or not (PC=0) with $T_{f_{r=0}} = 1600$ K and $T_{f_{r=R_e}} = 300$ K.

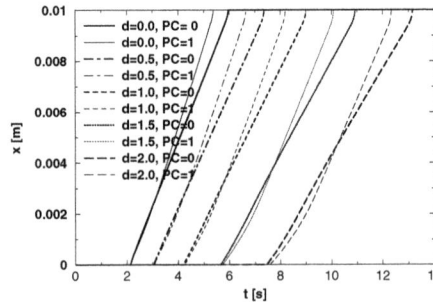

Figure 29. Time evolution of the front's position $x_{\zeta=1/2}(t)$ for different values of exponent d when phase change is taken into account (PC=1) or not (PC=0) with $T_{f_{r=0}} = 1600$ K and $T_{f_{r=R_e}} = 300$ K.

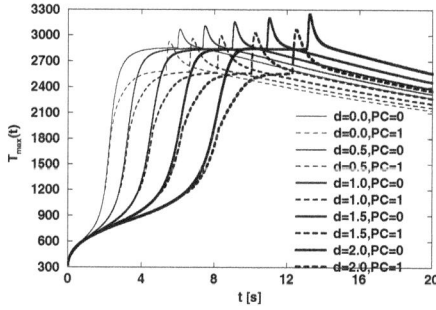

Figure 30. Time evolution of the front's maximum temperature $T_{max}(t)$ for different values of exponent d when phase change is taken into account (PC=1) or not (PC=0) with $T_{f_{r=0}} = 1600$ K and $T_{f_{r=R_e}} = 300$ K.

combustion front inside the cylindrical sample. The boundary conditions are defined over $\partial\Omega = \Gamma_{r=0} \cup \Gamma_{r=R_e} \cup \Gamma_{z=0} \cup \Gamma_{z=H_e}$ by

$$\begin{cases} \forall (r,z) \in \Gamma_{r=0} : \left(-\lambda \dfrac{\partial T(0,z,t)}{\partial n^e_{r=0}} \right) = 0, \\[2mm] \forall (r,z) \in \Gamma_{r=R_e} : \left(-\lambda \dfrac{\partial T(R_e,z,t)}{\partial n^e_{r=R_e}} \right) = \varepsilon.\sigma.\left(T(R_e,z,t)^4 - \left(T_{f_{z,r=R_e}}(t) \right)^4 \right), \\[2mm] \forall (r,z) \in \Gamma_{z=0} : \left(-\lambda \dfrac{\partial T(r,0,t)}{\partial n^e_{z=0}} \right) = \varepsilon.\sigma.\left(T(r,0,t)^4 - \left(T_{f_{r,z=0}}(t) \right)^4 \right), \\[2mm] \forall (r,z) \in \Gamma_{z=H_e} : \left(-\lambda \dfrac{\partial T(r,H_e,t)}{\partial n^e_{z=H_e}} \right) = \varepsilon.\sigma.\left(T(r,H_e,t)^4 - \left(T_{f_{r,z=H_e}}(t) \right)^4 \right). \end{cases} \quad (28)$$

The choice of the temperature triple $\left(T_{f_{z,r=R_e}}(t),\ T_{f_{r,z=0}}(t),\ T_{f_{r,z=H_e}}(t) \right)$ characterizes the way the heat supplied on the exterior surface of the cylinder (radius $R_e = 2$cm, height $H_e = 2$cm) will induce the propagation of the combustion front inside the cylinder, initially at room temperature $T = T_a = 300$ K. Each computation will analyze the phenomena during $t = 10\,s$. The computational grid is uniform along z-axis and iso-volume along r-axis.

4.2.1. Reaction-diffusion vs thermal diffusion

We choose $\left(T_{f_{z,r=R_e}}(t),\ T_{f_{r,z=0}}(t),\ T_{f_{r,z=H_e}}(t) \right) = (300\,K, 1600\,K, 300\,K)$ for both simulations. The thermal diffusion case means that the kinetics is not active, i.e. $k_d = 0$, so the initial mixture of reactive powders is not transformed. At each point (r,z) of the sample and at each instant $t > 0$, we observe on Fig.(31) that $T_a \leq T(r,z,t) \leq T_f$, when the exothermic kinetics is not taken into account, moreover the numerical solution fulfills a discrete maximum principle. A similar principle is observed when the kinetics is taken into account, but the time evolution is significantly different because of the sharp rise in temperature when the synthesis starts.

Moreover $T(R_e,0,t) < T(0,0,t)$ because of the radiative heat losses at $r = R_e$.

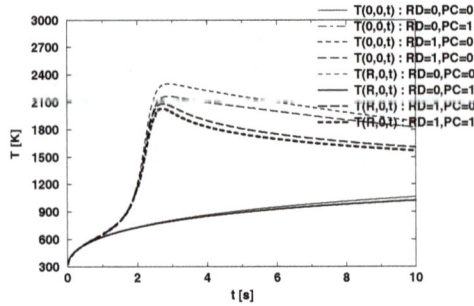

Figure 31. Time evolution of $T(0,0,t)$ et $T(R_e,0,t)$.

4.2.2. Single, dual longitudinal, dual longitudinal/single radial front

These three cases were considered and analyzed in detail in [2].

4.2.3. Dual radial-longitudinal front

We consider the case where $\left(T_{f_{z,r=R_e}}(t),\ T_{f_{r,z=0}}(t),\ T_{f_{r,z=H_e}}(t)\right) = (2400\,K, 2400\,K, 300\,K)$. At the same time a longitudinal front is moving from bottom to top, while a radial front moves from the exterior surface of the cylinder to the center, as seen on the temperature distribution at time $t = 3$s represented by Fig.(32). The conversion rate $\xi(r,z,3)$ has a similar shape. The energy released by the exothermic process is nearly uniformly distributed on Ω as seen on Fig.(33). A similar conclusion can be drawn upon the shape of the synthesis temperature $T_{syn}(r,z,5)$ represented on Fig.(34), except along the line where the two fronts are joining themselves.

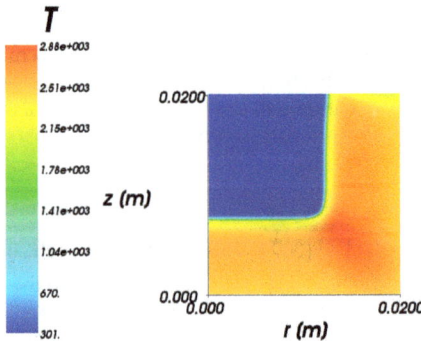

Figure 32. Temperature distribution $T(r,z,3)$ for $(r,z) \in \Omega$.

According to [2], in order to analyze the kind of propagation, we assume given a temperature $\Theta \in [300, 3000]$ and a simulation time $t_f > 0$. We define for each point $x \in \Omega$, the thermal history time $t_\Theta(x)$ which corresponds to the total time for which a given point $x \in \Omega$ remains

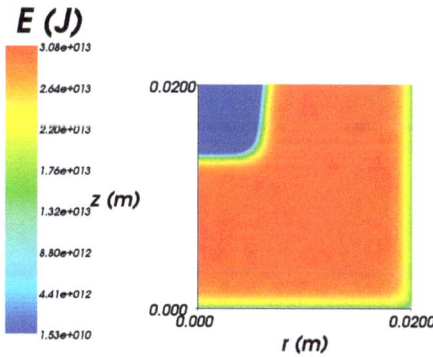

Figure 33. Integral of the energy released by the kinetics during the process at each $(r, z) \in \Omega$.

at a temperature greater or equal to Θ and gives an indication of the thermal history of the material as a function of the coupling between the exothermic reaction and the thermal diffusion. More precisely, $t_\Theta(r)$ determines if the energy involved in the reaction-diffusion process is uniformly distributed or not in Ω and what is the average temperature of the process. Fig.(35) represents such time distribution at $t = 5$s for $\Theta = 1800$K. Fig.(36) represents such time distribution at $t = 5$s for $\Theta = 2400$K. As in the previous figures, it is influenced by the propagation of the combustion wave. It is worth mentioning that on these two figures the time distribution is nearly equal, from up to 4.16s (resp. 4.70) for 1800K (resp. 2400K), and that they can be superposed.

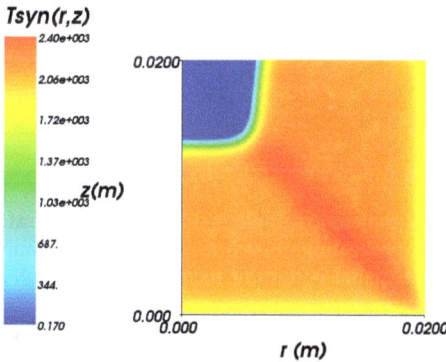

Figure 34. Synthesis temperature distribution $T_{syn}(r, z, 5)$ for $(r, z) \in \Omega$.

4.3. 3D numerical study

This subsection accounts for the heterogeneity of the radiative boundary conditions to analyze the ignition and propagation of the combustion front in a cubic sample. We assume that

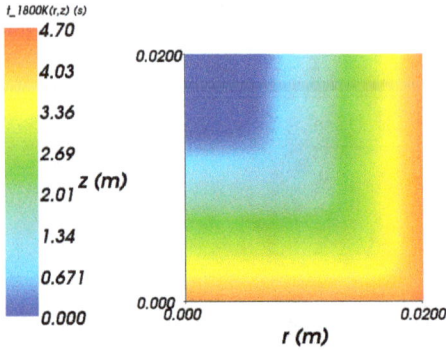

Figure 35. Time distribution of the thermal history time $T_{1800K}(r,z,5)$ for $(r,z) \in \Omega$.

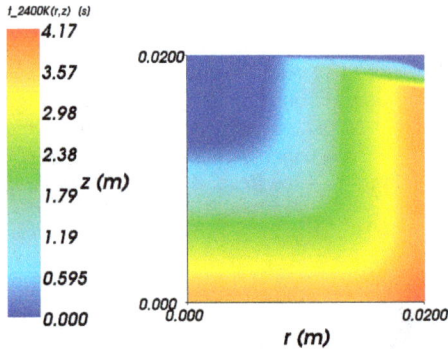

Figure 36. Time distribution of the thermal history time $T_{2400K}(r,z,5)$ for $(r,z) \in \Omega$.

the heat supplied to the six faces of the cubic sample is different, in the sense that the wall temperature that contributes to the preheating of each face of the cubic sample is different. This will induce a specific transient spatial pattern of the combustion front. A meaningful snapshot of the transient temperature profile is depicted in Fig.(37). It is clearly observed that the shape of the front's propagation inside the cube is influenced by the boundary conditions. If a regular propagation is required, providing heat supply over one single face of the cube is enough to obtain quickly titanium-carbide. So combining a different heat supply system for each face of the cube appears complicated from an experimental point of view. Numerical simulation show that the pattern of the propagation in this case is more complicated than in the single heat supply case and requires more computational power to get finely resolved. An analysis, not presented here due space constraints, and using the same methodology as in the previous 2D case show that the same conclusions can be drawn for $T_{2400K}(x,y,z,.)$, $T_{syn}(x,y,z,.)$.

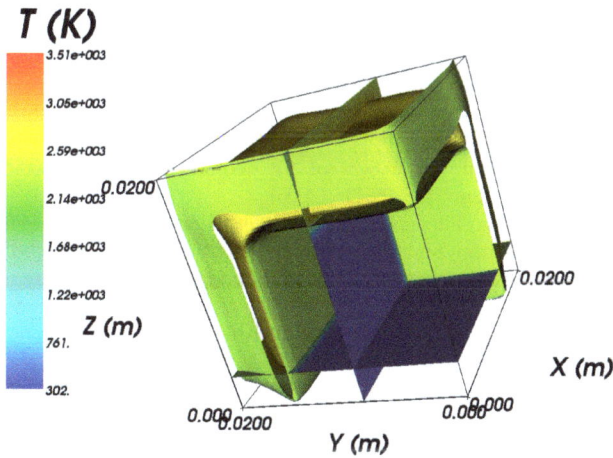

Figure 37. Temperature distribution $T(x,y,z,3)$ for $(x,y,z) \in \Omega$.

5. Conclusions and perspectives

In this book chapter, we have presented a multidimensional modelling for the numerical computation of ignition and propagation of combustion fronts during combustion synthesis of ceramic materials such as titanium-carbide. A detailed computational study was done in 1D slab/cylindrical/spherical geometry, 2D cylindrical/cartesian and 3D cartesian geometry to analyze the influence of the radiative boundary conditions over the induction time. The radiation contribution to the thermal conductivity was taken into account and the sensitivity of the induction time to several parameters such as the kinetics, wall temperature, phase change was carefully analyzed. Our numerical software Hephaïstos was presented. It implements an implicit finite-volume scheme for which error estimates, discrete maximum principles were reported and used to ensure the consistency of the numerical results. This modelling study was done for titanium carbide TiC. It can be applied to other ceramic materials such as silicium carbide-SiC-. Moreover in order to analyze the combustion front propagation for high Zeldovich number cases, an adaptive finite-volume scheme is required. A new monotonicity preserving refinement/derefinement conservative algorithm has been designed for multidimensional computations in various coordinate systems. This algorithm maintains the structured topology of the mesh. It is currently under implementation.

Author details

A. Aoufi
SMS/RMT, LCG, UMR CNRS 5146, Ecole des Mines de Saint-Etienne, 158 cours Fauriel, 42023 Saint-Etienne Cedex, France

G. Damamme
CEA, Centre DAM-Île de France, Bruyères le Châtel, 91297 Arpajon cedex, France

6. References

[1] A. Aoufi, "A finite volume scheme for a nonlinear reaction-Diffusion PDE arising in combustion", in: C. Constanda, M. Ahues, A. Largillier (Eds.), Integral Methods in Science and Engineering, Birkhauser, Basel, 2004, pp. 1-6.

[2] A.Aoufi, G.Dammame, "An implicit finite-volume discretization for the multidimensional numerical simulation of TiC combustion synthesis", Applied Mathematics and Computation, 176, 99-116 (2006).

[3] A.Aoufi, G.Dammame, "Two-Fluxes and Reaction-Diffusion Computation of Initial and Transient Secondary Electron Emission Yield by a Finite Volume Method", 89-108, in Numerical Simulations -Applications, Examples and Theory, Edited by Prof. Lutz Angermann, January 2011, InTech.

[4] A.Aoufi, "Numerical computation of multidimensional TiC combustion synthesis problems using openmp and cuda", To be submitted in (2012).

[5] Lakshmikantha M., Bhattacharya A., Sekhar J. A., "Numerical Modeling of Solidification Combustion Synthesis", Metallurgical Transactions A, 23A, 23-34, (1992).

[6] H.P.Li, J.A. Serkhar, "Dimensional Changes during micropyrectic synthesis", Material Science and Engineering, Vol A 1690 (1993), pp. 221-227.

[7] A.G. Merzhanov, The theory of stable homogeneous combustion of condensed systems, Combustion and Flame 13 (1969) 143-156.

[8] A.G. Merzhanov, Solid flames: discoveries, concept, and horizons of cognition, Combustion Science and Technology 98 (1994) 307-336.

[9] S.V. Patankar, "Numerical Heat Transfer and Fluid Flow", McGRAW-HILL BOOK COMPANY, 1980.

[10] Wei Shyy, H.S. Udaykumar, MadhuKar, M. Rao, Richar W. Smith, "Computational Fluid Dynamics with Moving Boundaries", Taylor and Francis, (1996).

[11] D.R. Stull, H. Prophet et al., "Janaf Thermochemical Tables Second Edition", NSRDS-NBS 37, Editor, US Department of Commerce, National Bureau of Standards, (1971).

[12] C.L. Tien, "Thermal Radiation in Packed Beds and Fluidized Beds", Transactions of the ASME, Vol. 110, November 1988, pp. 1230-1242.

[13] Y.S. Touloukian, D.P. Dewitt, "Thermal Radiative Properties, Nonmetallic Solids", The TPRC Data Series, Vol. 8, IFI Plenum, New-York, 1972.

[14] E. Tsotsas, H. Martin, "Thermal Conductivity of Packed Beds: A Review", Chem. Eng. Process. 22 (1987), pp. 19-37.

[15] Viljoen H.J. and Hlavacek V., "Effect of Radiation on the Combustion Rate in a Condensed Phase", AIChE J., 37, 10, 1595-1597, (1991).

[16] Transactions of mathematical monographs Volume 140 Travelling wave solutions of parabolic systems American Mathematical Society Providence Rhode Island Aizik I. Volpert, Vitaly A. Volpert, Vladimir A. Volpert 2000.

[17] D. Vrel, J.-M. Lihrmann and P. Tobaly, "Contribution of Solid-State Diffusion to the Formation of Titanium Carbide by Combustion Synthesis, Journal of Materials Synthesis and Processing, Vol. 2, No. 3, 1994, pp 179-187.

[18] N. Wakao, K. Kato, "Effective Thermal Conductivity of Packed Beds", Journal of Chemical Engineering of Japan, Vol. 2, No 1, 1969, pp. 24-32.

Chapter 10

3D Finite Element Simulation of T-Joint Fillet Weld: Effect of Various Welding Sequences on the Residual Stresses and Distortions

Nur Syahroni and Mas Irfan Purbawanto Hidayat

Additional information is available at the end of the chapter

1. Introduction

Due to the nature of welding process involving localized heat generation from moving heat source (s), rapid heating in the welded structures, and subsequent rapid cooling, problems such as residual stresses and distortions of welded structures remain great challenges to welding practitioners, designers and modeler. From modeling point of view, it will be very useful if the parameters of interest which contribute to the residual stresses and distortions in various types of welded joint and structure application can be simulated numerically so that welding performance with respect to the various aspects could be assessed and evaluated in an efficient manner (Goldak & Akhlagi, 2005; Lindgren, 2006; and Zacharia et al., 1995). Thorough consideration and assessment of the welding quality could then also be performed in earlier stage in a virtual environment. Moreover, dimensional inaccuracies due to the welding deformation giving rise problems in subsequent assembly and fabrication processes could also be predicted along with the necessary justification needed.

In recent years, various aspects and interests in the numerical modeling of welding residual stresses and distortions, mostly using finite element method, have been elaborated by researchers. Teng & Lin (1998) predicted the residual stresses during one-pass arc welding in steel plate using ANSYS software and discussed the effects of travel speed, specimen size, external mechanical constraints and preheating on the residual stresses. Tsai et al. (1999) studied the distortion mechanisms and the effect of welding sequence on panel distortion and utilized 2D finite element model. Residual stresses and distortions in T-joint fillet welds with the effects of flange thickness, welding penetration depth and restraint condition of welding was simulated by Teng et al. (2001) using thermal elasto-plastic finite element techniques. Further, effect of welding sequences on residual stresses of multi-pass butt-welds and circular patch welds was also investigated by Teng et al. (2003). Moreover, Chang & Lee (2009) performed the finite element analysis of the residual stresses in T-joint fillet welds made of similar and dissimilar steels.

The present study extends the previous work of Teng et al. (2001) and focuses on numerical simulation of welding sequence effect on temperature distribution, residual stresses and distortions of T-joint fillet welds. Several welding sequences were considered and the resulted distribution of welding temperature, longitudinal and transverse residual stresses and angular distortions were simulated utilizing three dimensional finite element models. Four welding sequences considered were one direction welding, contrary direction welding, welding from centre of one side and welding from centres of two sides. Further, a welding sequence producing the smallest residual stress, distortion as well as distortion difference between both flanges was then investigated. The numerical simulation was done in ANSYS environment.

2. Theoretical background

Basic mechanisms of welding residual stress and distortion together with the finite element formulations used in the 3D numerical simulation are described in the following sub-sections.

2.1. Basic mechanism of welding residual stresses

Complex heating and cooling cycles encountered in weldments lead to transient thermal stresses and incompatible strains produced in region near the weld. After heat cycles of welding diminished, the incompatible strains remain and provoking locked stresses or frequently termed as welding residual stresses. In general, term of residual stress deal with those remaining stress in a structure even though no external load applied (Masubuchi, 1980). Several terms having similar meaning with residual stress were found in some literatures, namely: internal stress, initial stress, inherent stress, reaction stress, lock-in stress, etc. In term of welding process, residual stress are the remaining internal stresses after welding and cooling down to room temperature.

There are two basic mechanisms to explain how residual stress produced by welding process, namely: the structural mismatch and the uneven distribution of non-elastic strain composed by plastic and thermal strains.

2.1.1. Residual stress due to mismatch

The residual stress mechanism due to mismatch may be simply illustrated in Fig. 1. Consider three carbon-steel bars of equal length and cross section connected together with two rigid blocks at the ends. The middle bar is heated up to 600ºC and then cooled to room temperature while no applied heating on the other two bars. Since the expansion of the

3D Finite Element Simulation of T-Joint Fillet Weld: Effect of Various Welding Sequences on
the Residual Stresses and Distortions

281

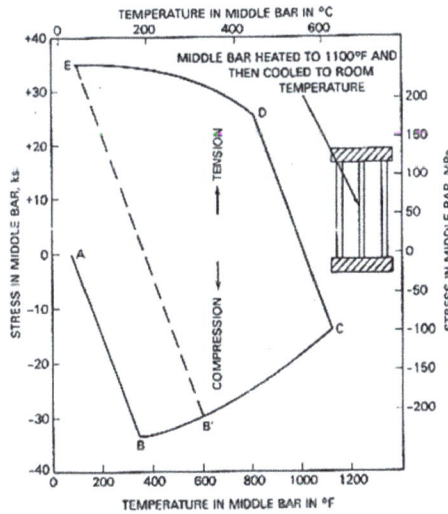

Figure 1. Illustration of residual stress mechanism in welding (source: Masubuchi, 1980)

middle bar is restricted by other bars, compressive stress is encountered at the middle bar
and the two side bars are subjected to opposite tensile stress. The compressive stress on
middle bar, increases in linear elastic manner when it is heated (AB curve) until the yield
stress of material in particular temperature reached, then plastic deformation is encountered
which affects in decreasing compressive stress (BC curve). During cooling stage, the stress
sign in middle bar is dramatically changed from compressive to tension stress and increases
in linear elastic way (CD curve) up to the yield stress at point D. Then, non-linear plastic
behaviour takes place (DE curve) in room temperature resulting in a tensile residual stress
in the middle bar and contrary a compressive residual stress in both side bars which are
equal to one-half of tensile stress in the middle bar.

2.1.2. Residual stress due to uneven distribution of non-elastic strains

When a metal bar is subjected to a uniform heat, it produces a uniform expansion lead to no
thermal stresses. However, when it is subjected to non-uniform heat as the case of welding,
thermal stresses and strains will be formed. Residual stress field in plane stress condition (σ_z
= 0) can be expressed by the following formulas:

- Elastic and plastic strains:

$$\varepsilon_x = \varepsilon_x{}' + \varepsilon_x{}'' ,$$

$$\varepsilon_y = \varepsilon_y{}' + \varepsilon_y{}'' , \qquad (1)$$

$$\gamma_{xy} = \gamma_{xy}{}' + \gamma_{xy}{}'' .$$

where:

$\varepsilon_x, \varepsilon_y, \gamma_{xy}$ is components of the total strain,

$\varepsilon'_x, \varepsilon'_y, \gamma'_{xy}$ is components of the elastic strains,

$\varepsilon''_x, \varepsilon''_y, \gamma''_{xy}$ is components of the plastic strains.

- Relationships of stress vs. elastic strain by Hooke's law:

$$\varepsilon_x' = \frac{1}{E}\left(\sigma_x - v\sigma_y\right),$$
$$\varepsilon_y' = \frac{1}{E}\left(\sigma_y - v\sigma_x\right), \tag{2}$$
$$\gamma_{xy}' = \frac{1}{G}\tau_{xy}.$$

- The stress must satisfy the equilibrium conditions:

$$\frac{\partial \sigma_x}{\partial x} + \frac{\partial \tau_{xy}}{\partial y} = 0,$$
$$\frac{\partial \tau_{xy}}{\partial x} + \frac{\partial \sigma_y}{\partial y} = 0. \tag{3}$$

- The total strain must satisfy the conditions of compatibility:

$$\left[\frac{\partial^2 \varepsilon'_x}{\partial y^2} + \frac{\partial^2 \varepsilon'_y}{\partial x^2} - \frac{\partial^2 \gamma'_{xy}}{\partial x \partial y}\right] + \left[\frac{\partial^2 \varepsilon''_x}{\partial y^2} + \frac{\partial^2 \varepsilon''_y}{\partial x^2} - \frac{\partial^2 \gamma''_{xy}}{\partial x \partial y}\right] = 0. \tag{4}$$

The second term of Eq. (4), which is called the incompatibility term, R, is determined by plastic strain. When the value of R is not zero, thus residual stresses will exist in the weld joint.

$$R = -\left[\frac{\partial^2 \varepsilon''_x}{\partial y^2} + \frac{\partial^2 \varepsilon''_y}{\partial x^2} - \frac{\partial^2 \gamma''_{xy}}{\partial x \partial y}\right]. \tag{5}$$

More realistic illustration of the residual stress mechanisms during welding in typical plate joints is shown in Fig. 2. Welding bead is made along x-axis on the plate. Welding is carried out by moving the welding arc at speed v, and presently it is located at the origin O, as illustrated in Fig. 2a. Temperature distributions along particular points at weldline are shown in Fig. 2b, while stress resulted in the respect points are shown in Fig. 2c.

Along point A-A which is located ahead of the welding arc is not affected by heat yet. Section B-B experiences highest heat distribution (Fig. 2b. 2) which results in compressive stresses at just besides of weldline and surrounded by opposite tensile stresses in the side far

3D Finite Element Simulation of T-Joint Fillet Weld: Effect of Various Welding Sequences on
the Residual Stresses and Distortions

283

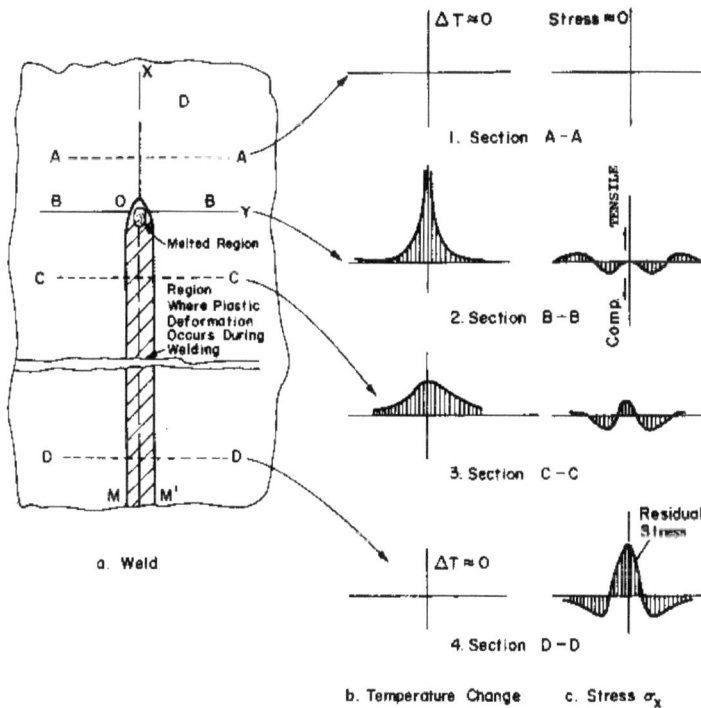

Figure 2. Schematic illustrations of heat cycles in welding and residual stress results (source:
Masubuchi, 1980)

away from weld line, whilst at weldment has zero stress due to metal melted (Fig. 2c. 2).
Section C-C which is located at some distance behind welding arc is subjected to moderate
heat (Fig. 2b. 3) due to cooling stage started in this section in which the condition at this
section is similar to those CD curve in Fig. 1. Some distance far away from heat source,
cooling down into room temperature is achieved which results in residual stresses in similar
way to those in the end of DE curve in Fig. 1.

Furthermore, typical distributions of butt joints in plate are presented in Fig. 3. Components
of residual stress are categorized into transverse and longitudinal, designated as σ_x and
$\sigma_y$respectively (Fig. 3a). Across the weldline, tensile residual stress in longitudinal direction
parallel to the weldline is found in the weldment region and compressive residual stresses
occur in the others region away from weldline (Fig. 3b). Transverse residual stresses
distributions along weldline are typically compressive part in the ends of plate, otherwise
are tensile part with magnitude of stresses is lower than longitudinal residual stress (Fig.
3c). Masubuchi & Martin (Masubuchi, 1980) have developed the distribution of longitudinal
residual stress σ_x which can be estimated as follows:

$$\sigma_x = \sigma_m \left\{ 1 - \left(\frac{y}{b} \right)^2 \right\} e^{-\frac{1}{2}(y/b)^2}.$$ (6)

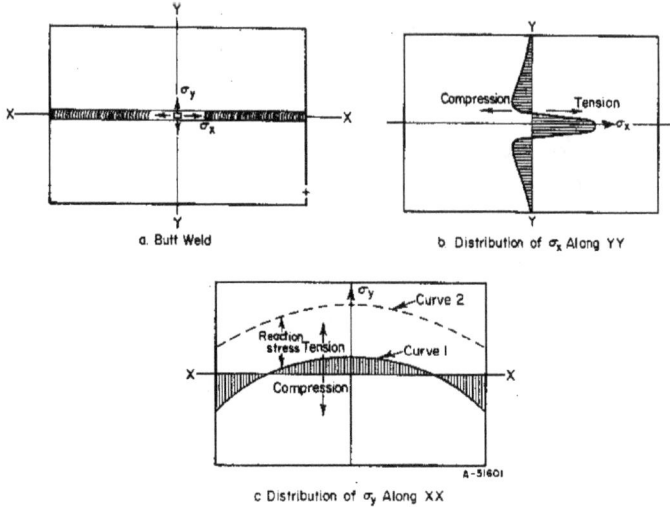

a. Butt Weld   b Distribution of σ_x Along YY

c Distribution of σ_y Along XX

Figure 3. Typical distributions of residual stress in plate butt joints (source: Masubuchi, 1980)

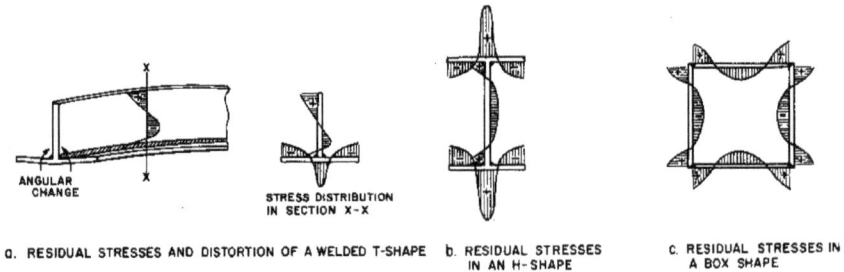

a. RESIDUAL STRESSES AND DISTORTION OF A WELDED T-SHAPE   b. RESIDUAL STRESSES   c. RESIDUAL STRESSES IN
                                                              IN AN H-SHAPE   A BOX SHAPE

Figure 4. Typical residual stresses in welded structural profiles (source: Masubuchi, 1980)

Fig. 4a shows residual stresses produced in welded T-shape and the residual stresses distributions. As can be further seen, high tensile residual stresses parallel to the axis are produced in areas near the weld in section away from the end of the column. In addition, stresses in the flange are tensile near the weld and compressive away from the weld. The tensile stresses near the upper edge of web are due to longitudinal bending distortion caused by longitudinal shrinkage. Furthermore, Figs. 4b and 4c show the typical distribution of residual stress in an H-shape and a box shape, respectively, particularly the distributions of residual stresses parallel to the weld line, in which the residual stresses are tensile in areas near the welds and compressive in area away from the welds.

2.2. Welding distortions

Distortion is closely related to the amount of residual stress and the degree of joint restraint during welding process. The correlation between distortion and residual stress is illustrated in Fig. 5. As rule of thumb, the welded joint with lower degree of restraint has an advantage due to less residual stress but it tends to get higher distortion. Conversely, the welded joint with higher degree of restraint has less distortion but it will further result in higher residual stress.

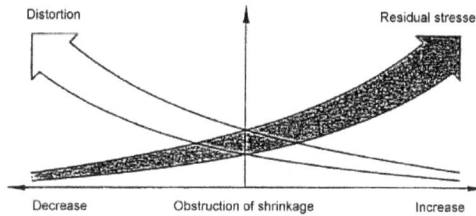

Figure 5. Welding residual stress and distortion correlation (source: Bette, 1999)

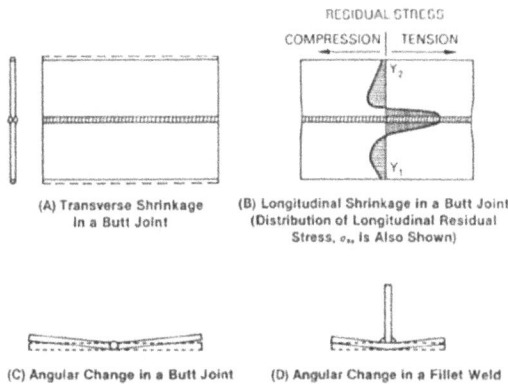

Figure 6. Three basic dimensional changes during welding (source: AWS Welding Handbook, 1987)

There are three basic dimensional changes during welding process with which we can easily understand the mechanism of distortion, namely:

- Transverse shrinkage, Fig. 6A, is a distortion perpendicular to the weld line
- Longitudinal shrinkage, Fig. 6B, is a distortion parallel to the weld line
- Angular change, in butt joint and T joint fillet weld, as shown in Figs. 6C and 6D, respectively, deformation in rotation form around the weld. It happens when the transverse shrinkage is not uniform in the thickness direction

In actual structures, the welding distortions are frequently more complex than these basic distortions or taking place with some conditions. For examples, pure transverse or longitudinal shrinkage will only take place when the following conditions apply, i. e.

thickness of member is large enough and centre of gravity of the welds is in line with the neutral axis of the components. When it is not the case, the rotational deformations such as the angular, bending and buckling distortion may be happened.

The empirical formula to estimate the quantity of transverse shrinkage of carbon and low alloy steel butt welds can be found in American Welding Society (AWS) Welding Handbook (1987) as follows:

$$S = 0.2\frac{A_w}{t}0.05d. \tag{7}$$

where:

S is transverse shrinkage, in,
A_w is cross sectional area of weld, in²,
t is thickness of plate, in,
d is root opening, in.

In fillet weld, the amount of transverse shrinkage is less than that happened in butt weld. The transverse shrinkage in fillet weld may be expressed by the following formulas found in AWS Welding Handbook (1987):

- For T-joint with two continuous fillet welds:

$$S = C_1\left(\frac{D_f}{t_b}\right). \tag{8}$$

where:

S is transverse shrinkage, in. or mm,
D_f is fillet leg length, in. or mm,
t is bottom plate thickness, in. or mm,
C_1 is 0. 04 or 1. 02 when using unit in. or mm, respectively.

- For lap joint with two fillet welds (the thickness of two plates are equal):

$$S = C_2\left(\frac{D_f}{t}\right). \tag{9}$$

where:

S is transverse shrinkage, in. or mm,
D_f is fillet leg length, in. or mm,
t is plate thickness, in. or mm,
C_2 is 0. 06 or 1. 52 when using unit in. or mm, respectively.

Compared to transverse shrinkage, the quantity of longitudinal shrinkage for butt joint is much less, approximately 1/1000 of the weld length. King, 1944 (as cited in AWS Welding

Handbook, 1987) proposed a formula to estimate the longitudinal shrinkage of butt joint as follows:

$$\Delta L = \frac{C_3 I L}{t} 10^{-7}.$$ (10)

where:

ΔL is longitudinal shrinkage, in. or mm,
I is welding current, A,
L isweld length, in. or mm,
t is plate thickness, in. or mm,
C_1 is 12 or 305 when using unit in. or mm, respectively.

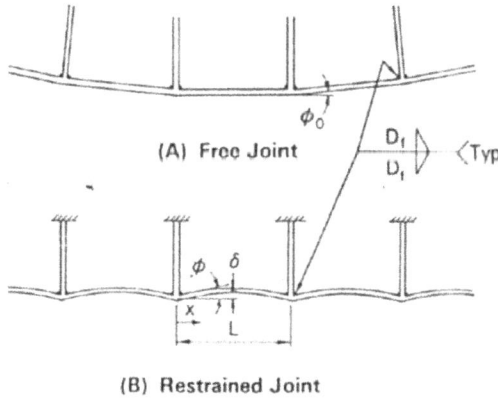

(A) Free Joint

(B) Restrained Joint

Figure 7. Angular change in T-joint fillet weld, (A) free restrained stiffeners, (B) restrained stiffeners

The primary source of angular change is due to non-uniform of transverse shrinkage in thickness direction. Fig. 7a shows angular change of the free restraint T-joint fillet weld. When the stiffeners are prevented from moving, a wavy distortion occurs as can be seen in Fig. 7b. Masubuchi et al., 1956 (as cited in AWS Welding Handbook, 1987) established a relationship between angular change and distortion at fillet weld using a rigid frame analysis in the following expression:

$$\frac{\delta}{L} = 0.25\phi - \left[\left(\frac{x}{L}\right) - 0.5\right]^2 \phi.$$ (11)

where:

δ is distortion,
L is length of stiffener spacing,
ϕ is angular change,
x is distance from centreline of frame to the point at which δ is measured, Fig. 7b.

To summary this section, many factors affect the welding process, thus the produced residual stresses and distortions, such as types of material, types of welded joints, structure thickness, joint restraint, heat input as well as welding sequence, which is the subject of the present study.

2.3. Thermal and Mechanical Finite Element Equations

The corresponding finite element equations of thermal and mechanical are obtained by choosing a form of interpolation function representing the variation of the field variables, namely temperature, T and displacement, U, within the corresponding finite elements of the structural model and by applying further the weighted-residual or variational argument to the mathematical models. Furthermore, with imposing the boundary and initial conditions, the discritized equations obtained are solved by finite element techniques through which the approximated solution over the finite element model considered could then be obtained.

The thermal finite element equation including boundary condition may be written as follows:

$$[C]\{\dot{T}\} + [K]\{T\} = \{F_T\}, \tag{12}$$

in which:

$$[C] = \int_V \varrho c [N]^T [N] dV, \tag{13}$$

$$[K] = \int_V k [B]^T [B] dV + \int_S h_f [N]^T [N] dS, \tag{14}$$

$$\{F_T\} = \int_V Q [N]^T dV + \int_S h_f T_{ref} [N]^T dS. \tag{15}$$

where:

ρ is the density (kg/m³),
c is the specific heat (J/kg. K),
k isthe conductivity (W/m. K),
h_f is the convective heat transfer coefficient (W/m². K),
Q isthe rate of internal heat generation per unit volume (W/m³),
[N]   is the matrix of element shape functions,
[B]   is the matrix of shape functions derivative, and
$\{T\}$ is the vector of nodal temperature.

The results of temperature distribution and history obtained from Eq. (12) are then inserted into the mechanical model in the form of thermal load. Incorporating the elasto-plasticity analysis, the mechanical finite element equation may be written in the form of incremental as:

$$^{i+1}\left[K_1\right]\{\Delta U\} - {}^{i+1}\left[K_2\right]\{\Delta T\} = {}^{i+1}\{R\} - {}^{i}\{R\}, \tag{16}$$

in which:

$$\left[K_1\right] = \int_V \left[B\right]^T \left[D^{ep}\right]\left[B\right]dV, \tag{17}$$

$$\left[K_2\right] = \int_V \left[B\right]^T \left[C^{th}\right]\left[M\right]dV, \tag{18}$$

$$\{R\} = \int_S \left[N\right]^T\{p\}dS + \int_V \left[N\right]^T\{f\}dV, \tag{19}$$

$$\left[D^{ep}\right] = \left[D^e\right] + \left[D^p\right]. \tag{20}$$

where:

$\{\Delta U\}$ is the incremental of nodal displacement,
$\{\Delta T\}$ is the incremental of nodal temperature,
[B] is the matrix of strain-displacement,
$[D^e]$ is the matrix of elastic stiffness,
$[D^p]$ is the matrix of plastic stiffness,
$[C^{th}]$ is the matrix of thermal stiffness,
[M] is the temperature shape function,
$\{p\}$ is the vector of traction or surface force,
$\{f\}$ is the vector of body force, and
i is the current step of analysis.

The vector of nodal displacement at the next step of analysis, $^{i+1}\{U\}$ could be obtained from:

$$^{i+1}\{U\} = {}^{i}\{U\} + \{\Delta U\}. \tag{21}$$

Furthermore, the updated condition of stress in the structure could be obtained from the following stress-strain relation:

$$^{i+1}\{\sigma\} = {}^{i}\{\sigma\} + \{\Delta\sigma\}, \tag{22}$$

$$\{\Delta\sigma\} = \left[D^{ep}\right]\lfloor B\rfloor\{\Delta U\} + \left[C^{th}\right]\left[M\right]\{\Delta T\}. \tag{23}$$

Commonly, the iterative method of Newton-Raphson is employed in the finite element solver to solve the nonlinear equations. For further treatment, see (Bathe, 1996). Note also that from the thermal analysis results, the updated stress and displacement conditions are now obtained.

3. Material and methods

In this study, material used for the welding simulation was SAE 1020 with the material properties vary according to the temperature history (Teng et al, 2001 and ASM, 1990). In addition, the welding parameters used in this analysis were as follows: single pass GTAW welding method, welding current, $I = 260$ A, welding voltage, $V = 20$ V, and welding speed, $v = 5$ mm/s.

3.1. The variations of welding sequence

Several welding sequences (WS) were considered in this study and the numerical investigation of the resulted temperature distribution, longitudinal and transverse residual stresses and angular distortions due to the welding sequences was then carried out. Four welding sequences considered were the one direction welding (WS-1), the contrary direction welding (WS-2), the welding from centre of one side (WS-3), and the welding from centres of two sides (WS-4), which are illustrated in Fig. 8.

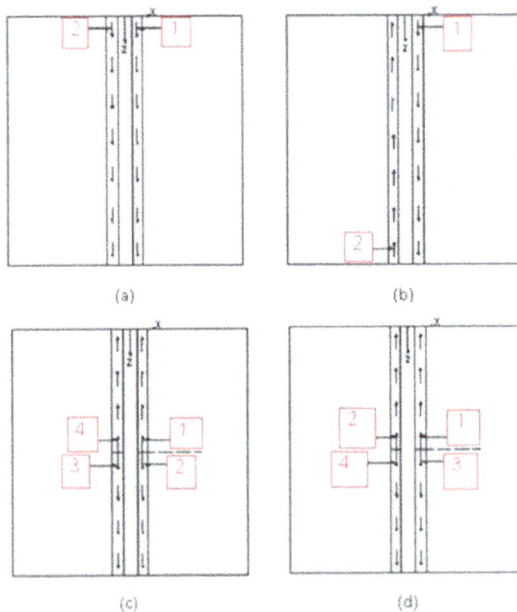

(a)   (b)

(c)   (d)

Figure 8. Variation of welding sequence employed in this study: (a) the one direction welding (WS-1), (b) the contrary direction welding (WS-2), (c) the welding from centre of one side (WS-3), and (d) the welding from centres of two sides (WS-4).

3. 2. Finite element simulation of welding

In the present study, a thermal elasto-plastic finite element procedure was employed to simulate the thermo-mechanical response of welding problem. In the procedure, two sequenced thermal and mechanical analyses were carried out independently (uncoupled) to obtain the total or desired response of the welding structure modelled.

A transient thermal analysis of heat conduction was carried out in the first step to obtain temperature distribution histories over the structural model. In the thermal analysis, the welding heat input, Q_a was calculated according to Masubuchi (1980) and the arc efficiency, η_a for GTAW was assumed to be 0. 60 (Grong, 1994). Also, the values of convective heat transfer coefficient, h_f and reference temperature were taken, respectively, to be 15 W/m². K and 25°C (298. 15 K).

In the next step, a structural analysis was carried out to now obtain the mechanical response of the structural model, where the temperature history obtained from the first step was employed as a thermal load in the analysis. The material model of elasto-plastic based on the von Mises yield criterion and isotropic strain hardening rule was chosen, in which its response over the history was determined by the temperature-dependent material properties inputted. The boundary condition or constraint on the structural model needs also to be assigned accordingly.

Fig. 9 represents the mesh of T-joint fillet weld employed in this study along with the position of constraint assigned on the finite element model. The total number of nodes and elements utilized for the 3D model were 3654 and 2961, respectively. The analyses were implemented in ANSYS environment utilizing the element type of SOLID70 for the thermal analysis and that of SOLID45 for the structural analysis.

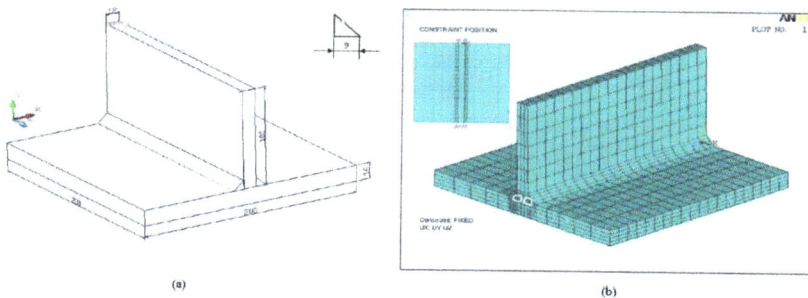

(a)   (b)

Figure 9. (a) Geometry of T-joint fillet welds, (b) Mesh of T-joint fillet weld along with its constraint position.

4. Results and discussion

With the finite element procedures described in the previous section, results on the problem considered are presented in this section. The finite element simulation for all the variation of welding was completed in 45 load-steps (LS). During the number of load-steps, the welding process took for 40 load-steps, while the cooling one took for the rest of the LS. For the presentation of welding simulation, the results of the LS which respectively represent the conditions of the peak temperature and the beginning of cooling processes were taken and plotted. Note that the temperature went down towards the reference (room) temperature after the LS of 41. Accordingly, the longitudinal and transverse residual stresses and the distortions occurred due to the welding sequences were presented and discussed.

4.1. Welding simulations and temperature distributions

First, thermal profile produced during welding as the heat source travels is presented as shown in Fig. 10. Fig. 10 represents the thermal profiles on several selected nodes along one fillet weld taken from WS-1 simulation results. It was shown that heat was moving as the welding heat source travelled. This can also be seen from the high temperature of the next adjacent node when the previous node has achieved its peak temperature. In addition, the next adjacent node's peak temperature was higher than that of the previous one, which also indicated that heat was accumulated. Subsequently, it has been distributed through the welding structure and the heat release to the surroundings was due to convective heat transfer.

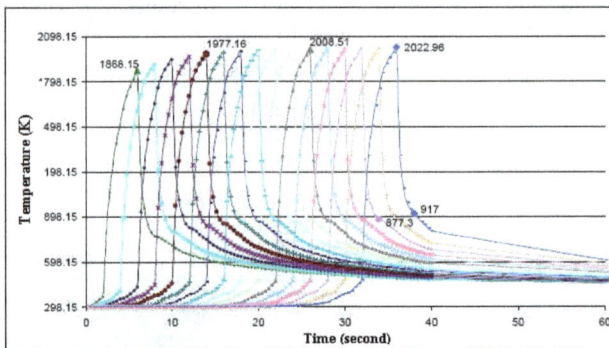

Figure 10. Thermal profiles on several selected nodes along the fillet weld.

Figs. 11 - 14 illustrate the welding simulation showing the peak temperature for each welding sequence and the temperature distribution after welding towards the room temperature. From the temperature distributions, it is clear that the peak temperature achieved in the welding was greatly affected by the welding sequence. The welding sequences produced different interaction between the current step and the accumulation of heat carried out from the previous steps due to the sequential path followed.

3D Finite Element Simulation of T-Joint Fillet Weld: Effect of Various Welding Sequences on
the Residual Stresses and Distortions

293

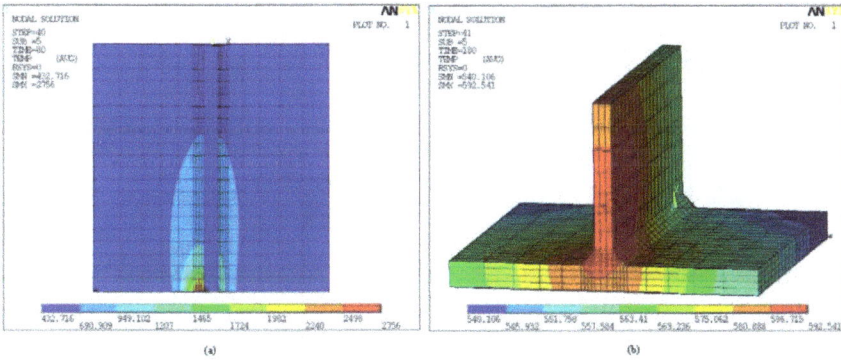

Figure 11. The welding simulation for WS-1: (a) the peak temperature achieved at the LS of 40, and (b) the temperature distribution after the welding process at the LS of 41.

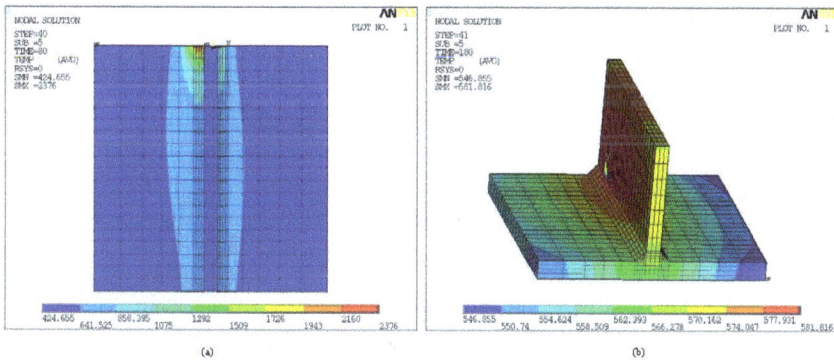

Figure 12. The welding simulation for WS-2: (a) the peak temperature achieved at the LS of 40, and (b) the temperature distribution after the welding process at the LS of 41.

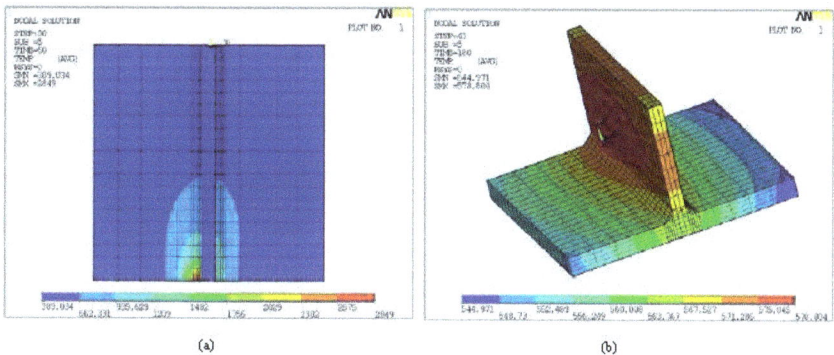

Figure 13. The welding simulation for WS-3: (a) the peak temperature achieved at the LS of 30, and (b) the temperature distribution after the welding process at the LS of 41.

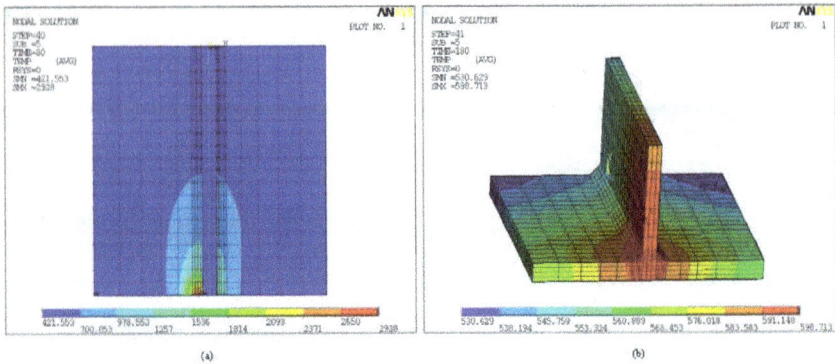

Figure 14. The welding simulation for WS-4: (a) the peak temperature achieved at the LS of 40, and (b) the temperature distribution after the welding process at the LS of 41.

The peak temperature achieved for each welding sequence as well as the peak temperature difference between WS were summarized in Table 1, in which the highest peak temperature of 2928 K belongs to WS-4 having the highest heat accumulation at the end of the welding process. The shapes of the temperature profile at the fillet welds during welding were depicted in Fig. 15.

From Fig. 15, it can be seen the differences of the temperature profile at the fillet welds during different WS. It is interesting to note that in general the temperature profiles of WS-1 and WS-2 tend to be similar. In a less extent, it also happened for those of WS-3 and WS-4, as the peak temperature of WS-3 was achieved at the LS of 30. Nevertheless, the peak temperature achieved was very different, even for the WS having similar temperature profiles such as WS-1 and WS-2. This verified again that the peak temperature achieved in the welding was greatly affected by the welding sequence.

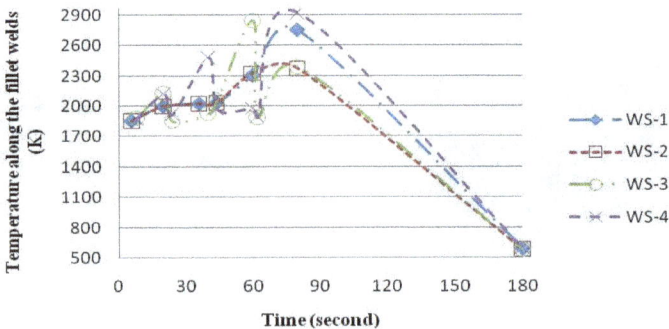

Figure 15. Peak temperature for each welding sequence.

Table 1 describes further the peak temperature achieved in a WS and the peak temperature difference between WS, in which the smallest and largest peak temperature differences between WS were 79 and 552 K, respectively.

Moreover, it may be also interesting to note how the peak temperature achieved in a WS may be related to the corresponding residual stresses and angular distortions produced.

Welding sequence (WS)	Load-step (LS)	The peak temperature achieved [K]	The peak temperature difference between WS [K]
4	40	2928	-
3	30	2849	79
1	40	2756	93
2	40	2376	380

Table 1. The peak temperature achieved for each welding sequence.

4. 2. Residual stress distributions

Fig. 16 and 17 shows respectively the simulated distributions of longitudinal and transverse residual stresses for each welding sequence investigated in this study. It is seen from Fig. 16 and 17, the maximum values of the longitudinal and transverse residual stresses occurred in the weld bead region for all the welding sequences. Note also that the distribution of the residual stresses produced from each of the welding sequences.

It can be seen that the smallest longitudinal and transverse residual stresses occurred in WS-2. It is interesting to note that the welding sequence also had the lowest peak temperature as indicated in Table 1. Also, for longitudinal residual stresses, their distributions due to the welding sequences tend to be similar. For transverse ones, the distributions were different. It seems that for the later, it could be related to the way of the welding had been performed.

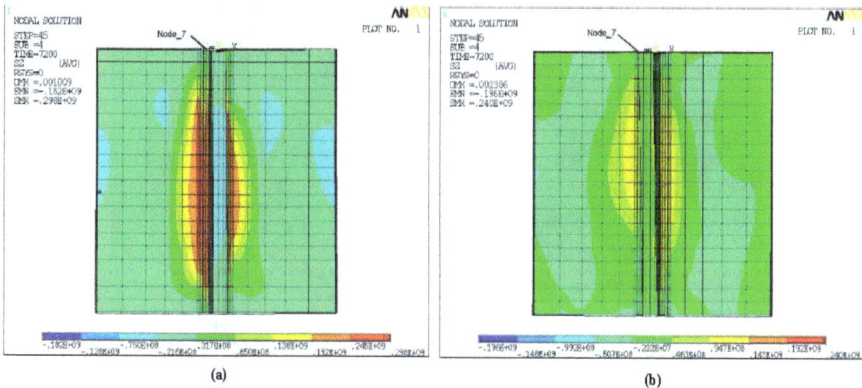

(a)   (b)

(c)   (d)

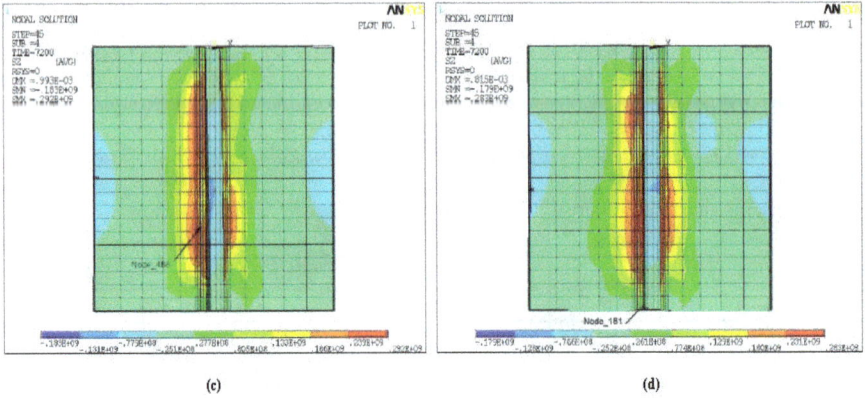

Figure 16. Simulated distributions of longitudinal residual stresses for: (a) WS-1, (b) WS-2, (c) WS-3, and (d) WS-4.

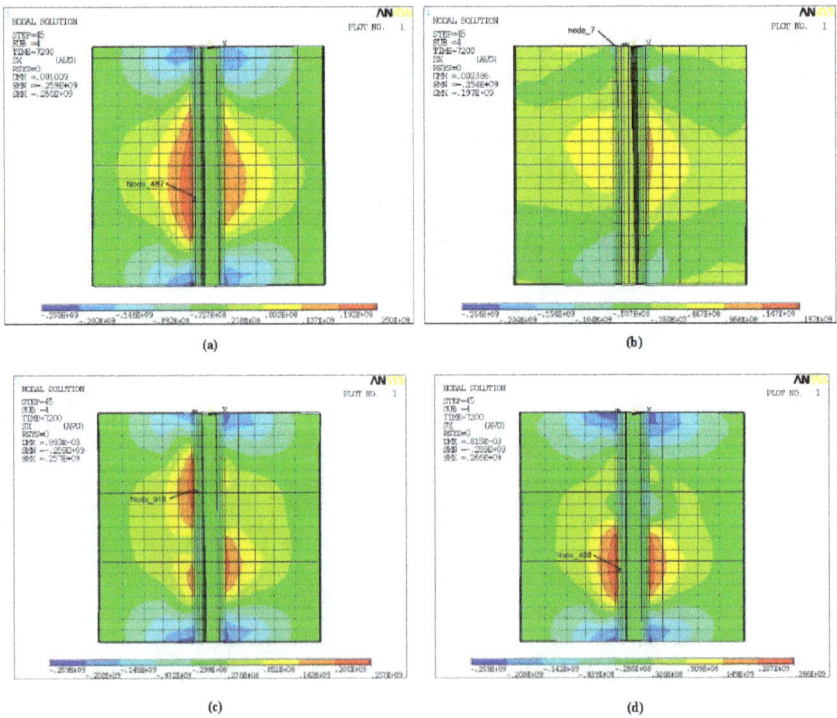

(a)   (b)

(c)   (d)

Figure 17. Simulated distributions of transverse residual stresses for: (a) WS-1, (b) WS-2, (c) WS-3, and (d) WS-4.

Fig. 18 describes the transverse residual stress distribution along the fillet weld for each WS. The maximum values of longitudinal and transverse stresses as well as von Mises stress for each welding sequence were summarized in Table 2. The ratio between the longitudinal and the transverse residual stress values for the problem considered varies from 1. 06 to 1. 22.

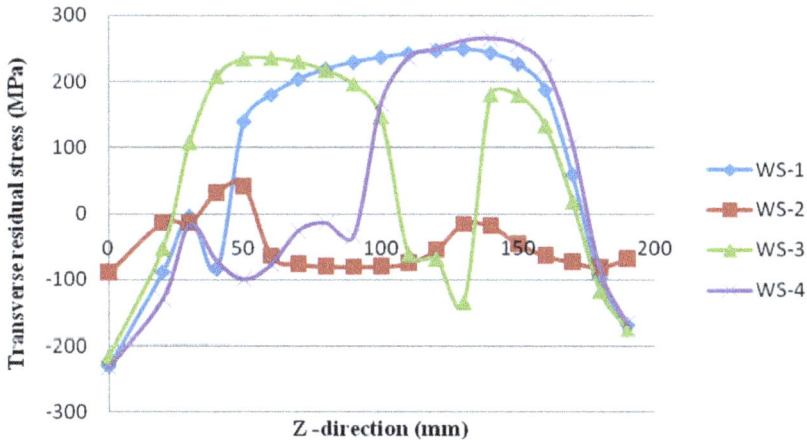

Figure 18. Distribution of transverse residual stress along the fillet weld for each welding sequence.

Observing further Fig. 18, it is also interesting to note the consistency of trends of the transverse residual stresses distributions produced by the WS simulated in the present study. It can be clearly observed that the distributions of transverse residual stresses produced by WS-3 and WS-4 and WS-1 and WS-2, respectively, are in consistent nature with respect to the welding sequences.

Welding sequence (WS)	The maximum longitudinal stress value [MPa]	The maximum transverse stress value [MPa]	The maximum von Mises stress value [MPa]
2	240	197	117
4	283	266	251
3	292	257	249
1	298	250	250

Table 2. The maximum longitudinal and transverse stress values for each welding sequence.

4.3. Distortions

Fig. 19 illustrates the distortions of the welding structure due to the welding sequences. The initial undeformed configurations were also shown. From Fig. 19, it can be seen the angular distortions occurred in both flanges. It can be further revealed that there was the difference of distortion between the flanges showing that the distortion was unsymmetrical. The maximum value of angular distortion took place on the right flange for all the welding sequences, unless that of WS-2 which took place on the left one. The simulation results obtained also clearly indicate the influence of the welding sequences examined in the present study to the angular distortions of the T-joint fillet weld considering the same boundary conditions appliedin the corresponding FEM models of the T-joint fillet weld.

Furthermore, Table 3 summarizes the vertical displacements and the angular distortions of both flanges due to the welding sequences. The angular distortion differences were also shown in Table 3.

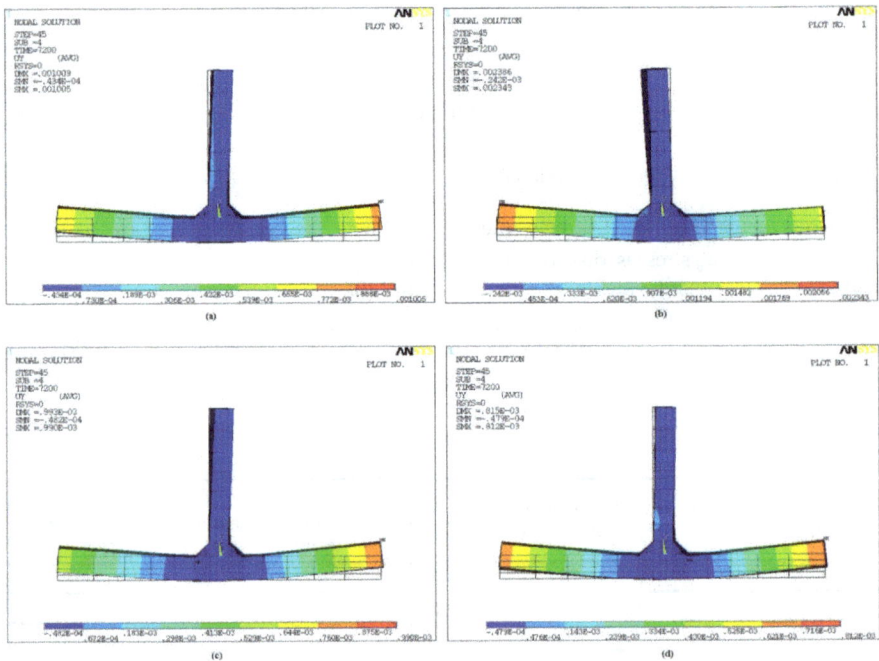

Figure 19. Distortions of the welding structure due to the welding sequences: (a) WS-1, (b) WS-2, (c) WS-3, and (d) WS-4.

3D Finite Element Simulation of T-Joint Fillet Weld: Effect of Various Welding Sequences on
the Residual Stresses and Distortions

299

Welding sequence (WS)	The left flange (X = -100 mm)		The right flange (X = 100 mm)		Angular distortion difference [rad]
	Uy [mm]	Angular distortion [rad]	Uy [mm]	Angular distortion [rad]	
1	0. 897	0. 0090	1. 005	0. 0101	0. 0011
3	0. 755	0. 0076	0. 990	0. 0099	0. 0023
2	2. 344	0. 0240	0. 897	0. 0090	0. 0150
4	0. 783	0. 0078	0. 812	0. 0081	0. 0003

Table 3. The vertical displacements and the angular distortions of both flanges due to the WS.

4.4. Discussions and recommendation for further research

From the results, it seems that, for the problem considered in this numerical study, two welding sequences, namely WS-2 and WS-4, have taken the attention. The WS-2, which is called as simple alternating welding, has produced the lowest peak temperature and the smallest longitudinal and transverse residual stresses as well. Meanwhile the WS-4, which is called as multiple crossing welding, has produced the smallest angular distortion and angular distortion difference, although it produced the highest peak temperature.

The information appears to be consistent with respect to the welding sequences performed. The corresponding value of the von Mises stress and the distortion difference produced as shown respectively in Table 2 and 3 indicated this as well. In particular, the results were also in contrast to those of WS-1 and WS-3. Not only did the welding sequences produce high angular distortions, but also they resulted in relatively high values of the von Mises stresses. Furthermore, the distortion results obtained appears to be match with the ones usually found in the welding practice incorporating alternating welding.

Also, considering limited literatures concerning welding simulation of T-joint fillet welds in 3D (Chang & Lee, 2009 and Deng et al., 2007), the results obtained would be very valuable and useful to welding designers and practitioners, because the results have been describing the predicted or anticipated residual stresses and distortions with respect to the welding sequences, varied from simple to multiple crossing welding. In addition, the assessment of welding performance which can be taken in an efficient and fast manner allows the designers to integrate it in their subsequent design plans.

Furthermore, the 3D simulation results of T-joint fillet weld may be further used as validation model for 3D welding simulations as well as for other numerical technique implementations such as mesh-less techniques, where no predefined mesh is required to build interpolation of the potential field variables investigated thus reducing cumbersome mesh preparation and increasing the related simulation time.

Moreover, the relationship between the input and output variables of the welding process may be further investigated and optimized using techniques from artificial intelligence (AI) family, such as neural networks and genetic algorithm. For examples, in the single pass GTAW welding method presented in this study, the variables of welding current, voltage, welding speed and welding sequences have been examined, in which more output variables may be also examined, such as the nature and dimensions of weld bead. Thus, much more information and insights can be revealed in such a welding process, which is in turn very useful to optimize the welding process.

It is noted here that the aspects of shrinkage were not discussed in the present paper. The aspects could be also related to the variation of welding speed. Also, it may be interesting if some welding paths in one WS are performed and simulated simultaneously thus allowing the exploitation of symmetry and anti-symmetry boundary conditions in the finite element model. The aforementioned aspects would be the subjects of further investigations.

5. Conclusions

Welding sequences effect on temperature distribution, residual stresses and distortions of T-joint fillet welds has been studied numerically in this paper. The simulation results revealed that peak temperature achieved in the welding was greatly affected by the WS and residual stress and angular distortion produced cannot both hold in minimum for a WS. The smallest longitudinal and transverse residual stresses occurred in WS-2, while the smallest angular distortion and difference in WS-4. The distributions of temperature, longitudinal and transverse residual stresses as well as angular distortions were also presented.

Investigating the aspects of shrinkage and simultaneous welding as well as the implementations of other related numerical techniques for further and better understanding of the welding process and its optimization would be the subjects of further publication in the future time.

Author details

Nur Syahroni
*Department of Ocean Engineering, Institut Teknologi Sepuluh Nopember (ITS),
Surabaya, Indonesia*

Mas Irfan Purbawanto Hidayat
*Department of Materials and Metallurgical Engineering, Institut Teknologi Sepuluh Nopember
(ITS), Surabaya, Indonesia*

Acknowledgement

Funding provided by Institut Teknologi Sepuluh Nopember (ITS) Surabaya is gratefully acknowledged.

6. References

ASM (1990). Metals Handbook Vol. 1, 10th ed. , Properties and Selection: Irons, Steels, and High-Performance Alloys, ASM International USA.

Bathe, K. J. (1996). Finite Element Procedures, Prentice Hall, Inc. , New Jersey.

Bette (1999). Fabrication, Applications Engineering, in: The Welding Engineer's Current Knowledge ed. 2000, EWE-3/4. 12 pp. 8-11, SLV Duisburg Gmbh, Germany.

Chang, K. H. & Lee, C. H. (2009). Finite Element Analysis of the Residual Stresses in T-joint Fillet Welds Made of Similar and Dissimilar Steels, Int. J. Adv. Manuf. Technol. , Vol. 41, 250-258.

Deng, D. , Murakawa, H. & Liang, Wei (2007). Numerical Simulation of Welding Distortion in Large Structures, Comput. Methods Appl. Mech. Eng. , Vol. 196, 4613-4627.

Goldak, J. A. & Akhlaghi, M. (2005). Computational Welding Mechanics, Springer, Inc. , New York.

Grong, O. (1994). Metallurgical Modelling of Welding, The Institute of Materials, Cambridge.

Lindgren, L. E. (2006). Numerical Modelling of Welding, Comput. Methods Appl. Mech. Eng. , Vol. 195, 6710-6736.

Masubuchi, K. (1980). Analysis of Welded Structures: Residual Stress, Distortion, and their Concequences, Pergamon Press Ltd, London.

Teng, T. L. & Lin, C. C. (1998). Effect of Welding Conditions on Residual Stresses due to Butt Welds, International Journal of Pressure Vessels and Piping, Vol. 75, 857-864.

Teng, T. L. , Fung, C. P. , Chang, P. H. & Yang, W. C. (2001). Analysis of Residual Stresses and Distortions in T-joint Fillet Welds, International Journal of Pressure Vessels and Piping, Vol. 78, 523-538.

Teng, T. L. , Chang, P. H. & Tseng, W. C. (2003). Effect of Welding Sequences on Residual Stresses, Computers and Structures, Vol. 81, 273-286.

Tsai, C. L. , Park, S. C. & Cheng, W. T. (1999). Welding Distortion of a Thin-Plate Panel Structure, Welding Research Supplement, 156-165.

Zacharia, T. , Vitek, J. M. , Goldak, J. A. , DebRoy, T. A. , Rappaz, M. & Bhadeshia, H. K. D. H. (1995). Modeling of Fundamental Phenomena in Welds, Modelling Simul. Mater. Sci. Eng. , Vol. 3, 265-288.

Permissions

The contributors of this book come from diverse backgrounds, making this book a truly international effort. This book will bring forth new frontiers with its revolutionizing research information and detailed analysis of the nascent developments around the world.

We would like to thank Mykhaylo I. Andriychuk, for lending his expertise to make the book truly unique. He has played a crucial role in the development of this book. Without his invaluable contribution this book wouldn't have been possible. He has made vital efforts to compile up to date information on the varied aspects of this subject to make this book a valuable addition to the collection of many professionals and students.

This book was conceptualized with the vision of imparting up-to-date information and advanced data in this field. To ensure the same, a matchless editorial board was set up. Every individual on the board went through rigorous rounds of assessment to prove their worth. After which they invested a large part of their time researching and compiling the most relevant data for our readers. Conferences and sessions were held from time to time between the editorial board and the contributing authors to present the data in the most comprehensible form. The editorial team has worked tirelessly to provide valuable and valid information to help people across the globe.

Every chapter published in this book has been scrutinized by our experts. Their significance has been extensively debated. The topics covered herein carry significant findings which will fuel the growth of the discipline. They may even be implemented as practical applications or may be referred to as a beginning point for another development. Chapters in this book were first published by InTech; hereby published with permission under the Creative Commons Attribution License or equivalent.

The editorial board has been involved in producing this book since its inception. They have spent rigorous hours researching and exploring the diverse topics which have resulted in the successful publishing of this book. They have passed on their knowledge of decades through this book. To expedite this challenging task, the publisher supported the team at every step. A small team of assistant editors was also appointed to further simplify the editing procedure and attain best results for the readers.

Our editorial team has been hand-picked from every corner of the world. Their multi-ethnicity adds dynamic inputs to the discussions which result in innovative outcomes. These outcomes are then further discussed with the researchers and contributors who give their valuable feedback and opinion regarding the same. The feedback is then collaborated with the researches and they are edited in a comprehensive manner to aid the understanding of the subject.

Apart from the editorial board, the designing team has also invested a significant amount of their time in understanding the subject and creating the most relevant covers. They scrutinized every image to scout for the most suitable representation of the subject and create an appropriate cover for the book.

The publishing team has been involved in this book since its early stages. They were actively engaged in every process, be it collecting the data, connecting with the contributors or procuring relevant information. The team has been an ardent support to the editorial, designing and production team. Their endless efforts to recruit the best for this project, has resulted in the accomplishment of this book. They are a veteran in the field of academics and their pool of knowledge is as vast as their experience in printing. Their expertise and guidance has proved useful at every step. Their uncompromising quality standards have made this book an exceptional effort. Their encouragement from time to time has been an inspiration for everyone.

The publisher and the editorial board hope that this book will prove to be a valuable piece of knowledge for researchers, students, practitioners and scholars across the globe.

List of Contributors

Takaaki Uda
Public Works Research Center, Taito, Tokyo

Masumi Serizawa and Shiho Miyahara
Coastal Engineering Laboratory Co. Ltd., Shinjuku, Tokyo

Julien Touboul
Mediterranean Institute of Oceanography (MIO), Aix-Marseille Univ., Université du Sud Toulon-Var, CNRS/INSU, UMR 7294, IRD, UMR235, France

Christian Kharif
Institut de Recherche sur les phénomènes hors équilibre (IRPHE), Aix-Marseille Univ., Ecole Centrale Marseille, CNRS/INSIS UMR 7342, France

Manabendra Pathak
Department of Mechanical Engineering, Indian Institute of Technology Patna, India

Mohamed Riahi and Taieb Lili
Faculté des Sciences de Tunis, Département de Physique, Laboratoire de Mécanique des Fluides, Campus Universitaire, Manar II, Tunis, Tunisia

Hossein Jalalifar
Shaihid Bahonar University of Kerman-Iran

Naj Aziz
Wollongong University- Australia

Dumitru Toader, Stefan Haragus and Constantin Blaj
"Politehnica" University of Timisoara, Romania

Masoud Ziabasharhagh and Arash Mohammadi
K. N. Toosi University of Technology, Iran

Jian-Xun Fu and Weng-Sing Hwang
Research Center for Energy Technology and Strategy & Department of Materials Science and Engineering, National Cheng Kung University, Tainan 701, Taiwan

A. Aoufi
SMS/RMT, LCG, UMR CNRS 5146, Ecole des Mines de Saint-Etienne, 158 cours Fauriel, 42023 Saint-Etienne Cedex, France

G. Damamme
CEA, Centre DAM-Île de France, Bruyères le Châtel, 91297 Arpajon cedex, France

Nur Syahroni
Department of Ocean Engineering, Institut Teknologi Sepuluh Nopember (ITS), Surabaya, Indonesia

Mas Irfan Purbawanto Hidayat
Department of Materials and Metallurgical Engineering, Institut Teknologi Sepuluh Nopember (ITS), Surabaya, Indonesia